Progress in Mathematics

Volume 124

Series Editors

J. Oesterlé
A. Weinstein

B. Aebischer
M. Borer
M. Kälin
Ch. Leuenberger
H. M. Reimann

Symplectic Geometry

An Introduction based on the Seminar in Bern, 1992

Springer Basel AG

Authors:
Institute of Mathematics
Sidlerstrasse 5
3012 Bern
Switzerland

A CIP catalogue record for this book is available from the Library of Congress,
Washington D.C., USA

Deutsche Bibliothek Cataloging-in-Publication Data

Symplectic geometry: an introduction based on the seminar in
Bern in 1992 / B. Aebischer ... – Basel ; Boston ; Berlin :
Birkhäuser, 1994
  (Progress in mathematics ; Vol. 124)
  ISBN 978-3-0348-7514-1   ISBN 978-3-0348-7512-7 (eBook)
  DOI 10.1007/978-3-0348-7512-7
NE: Aebischer, Beat; GT

Camera-ready copy prepared by the authors
Printed on acid-free paper produced of chlorine-free pulp

9 8 7 6 5 4 3 2 1

Contents

Preface

The seminar *Symplectic Geometry* at the University of Berne in summer 1992 showed that the topic of this book is a very active field, where many different branches of mathematics come together: differential geometry, topology, partial differential equations, variational calculus, and complex analysis. As usual in such a situation, it may be tedious to collect all the necessary ingredients. The present book is intended to give the nonspecialist a solid introduction to the recent developments in symplectic and contact geometry.

Chapter 1 gives a review of the symplectic group $Sp(n, \mathbf{R})$, symplectic manifolds, and Hamiltonian systems (last but not least to fix the notations). The Maslov index for closed curves as well as arcs in $Sp(n, \mathbf{R})$ is discussed. This index will be used in chapters 5 and 8.

Chapter 2 contains a more detailed account of symplectic manifolds starting with a proof of the Darboux theorem saying that there are no local invariants in symplectic geometry. The most important examples of symplectic manifolds will be introduced: cotangent spaces and Kähler manifolds. Finally we discuss the theory of coadjoint orbits and the Kostant-Souriau theorem, which are concerned with the question of which homogeneous spaces carry a symplectic structure.

Chapter 3 is devoted to the theory of generating functions, an important tool for dealing with Lagrange submanifolds and their intersections. The intersection problem for Lagrange manifolds is a generalization of the problem of finding (or counting) the fixed points of a symplectomorphism. Here the minimax principle and Lusternik-Schnirelman theory can be used and will be explained. Generating functions also lead to a symplectic invariant, called the Viterbo capacity.

In Chapter 4 symplectic capacities will be discussed. For this one first has to deal with periodic trajectories in Hamiltonian systems. The variational method makes it possible to identify these closed trajectories with the critical points of the action functional defined on a loop space. Leray-Schauder degree theory will then be applied to prove an existence theorem for closed trajectories. This theorem (or rather its proof) is then used to show that the Hofer-Zehnder capacity has all the properties required of a symplectic capacity. Roughly, this capacity measures how large a positive Hamiltonian which is zero on an open set can get without creating nonconstant periodic solutions with period less than one. Gromov's squeezing theorem and some rigidity results for symplectomorphisms will be derived from the existence of a symplectic capacity.

In Chapter 5 we discuss Floer's proof of the Arnol'd conjecture. In order to understand this, one first has to study Morse homology. Here one constructs a differential complex using the critical points of a Morse function $f : M \to \mathbf{R}$ and their connecting trajectories along the gradient flow of f. The homology of this complex is called Morse homology and it is isomorphic to the (singular) homology of the manifold M. This will be shown with the aid of the Conley index of isolated invariant sets (in a gradient dynamical system). As a corollary the classical Morse inequalities evolve. They imply that the number of critical points of f is at least the sum of the Betti numbers of M.

Arnol'd conjectured that every exact symplectomorphism of a symplectic manifold M has at least as many fixed points as a smooth function on M has critical points. An exact symplectomorphism is the time-1 map of a 1-periodic time-dependent Hamiltonian vector field. Its fixed points therefore correspond to 1-periodic solutions of the Hamiltonian equation. Since by the variational method these periodic solutions are identified with the critical points of the action functional, one has to count critical points, as in Morse theory. But instead of a Morse function one now has a functional on an infinite dimensional space, the loop space. In spite of the mathematical difficulties Floer was able to construct an analog of the Morse complex in many cases. Its homology is now called Floer homology. Using a deformation to a time independent Hamiltonian one then shows that Floer homology agrees with Morse homology. This proves the Arnol'd conjecture.

Chapter 6 is an introduction to pseudoholomorphic curves in a symplectic manifold. To a symplectic form on a manifold a natural family of almost complex structures is associated, imitating the relationship between hermitian forms and Kähler forms on Kähler manifolds. Although these almost complex structures in general do not define an honest complex structure of the manifold, one-dimensional complex submanifolds exist. The study of these (pseudo-) holomorphic curves yields a powerful tool for proving theorems on symplectic manifolds. We try to give a self-contained presentation of the fundamental construction of the moduli space of parametrized closed holomorphic curves and its simplest compactness properties. The moduli-space is constructed with the Sard-Smale theorem and the compactness is investigated via a differential geometric approach relying on a Sachs-Uhlenbeck argument and a Morrey inequality. At the end we show how pseudoholomorphic curves can be used to prove the so-called squeezing theorem.

In Chapter 7 the compactness theorem for pseudoholomorphic curves is taken up again, but this time from a more geometrical point of view. It can be read independently of Chapter 6. Essentially, Gromov's original proof as worked out by P. Pansu and C. Hummel will be sketched. This approach offers the advantage of being geometrically intuitive and displays the analogy to function theoretic methods. Pseudoholomorphic analogs of the Schwarz lemma, of the Weierstrass theorem, and of a well-known monotonicity property will be formulated.

To make the notion of cusp-curves precise, a delicate notion of convergence will be studied. It is closely related to convergence of hyperbolic structures on Riemann surfaces (using Fenchel-Nielsen parameters). For this, deformations of Riemann surfaces are studied.

Contact structures are the odd-dimensional counterpart to symplectic structures. They naturally appear in many ways in symplectic geometry: first of all as certain integral surfaces of Hamiltonian systems, and then, in the context of complex analysis, as strictly pseudoconvex hypersurfaces. The process of symplectification provides a direct link between contact and symplectic geometry. In Chapter 8 we consider contact structures, first from a general point of view and then with special emphasis on complex analytic methods in 3-dimensional contact geometry. We study topological invariants of two-dimensional manifolds embedded in 3-dimensional contact or 4-dimensional symplectic manifolds and we present Eliashberg's classification results.

The following lectures were given at the Symplectic Geometry seminar in Berne, 1992:

K. Cieliebak, *Symplectic spaces: introduction*

H. Hofer, *Symplectic spaces: survey*

A. Künzle, *Capacities, rigidity, and periodic solutions I*

T. Wurms, *Capacities, rigidity, and periodic solutions II*

C. Viterbo, *Generating functions and symplectic invariants I and II*

H. Hofer, *Elliptic methods in symplectic geometry I and II*

M. Schwarz, *Floer homology and Arnol'd conjectures I and II*

P. Pansu, *Holomorphic curves and symplectic geometry I and II*

Y. Eliashberg, *Contact geometry and lower dimensional topology I and II*

Y. Eliashberg, *Holomorphic discs I and II*

P. Pansu, *Holomorphic curves*

D. Salamon, *Floer homology and Novikov rings*

H. Hofer, *Symplectic homology I and II*

The topics presented in this book reflect the main lines exposed in the lectures, but they should not be understood as a reproduction of the content of these talks. We rather conceived the present notes as a beginner's introduction and an initiation to a fascinating field of mathematics. In symplectic geometry many disciplines come together and contribute with their results and methods. Our aim is to introduce the relevant techniques and to explain how they interact.

Our thanks go to all the speakers in the seminar and to all those who participated in the discussions and in the workshop. We also are indebted to Ch. Riedtmann and E. Zehnder for their advice and to N. A'Campo and E. Zehnder for their help in organizing the Berne seminar. Last but not least, we thank J. Giger for expertly typing large parts of the manuscript.

This project has been made possible by the support of the Swiss National Science Foundation.

Berne, December 1993

B. Aebischer, M. Borer, M. Kälin,
Ch. Leuenberger, H.M. Reimann

1 Introduction

1.1 Symplectic Linear Algebra

An even dimensional real vector space V is called *symplectic* if on V we have a skew-symmetric and non-degeneate bilinear form ω. More precisely, we say that ω is non-degenerate if $\omega(x, y) = 0$ for all $y \in V$ implies $x = 0$. A linear map $f : V_1 \to V_2$ between two symplectic vector spaces (V_1, ω_1) and (V_2, ω_2) is *symplectic* if $f^* \omega_2 = \omega_1$, i.e. if

$$\omega_1(x, y) = \omega_2(f(x), f(y)) \text{ for all } x, y \in V_1.$$

The standard example of a symplectic vector space is $V = \mathbf{R}^{2n}$ with the canonical symplectic form given by $\omega(x, y) = \langle Jx, y \rangle$ where $\langle \cdot, \cdot \rangle$ is the Euclidean inner product of $\mathbf{R}^{2n}$ and J is the $2n \times 2n$-matrix

$$J = \begin{pmatrix} 0 & I_{n \times n} \\ -I_{n \times n} & 0 \end{pmatrix}.$$

Thus if $x = (x_1, ..., x_{2n}), y = (y_1, ..., y_{2n}) \in V$ we get

$$\omega(x, y) = \sum_{i=1}^{n} x_{i+n} y_i - x_i y_{i+n}.$$

Moreover, this is essentially the only example of a symplectic vector space. More precisely, if (V, ω) is symplectic we can always find a canonical basis $e_1, ..., e_n, f_1, ..., f_n$ of V such that $\omega(e_i, e_j) = \omega(f_i, f_j) = 0$, $\omega(e_i, f_j) = \delta_{ij}$. The standard basis $e_1 = (0, ..., 0, 1, 0, ..., 0), ..., f_1 = (1, 0, ..., 0, ..., 0), ...$ satisfies these conditions in the case of $V = \mathbf{R}^{2n}$.

It is now clear that two symplectic vector spaces V_1, V_2 of the same dimension $2n$ are isomorphic: Choosing symplectic bases $e_1^i, .., e_n^i, f_1^i, ..., f_n^i$ in V_i, $i = 1, 2$. we get a symplectic isomorphism S between V_1 and V_2 by setting $Se_j^1 = e_j^2, Sf_j^1 = f_j^2, j = 1, ..., n$.

For any subspace W of a symplectic vector space V denote by $W^\perp$ the ω-orthogonal subspace of W

$$W^\perp = \{ x \in V : \omega(x, y) = 0 \quad \forall y \in W \}.$$

A subspace is called *isotropic* if $W \subset W^\perp$, *coisotropic* if $W^\perp \subset W$ and *Lagrangian* if $W = W^\perp$. An isotropic subspace is Lagrangian if and only if its dimension is maximal, i.e. if $\dim W = n$.

The set of all symplectic transformations of $\mathbf{R}^{2n}$ forms a group under composition. the *symplectic group* $Sp(n, \mathbf{R})$. In particular, $J \in Sp(n, \mathbf{R})$. For a $2n \times 2n$-matrix A with respect to a canonical basis $\{e_i, f_i\}$ of $\mathbf{R}^{2n}$ the condition for $A \in Sp(n, \mathbf{R})$ reads

$$A^T J A = J.$$

Writing $A = (A_{ij})$, where A_{ij} $(j = 1, 2)$ are $n \times n$-matrices, one has $A \in Sp(n, \mathbf{R})$ if and only if $A_{11}^T A_{21}$ and $A_{12}^T A_{22}$ are symmetric and $A_{11}^T A_{22} - A_{21}^T A_{12} = I_{n \times n}$.

From $A^T J A = J$ one concludes $\det A = \pm 1$. In fact, all symplectic matrices have determinant one. To see this, consider the exterior power $\Omega = \omega \wedge \ldots \wedge \omega$ with n factors. Since ω is nondegenerate, Ω is a volume form. Now if $A^* \omega = \omega$ one also has $A^* \Omega = \Omega$, which implies $\det A = 1$.

We have the following

Theorem 1.1 (Symplectic eigenvalue theorem) *If λ is an eigenvalue of a symplectic matrix A with multiplicity k, then $1/\lambda, \bar{\lambda}, 1/\bar{\lambda}$ are also eigenvalues of A; especially $1/\lambda$ occurs with multiplicity k.*

Proof. Let $p(\lambda) = \det(A - \lambda I)$ be the characteristic polynomial of A. From $JAJ^{-1} = (A^{-1})^T$ and $J^{-1} = -J$ we conclude

$$p(\lambda) = \det(J(A - \lambda I)J^{-1}) = \det(A^{-1} - \lambda I) = \det(A^{-1}(I - \lambda A)) =$$

$$= \det(I - \lambda A) = \lambda^{2n} \det(\frac{1}{\lambda}I - A) = \lambda^{2n} p(\frac{1}{\lambda}).$$

As 0 is not an eigenvalue of A it follows that λ^{-1} and $\bar{\lambda}$ are also eigenvalues of A. Suppose now λ_0 occurs with multiplicity k. Then $p(\lambda) = (\lambda - \lambda_0)^k p_1(\lambda)$ with $p_1(\lambda_0) \neq 0$. The identity

$$p(\lambda) = \lambda^{2n} p(\frac{1}{\lambda}) = \lambda^{2n} (\frac{1}{\lambda} - \lambda_0)^k p_1(\frac{1}{\lambda}) = \lambda_0^k \lambda^{2n-k} (\frac{1}{\lambda_0} - \lambda)^k p_1(\frac{1}{\lambda})$$

shows that $\frac{1}{\lambda_0}$ has multiplicity k, since $\lambda^{2n-k} p_1(\frac{1}{\lambda})$ does not vanish at $\lambda = \frac{1}{\lambda_0}$.
$\square$

1.2 Symplectic Manifolds

A *symplectic manifold* is a smooth manifold M on which a closed and non-degenerate 2-form ω is given. More precisely, ω must satisfy

1. $\mathbf{d}\omega = 0$ and

2. on any tangent space $T_p M, p \in M$, we have: if $\omega|_{T_p M}(X, Y) = 0$ for all $Y \in T_p M$, then $X = 0$.

M must have even dimension since any tangent space at a point of M is a symplectic vector space whose bilinear form is given by restriction of the symplectic form. In the next chapter we will see that locally all symplectic manifolds look the same. This is in sharp contrast to Riemannian geometry. Symplectic geometry is essentially a global theory.

Not every manifold admits a symplectic structure. All manifolds of odd dimension are counter examples. Another example is $M = S^4$. Assume ω is a closed and non-degenerate 2-form on S^4. Then ω is exact, i.e. there exists a 1-form α with $\omega = \mathbf{d}\alpha$. This follows from the well-known fact that the second de Rham cohomology group of S^4 vanishes, i.e. all closed 2-forms on S^4 are exact. But then also the volume form $\Omega = \omega \wedge \omega$ is exact:

$$\mathbf{d}(\omega \wedge \alpha) = \mathbf{d}\omega \wedge \alpha + \omega \wedge \mathbf{d}\alpha = \omega \wedge \omega = \Omega.$$

By Stokes' theorem we have

$$\int_{S^4} \Omega = \int_{\partial S^4} \omega \wedge \alpha = 0,$$

which is is impossible for a volume form. So we see that on S^4 we cannot impose a symplectic form.

Recall that to any smooth map between two manifolds M and N we can associate the tangential map $Tf : TM \to TN$. If β is a k-form on N then the pullback of β to M is defined by

$$f^*\beta_p(X_1, ..., X_k) = \beta_{f(p)}(Tf(X_1). .... Tf(X_k)),$$

where $p \in M$ and $X_1, ..., X_k \in T_pM$. Recall also the definition of the *interior product*: Let β be a k-form on M and X a vector field. Then $i_X\beta = X \lrcorner \beta$ (we will use both notations later) is defined to be the $k - 1$-form

$$X \lrcorner \beta(X_1, ..., X_{k-1}) = \beta(X, X_1. ..., X_{k-1}).$$

We are now able to give the global analog of a symplectic linear map:

Definition 1.2 *Let* (M, ω) *and* (N, ρ) *be symplectic manifolds. A smooth mapping* $f : M \to N$ *is called* symplectic *if*

$$f^*\rho = \omega.$$

Note that if f is a symplectic diffeomorphism, then f^{-1} is also symplectic; in this case f is called a *symplectomorphism*. Most important are of course symplectomorphism from M to itself because they preserve the symplectic structure.

The classification of subspaces into isotropic, coisotropic and Lagrangian can be taken over to submanifolds of a symplectic manifold. For example, a submanifold N of (M, ω) is called *Lagrangian* if for any point $p \in N$ the tangent space T_pN is a Lagrangian subspace of $(T_pM, \omega|_{T_pM})$.

1.3 Hamiltonian Systems

The Hamiltonian vector field of a function H on a symplectic manifold is formed
in a manner analogous to the gradient in Riemannian geometry. However, the
skew-symmetry of the symplectic 2-form leads to conservative properties which
are totally different from the properties of gradient vector fields.

Definition 1.3 *Let (M, ω) be a symplectic manifold and $H : M \to \mathbf{R}$ a smooth
function. The vector field X_H on M determined by the condition*

$$\omega(X_H, \cdot) = \mathbf{d}H(\cdot) \tag{1.1}$$

is called the Hamiltonian vector field *with energy function H.*

The non-degeneracy of ω guarantees that X_H exists and is uniquely defined.

Written in local coordinates the condition (1.1) becomes more familiar —
at least for physicists. Let $(q_1, \ldots, q_n, p_1, \ldots, p_n)$ be canonical coordinates for
ω, i.e. $\omega = \sum dp_i \wedge dq_i$. By Darboux' theorem such coordinates can always be
found locally (cf. next chapter). Then $(q(t), p(t))$ is an integral curve for X_H
exactly if *Hamilton's equations* hold:

$$\dot{p}_i = -\frac{\partial H}{\partial q_i}, \quad \dot{q}_i = \frac{\partial H}{\partial p_i}.$$

Indeed, let $X_H = (-\frac{\partial H}{\partial p_i}, \frac{\partial H}{\partial q_i}) = -J\nabla H$. Then

$$
\begin{aligned}
X_H \lrcorner\, \omega &= \sum X_H \lrcorner (dp_i \wedge dq_i) \\
&= \sum [(X_H \lrcorner dp_i) \wedge dq_i - dp_i \wedge (X_H \lrcorner dq_i)] \\
&= \sum [\frac{\partial H}{\partial p_i} dp_i + \frac{\partial H}{\partial q_i} dq_i] = \mathbf{d}H.
\end{aligned}
$$

Hamiltonian systems are well-known from classical mechanics. Newton's
second law states that a particle of mass m moving in a potential $V(q)$, where
q is a point in configuration space $\mathbf{R}^3$, moves along a curve $q(t)$ such that
$m\frac{d}{dt}q = -\nabla V(q)$. If we introduce the momentum $p_i = m\dot{q}_i$, $i = 1, 2, 3$, and the
energy $H(p, q) = \frac{1}{2m}|p|^2 + V(q)$, then Newton's law is equivalent to Hamilton's
equations. The mechanical problem can now be investigated in the phase space
$T^*\mathbf{R}^3 = \{(q_1, q_2, q_3, p_1, p_2, p_3)\}$.

From physics we expect conservation of energy. Let $c(t)$ be an integral
curve for a Hamiltonian vector field X_H on a symplectic manifold. Then we
have

$$
\begin{aligned}
\frac{d}{dt}H(c(t)) &= \mathbf{d}H(c(t))\dot{c}(t) = \mathbf{d}H(c(t))X_H(c(t)) = \\
&= \omega(X_H(c(t)), X_H(c(t))) = 0,
\end{aligned}
$$

hence $H \circ c$ is constant. The integral trajectories of a Hamiltonian system lie on energy surfaces $H = \text{const}$. This shows for example that the dynamical system given by an irrational winding along the 2-dimensional torus—considered as a symplectic manifold—cannot be a Hamiltonian system. Nevertheless this system is *locally* Hamiltonian:

Definition 1.4 *A vector field X on a symplectic manifold (M, ω) is called locally Hamiltonian if at every point p on M there is a neighbourhood U of p such that X restricted to U is Hamiltonian.*

Proposition 1.5 *The following statements are equivalent:*

(i) X *is locally Hamiltonian.*

(ii) $X \lrcorner \omega$ *is a closed 1-form.*

(iii) *The flow of X consists of symplectic maps.*

Proof. X is locally Hamiltonian if and only if $X \lrcorner \omega$ is locally exact. By Poincaré's lemma we get the equivalence of **(i)** and **(ii)**. Let f_t denote the flow of X. Since $f_0^* \omega = \omega$, the maps f_t are symplectic if and only if $\frac{d}{dt} f_t^* \omega = 0$. We have

$$\frac{d}{dt} f_t^* \omega = f_t^* L_X \omega = f_t^*(X \lrcorner \mathbf{d}\omega + \mathbf{d}(X \lrcorner \omega)) = f_t^* \mathbf{d}(X \lrcorner \omega)$$

where we used Cartan's formula for the Lie derivative $L_X = i_X \circ \mathbf{d} + \mathbf{d} \circ i_X$ and $\mathbf{d}\omega = 0$. Thus $\frac{d}{dt} f_t^* \omega$ vanishes exactly if $\mathbf{d}(X \lrcorner \omega) = 0$, i.e. if X is locally Hamiltonian. $\quad\square$

As a corollary we get the fact that (locally) Hamiltonian vector fields preserve the phase volume $\Omega = \omega \wedge \ldots \wedge \omega$ (n times).

Of course all Hamiltonian systems are locally Hamiltonian. But the converse is not true as the following example shows: Take $M = \mathbf{R}^2 \setminus \{0\}$ with the symplectic form $\omega = r\, dr \wedge d\theta$ in polar coordinates. Then the vector field $X = \frac{1}{r}\frac{\partial}{\partial r}$ is locally but not globally Hamiltonian (in fact. $X \lrcorner \omega = d\theta$). Another example is the above mentioned irrational winding on the torus T^2.

Note, however, if Y, Z are locally Hamiltonian vector fields on a symplectic manifold, then

$$\begin{aligned}
[Y, Z] \lrcorner \omega &= L_Y(Z \lrcorner \omega) - Z \lrcorner (L_Y \omega) \\
&= Y \lrcorner \mathbf{d}(Z \lrcorner \omega) + \mathbf{d}(Y \lrcorner (Z \lrcorner \omega)) - Z \lrcorner \mathbf{d}(Y \lrcorner \omega) = \mathbf{d}(\omega(Z, Y)),
\end{aligned}$$

in other words $[Y, Z] = X_H$ for $H = \omega(Z, Y)$, so that we have proved

Proposition 1.6 *If Y and Z are locally Hamiltonian vector fields on a symplectic manifold (M, ω), then $[Y, Z]$ is Hamiltonian.*

The following observation is a simple but important one:

Proposition 1.7 *Let $H_1, H_2 : \mathbf{R}^{2n} \to \mathbf{R}$ be two smooth Hamiltonian functions on the standard symplectic manifold $(\mathbf{R}^{2n}, \omega_0)$ which possess the same energy hypersurface at the level a, i.e. $\mathcal{C} := H_1^{-1}(a) = H_2^{-1}(a)$. If the gradients of H_1 and H_2 nowhere vanish on $\mathcal{C}$, then the trajectories of the two Hamiltonian systems coincide on $\mathcal{C}$.*

Proof. Let $c(t)$ be an integral trajectory of H_1 on $\mathcal{C}$, i.e. a solution of the equation of motion
$$\dot{c}(t) + J\nabla H_1(c(t)) = 0.$$

As ∇H_1 and ∇H_2 are normal to $\mathcal{C}$ and nowhere zero, there is a function λ given by $\nabla H_2(c(t)) = \lambda(t)\nabla H_1(c(t))$. Because of $\lambda(t) \neq 0$ for all t we can define the function $h(s) := \int_0^s \frac{1}{\lambda(t)}\, dt$. Set $\tilde{c} = c \circ h^{-1}$. As

$$\begin{aligned}
\frac{d}{dt}\tilde{c} &= (\dot{c} \circ h^{-1}) \cdot (\frac{1}{h'} \circ h^{-1}) \\
&= -J\nabla H_1(c \circ h^{-1}) \cdot (\lambda \circ h^{-1}) \\
&= -J\nabla H_2(c \circ h^{-1}) = -J\nabla H_2(\tilde{c}),
\end{aligned}$$

we observe that $\tilde{c}(t)$ solves the equations of motion for H_2. $\square$

For two smooth functions on a symplectic manifold M we define the *Poisson bracket* to be

$$\{f, g\} = \omega(X_f, X_g) = \mathbf{d}f(X_g) = X_g(f) = -X_f(g).$$

The Poisson bracket defines a Lie algebra structure on $C^\infty(M)$. To see this one only has to check Jacobi's identity:

$$\{f, \{g, h\}\} + \{h, \{f, g\}\} + \{g, \{h, f\}\} = 0.$$

This immediately reduces to the Jacobi identity for the Lie bracket, since

$$X_{\{h,g\}} = [X_g, X_h],$$

as follows from the computation

$$\begin{aligned}
[X_g, X_h] \lrcorner \omega &= L_{X_g}(X_h \lrcorner \omega) - X_h \lrcorner L_{X_g}\omega = L_{X_g}(\mathbf{d}h) \\
&= \mathbf{d}(X_g \lrcorner \mathbf{d}h) = \mathbf{d}(\{h, g\}) = X_{\{h,g\}} \lrcorner \omega.
\end{aligned}$$

where in the first step we have used the identity $i_{[X,Y]} = L_X \circ i_Y - i_Y \circ L_X$.

Let $H, g \in C^\infty(M)$. Then if f_t denotes the flow of X_H we have

$$\frac{d}{dt}(g \circ f_t) = f_t^* X_H(g) = -f_t^*\{H, g\},$$

so that g is constant along the trajectories of the Hamiltonian vector field exactly if $\{H, g\} \equiv 0$. Such a g is called an *integral of motion*.

Generically, Hamiltonian systems do not allow an integral of motion which is independent of the Hamiltonian function. In the exceptional case that H on a compact $2n$-dimensional manifold possesses n independent integrals of motion $g_1 = H, g_2, ..., g_n$ which are pairwise in involution, i.e. $\{g_i, g_j\} = 0$, the trajectories of X_H move along n-dimensional tori as was first observed by Liouville. Such systems are called *completely integrable*. Kolmogorov-Arnold-Moser theory (KAM) studies the behaviour of Hamiltonian systems which are near an integrable one. We will not be concerned with KAM theory in this book.

1.4   The Maslov Index

The fundamental group of the symplectic group is infinite cyclic as we show below. Thus to any loop in $Sp(n, \mathbf{R})$ we can associate an integer, the *Maslov index* of the loop. In this paragraph we show how one can associate a Maslov index to paths in $Sp(n, \mathbf{R})$ with certain boundary conditions. We closely follow [91] where the full proofs are presented. Recently Robbin and Salamon [89] have generalized the Maslov index to arbitrary symplectic arcs.

First we compute the fundamental group of the symplectic group. An easy calculation shows that $A \in Sp(n, \mathbf{R}) \cap O(2n)$ if and only if A has the form

$$A = \begin{pmatrix} A_1 & -A_2 \\ A_2 & A_1 \end{pmatrix}$$

where $A_1^T A_2 - A_2^T A_1 = 0$ and $A_1^T A_1 + A_2^T A_2 = I_n$. This is equivalent to $A^c \in U(n)$, where $A^c = A_1 + iA_2$ is a complex $n \times n$-matrix. So we have proved

$$Sp(n; \mathbf{R}) \cap O(2n) \cong U(n).$$

From Bott's periodicity (see e.g. [77]) we conclude

$$\pi_1(GL(n, \mathbf{C})) \cong \mathbf{Z}$$

and since $U(n)$ is a deformation retract of $GL(n, \mathbf{C})$ (see below), it follows that

$$\pi_1(U(n)) = \pi_1(GL(n, \mathbf{C})) \cong \mathbf{Z}. \tag{1.2}$$

Since $Sp(n, \mathbf{R})/U(n)$ is contractible, we also have $\pi_1(Sp(n, \mathbf{R})) \cong \mathbf{Z}$. It is possible to find a continuous map $\varrho : Sp(n, \mathbf{R}) \to S^1$ which restricts to the determinant map

$$\det : U(n) \subseteq Sp(n, \mathbf{R}) \to S^1.$$

To see that $U(n)$ is a deformation retract of $GL(n, \mathbf{C})$ we explicitly construct a retraction map: let $A \in GL(n, \mathbf{C})$. Now let $B = (A^*A)^{-\frac{1}{2}}$. We connect the positive definite hermitian matrix B with the identity, i.e. we choose a path B_t satisfying $B_0 = I_n$ and $B_1 = B$. The retraction is now given by $A_t = AB_t$.

An alternative approach to prove (1.2) is the following: The map $\det : U(n) \to S^1$ has the homotopy lifting property with respect to any space X. i.e. if $f : X \to U(n)$, $F_t : X \times I \to S^1$ are continuous maps such that $F_0(x) = \det f(x)$, then there is a map $f_t : X \times I \to U(n)$ such that $f_0(x) = f(x)$ and $\det f_t = F_t$. This is equivalent to the existence of a map f_t making the following diagram commutative:

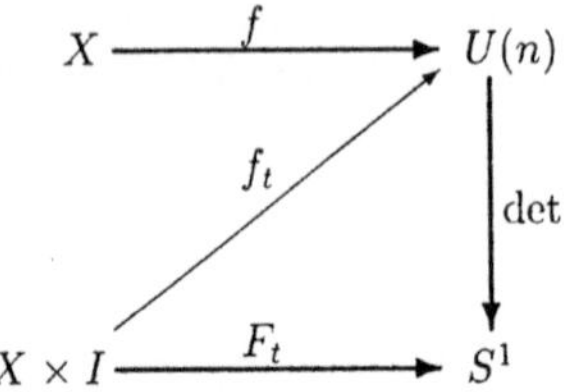

f_t can be given explicitly by $f_t(x) = (\frac{F_t(x)}{F_0(x)})^{1/n} f(x)$. This proves that $\det : U(n) \to S^1$ is a fibration. It is well-known that for fibrations the following homotopy sequence is exact (see e.g. [15, p. 209]):

$$\ldots \to \pi_2(S^1) \to \pi_1(SU(n)) \to \pi_1(U(n)) \to \pi_1(S^1) \to 0.$$

Note that $SU(n) = \det^{-1}(1) \hookrightarrow U(n)$, i.e. $SU(n)$ is the fiber over $1 \in S^1$. Since $SU(n)$ is simply connected we can read off $\pi_1(U(n)) = \mathbf{Z}$.

Recall that for a symplectic vector space (V, ω) we define $Sp(V, \omega)$ to be the group of automorphisms A satisfying $A^* \omega = \omega$. It can be shown (see [91]) that there exists for any symplectic vector space (V, ω) a unique continuous mapping

$$\varrho_V : Sp(V, \omega) \to S^1$$

which satisfies the following conditions:

(i) If $T : (V_1, \omega_1) \to (V_2, \omega_2)$ is a symplectic isomorphism, then

$$\varrho_{V_2}(TAT^{-1}) = \varrho_{V_1}(A), \quad A \in Sp(V_1, \omega_1).$$

(ii) If for $A_1 \in Sp(V_1, \omega_1)$, $A_2 \in Sp(V_2, \omega_2)$ we define $A \in Sp(V_1 \times V_2, \omega_1 \oplus \omega_2)$ to be $A(\xi_1, \xi_2) = (A\xi_1, A\xi_2)$, then

$$\varrho_{V_1 \times V_2}(A) = \varrho_{V_1}(A_1)\, \varrho_{V_2}(A_2).$$

(iii) For $A = \begin{pmatrix} A_1 & -A_2 \\ A_2 & A_1 \end{pmatrix} \in Sp(n, \mathbf{R}) \cap O(2n)$ (see above) we have

$$\varrho_{\mathbf{R}^{2n}}(A) = \det(A_1 + iA_2).$$

(iv) If A has no eigenvalues on the unit circle S^1, then

$$\varrho_V(A) = \pm 1.$$

For an explicit construction of ϱ_V see [91]. We denote with $Sp^*(n, \mathbf{R})$ the subset of symplectic matrices with no eigenvalue equal to 1, i.e.

$$Sp^*(n, \mathbf{R}) = \{A \in Sp(n, \mathbf{R}) : \det(I - A) \neq 0\}.$$

Lemma 1.8 ([91]) *(i) $Sp^*(n, \mathbf{R})$ has two connected components:*

$$Sp^*(n, \mathbf{R}) = Sp^+(n, \mathbf{R}) \cup Sp^-(n, \mathbf{R}),$$

where $Sp^\pm(n, \mathbf{R}) = \{A \in Sp(n, \mathbf{R}) : \pm \det(I - A) > 0\}$.

(ii) Any loop in $Sp^*(n, \mathbf{R})$ is contractible in $Sp(n, \mathbf{R})$.

We are now able to give a definition of the Maslov index for symplectic arcs: We call a path $\gamma : [0, 1] \to Sp(n, \mathbf{R})$ *admissible* if $\gamma(0) = I_{2n}$, $\gamma(1) \in Sp^*(n, \mathbf{R})$. The *Maslov index* is an integer $\mu(\gamma)$ assigned to an admissible γ in the following way: Choose a continuous function $\alpha : [0, 1] \to \mathbf{R}$ such that $\varrho(\gamma(t)) = e^{i\alpha(t)} \in S^1$ and set

$$\Delta(\gamma) = \frac{\alpha(1) - \alpha(0)}{\pi}.$$

Connect $\gamma(1) \in Sp^*(n, \mathbf{R})$ with W^+ or W^- using a path $\overline{\gamma}$ lying entirely in $Sp^*(n, \mathbf{R})$, i.e.

$$\overline{\gamma}(0) = \gamma(1), \quad \overline{\gamma}(1) = W^\pm \quad \text{and} \quad \overline{\gamma}(t) \in Sp^*(n, \mathbf{R}) \quad \forall t \in [0, 1].$$

Here W^+ and W^- are representatives from $Sp^+(n, \mathbf{R})$ and $Sp^-(n, \mathbf{R})$, namely

$$\begin{aligned}
W^+ &= -I_{2n} \in Sp^+(n, \mathbf{R}) \text{ and} \\
W^- &= \operatorname{diag}(2, -1, \ldots, -1, \tfrac{1}{2}, -1, \ldots, -1) \in Sp^-(n, \mathbf{R}).
\end{aligned}$$

Definition 1.9 *The Maslov index $\mu(\gamma)$ is defined by*

$$\mu(\gamma) = \Delta(\gamma) + \Delta(\overline{\gamma}).$$

It is shown in [91] that $\mu(\gamma) \in \mathbf{Z}$.

The Maslov index is closely related to the index of a certain Fredholm operator. We will present the main idea of this theorem of Salamon and Zehnder; for the details we refer to the original paper.

Recall that the Sobolev space $H^{1,2}(\mathbf{R} \times \mathbf{R}/\mathbf{Z}; \mathbf{R}^{2n})$ is the Banach space

$$\{\xi \in L^2(\mathbf{R} \times \mathbf{R}/\mathbf{Z}; \mathbf{R}^{2n}) : \frac{\partial \xi}{\partial s}, \frac{\partial \xi}{\partial t} \in L^2(\mathbf{R} \times \mathbf{R}/\mathbf{Z}; \mathbf{R}^{2n})\},$$

where the partial derviatives are understood in the weak sense and the norm is given by

$$\|\xi\|_{H^{1,2}}^2 = \|\xi\|_{L^2}^2 + \|\frac{\partial \xi}{\partial s}\|_{L^2}^2 + \|\frac{\partial \xi}{\partial t}\|_{L^2}^2.$$

$H^{1,2}(\mathbf{R} \times \mathbf{R}/\mathbf{Z}; \mathbf{R}^{2n})$ is even a Hilbert space with scalar product

$$(\xi, \eta)_{H^{1,2}} = \int_{\mathbf{R} \times \mathbf{R}/\mathbf{Z}} \left(\xi\eta + \frac{\partial \xi}{\partial s}\frac{\partial \eta}{\partial s} + \frac{\partial \xi}{\partial t}\frac{\partial \eta}{\partial t}\right) ds\, dt.$$

Obviously $(\cdot, \cdot)_{H^{1,2}}$ induces $\|\cdot\|_{H^{1,2}}$. We define the operator

$$F : H^{1,2}(\mathbf{R} \times \mathbf{R}/\mathbf{Z}; \mathbf{R}^{2n}) \longrightarrow L^2(\mathbf{R} \times \mathbf{R}/\mathbf{Z}; \mathbf{R}^{2n})$$

by

$$F\xi = \frac{\partial}{\partial s}\xi + J\frac{\partial}{\partial t}\xi + S\xi, \tag{1.3}$$

where $S(s, t)$ is a real symmetric $n \times n$ matrix continuous in (s, t) and 1-periodic in t. We further assume that $S(s, t)$ converges uniformly in t to the limits $S^+(t)$ or $S^-(t)$ as s tends to $+\infty$ or $-\infty$:

$$\lim_{s \to \pm\infty} S(s, t) \to S^\pm(t).$$

Recall the following

Definition 1.10 *A bounded linear operator* $F : X \to Y$, *where* X, Y *are Banach spaces, is called* Fredholm *if* $\ker F$ *and* $\operatorname{coker} F = Y/\operatorname{image}(F)$ *are finite dimensional. The index of* F *is given by*

$$\operatorname{ind} F = \dim \ker F - \dim \operatorname{coker} F.$$

If X, Y are Hilbert spaces we have an adjoint operator F^* and it is not difficult to show that F is Fredholm precisely if $\ker F$ and $\ker F^*$ are of finite dimension and $\operatorname{image}(F)$ is closed. The most important property of Fredholm operators is the invariance of their index under perturbations. Let $\mathcal{F}$ denote the set of Fredholm operators $F : X \to Y$ with the topology induced by the operator norm. Then we have the following

Theorem 1.11 (Dieudonné) *The index function* $\operatorname{ind} : \mathcal{F} \to \mathbf{Z}$ *is locally constant.*

However, $\dim \ker F$ need *not* be locally constant. For a proof see e.g. [60].

We will now show that the above defined operator F is indeed Fredholm. Observe that to F we can associate a path $\Psi(s,t)$ in $Sp(n,\mathbf{R})$ by the first order differential equation

$$\frac{\partial}{\partial t}\Psi(s,t) = JS(s,t)\,\Psi(s,t)$$

together with the initial condition $\Psi(s,0) = I$. As s tends to $\pm\infty$, $\Psi(s,t)$ converges to the matrices $\Psi^{\pm}(t)$ uniformly in t.

Theorem 1.12 (Salamon-Zehnder) *Let the operator F be given by (1.3). Assume further that the paths $\Psi^{\pm}(t)$ are admissible. Then F is Fredholm.*

To prove this we first need the following technical lemma:

Lemma 1.13 *Let X, Y, Z be Banach spaces, $F : X \to Y$ linear and $K : X \to Z$ compact. If for all $x \in X$ and a suitable $c > 0$ we have*

$$\|x\|_X \le c(\|Fx\|_Y + \|Kx\|_Z). \tag{1.4}$$

then $\dim \ker F < \infty$ and $\mathrm{image}\,(F)$ is closed.

Proof. (i) $\ker F$ is finite dimensional if and only if the unit ball $B = \{x \in \ker F : \|x\|_X \le 1\}$ is compact. Consider a sequence (x_k) in B. From (1.4) we conclude

$$\|x_k - x_l\|_X \le c\|K(x_k - x_l)\|_Z.$$

Because K is compact, i.e. (Kx_k) contains a convergent subsequence, (x_k) contains a convergent subsequence as well. Hence B is compact.

(ii) We write $X = X' \oplus \ker F$ and from the Hahn-Banach theorem we conclude that X' is closed. We choose a sequence (Fx_k) which is convergent in Y. We may assume $x_k \in X'$. If (x_k) were unbounded, then by (1.4) the normalized sequence $\frac{x_k}{\|x_k\|}$ would contain a subsequence (y_k) with

$$y_k \to y, \quad \|y_k\| = 1 \quad \text{and} \quad Fy = 0.$$

This is a contradiction to the closedness of X'. If (x_k) is bounded we get a convergent subsequence by the same argument as in part (i), i.e.

$$y = Fx \quad \text{for suitable } x \in X'.$$

Hence $\mathrm{image}\,(F)$ is closed in Y. $\qquad\qquad\square$

Proof of theorem 1.12. We follow [91] almost literally. We have to show that F satisfies the inequality (1.4) of the lemma, where we set $X = H^{1,2}(\mathbf{R} \times \mathbf{R}/\mathbf{Z}; \mathbf{R}^{2n})$, $Y = L^2(\mathbf{R} \times \mathbf{R}/\mathbf{Z}; \mathbf{R}^{2n})$ and $Z = L^2([-T,T] \times \mathbf{R}/\mathbf{Z}; \mathbf{R}^{2n})$ and choose a suitable operator K.

Suppose first that the function $\xi(s,t)$ is compactly supported: supp $\xi \subset [-T,T] \times \mathbf{R}/\mathbf{Z}$ for $T > 0$ large enough. By integration by parts we get

$$\|\nabla\xi\|_Y^2 = -\langle \xi, \Delta\xi \rangle = \|\frac{\partial\xi}{\partial s} + J\frac{\partial\xi}{\partial t}\|_Y^2.$$

Since $\|\xi\|_X \leq \|\nabla\xi\|_Y + \|\xi\|_Y$, it follows that

$$\|\xi\|_X \leq c_1 \left(\|F\xi\|_Y + \|\xi\|_Y \right), \tag{1.5}$$

where $c_1 = 1 + \sup |S(s,t)|$. So we have shown (1.4) for compactly supported ξ since the Sobolev embedding $K : X \hookrightarrow Y$ is compact.

Now assume that $S(s,t) = S(t)$ and $\Psi(s,t) = \Psi(t)$ are independent of s. Consider the operator $A : H^{1,2}(\mathbf{R}/\mathbf{Z}; \mathbf{R}^{2n}) \to L^2(\mathbf{R}/\mathbf{Z}; \mathbf{R}^{2n})$ defined by

$$(A\xi)(t) = J\frac{d}{dt}\xi(t) + S(t)\xi(t).$$

A is invertible if and only if 1 is not an eigenvalue of $\Psi(1)$. If this is the case there exists a constant $c_0 > 0$ such that

$$\|\xi\|_{H^{1,2}(\mathbf{R}/\mathbf{Z};\mathbf{R}^{2n})} \leq c_0 \|A\xi\|_{L^2(\mathbf{R}/\mathbf{Z};\mathbf{R}^{2n})}$$

(see e.g. [113]) and as A is self adjoint we have inequalities

$$|\omega|\,\|\xi\|_{L^2(\mathbf{R}/\mathbf{Z};\mathbf{R}^{2n})}^2 \leq |\langle \xi, i\omega\xi + A\xi \rangle| \leq \|\xi\|_{L^2(\mathbf{R}/\mathbf{Z};\mathbf{R}^{2n})}\,\|i\omega\xi + A\xi\|_{L^2(\mathbf{R}/\mathbf{Z};\mathbf{R}^{2n})}$$

and

$$\|A\xi\|_{L^2(\mathbf{R}/\mathbf{Z};\mathbf{R}^{2n})} \leq \|i\omega\xi + A\xi\|_{L^2(\mathbf{R}/\mathbf{Z};\mathbf{R}^{2n})}$$

for $\omega \in \mathbf{R}$. From this we conclude

$$|\omega|\,\|\xi\|_{L^2(\mathbf{R}/\mathbf{Z};\mathbf{R}^{2n})} \leq \|i\omega\xi + A\xi\|_{L^2(\mathbf{R}/\mathbf{Z};\mathbf{R}^{2n})}$$

and

$$\|\xi\|_{H^{1,2}(\mathbf{R}/\mathbf{Z};\mathbf{R}^{2n})} \leq c_0 \|i\omega\xi + A\xi\|_{L^2(\mathbf{R}/\mathbf{Z};\mathbf{R}^{2n})}.$$

Now let $\xi(s,t)$ be compactly supported and denote by $\hat{\xi}(\omega,t)$ its Fourier transform with respect to the variable s:

$$\hat{\xi}(\omega,t) = (2\pi)^{-\frac{1}{2}} \int_{\mathbf{R}} e^{-i\omega s}\,\xi(s,t)\,ds.$$

Using Plancherel's theorem we get

$$\|\xi\|_X^2 = \int_{\mathbf{R}} \left(\|\hat{\xi}\|_{H^{1,2}(\mathbf{R}/\mathbf{Z};\mathbf{R}^{2n})}^2 + |\omega|^2 \|\hat{\xi}\|_{L^2(\mathbf{R}/\mathbf{Z};\mathbf{R}^{2n})}^2 \right) d\omega$$

$$\leq (c_0^2 + 1) \int_{\mathbf{R}} \|i\omega\hat{\xi} + A\hat{\xi}\|_{L^2(\mathbf{R}/\mathbf{Z};\mathbf{R}^{2n})}^2 d\omega = (c_0^2 + 1) \|\frac{\partial\xi}{\partial s} + A\xi\|_Y^2.$$

This shows that $F = \frac{\partial}{\partial s} + A$ is invertible if S is independent of s and $\det(\Psi(1) - I) \neq 0$, i.e. if $\Psi(t)$ is an admissible path in $Sp(n, \mathbf{R})$.

Returning to the general case we see that the limit operators

$$F^{\pm} = \frac{\partial}{\partial s} + J\frac{\partial}{\partial t} + S^{\pm}$$

are invertible. Thus there are constants $T > 0$ and $c_2 > 0$ such that

$$\|\xi\|_X \leq c_2 \|F\xi\|_Y \tag{1.6}$$

for each ξ which vanishes in the interval $[-T + 1, T - 1]$.

In order to prove (1.4) we glue together the estimates (1.5) and (1.6) by choosing a smooth cut-off function $\alpha = \alpha(s)$ which is identically 1 in $[-T + 1, T - 1]$ and vanishes outside of $[-T, T]$. Writing $\xi = \alpha\xi + (1 - \alpha)\xi$ we compute

$$\begin{aligned}
\|\xi\|_X &\leq c_1 \left(\|F(\alpha\xi)\|_Y + \|\alpha\xi\|_Y \right) + c_2 \|F((1 - \alpha)\xi)\|_Y \\
&\leq c_3 \left(\|F\xi\|_Y + \|\xi\|_Z \right).
\end{aligned}$$

By Lemma 1.13 we see that F has a closed range and a finite dimensional kernel. By a similar argument for the formally adjoint operator

$$F^* = -\frac{\partial}{\partial s} + J\frac{\partial}{\partial t} + S$$

we conclude that the cokernel of F is finite dimensional as well, so that F is indeed Fredholm. $\qquad\qquad\square$

Once knowing that F is a Fredholm operator we can express its index in terms of the Maslov index. Note that we have assumed that the limit paths $\Psi^{\pm}(t)$ are admissible, so we can associate a Maslov index to them. We omit the proof of the following theorem which is proved in [91].

Theorem 1.14 *Under the conditions of the preceeding theorem the index of* $F = \frac{\partial}{\partial s} + J\frac{\partial}{\partial t} + S$ *is given by*

$$\mathrm{ind}\, F = \mu(\Psi^-) - \mu(\Psi^+).$$

We will now show, again omitting the proofs, how the Maslov index can be defined for certain periodic solutions of Hamiltonian systems. Let (M, ω) be a compact symplectic manifold and $H : \mathbf{R} \times M \to \mathbf{R}$ a smooth time-dependent Hamiltonian. Assume further that H is 1-periodic in time, i.e. $H(t + 1, x) = H(t, x)$. Recall that H gives rise to a 1-periodic Hamiltonian vector field X_H. We are interested in periodic solutions of the differential equation

$$\dot{x}(t) = X_H(t, x(t)). \tag{1.7}$$

The solutions of this equation give rise to a Hamiltonian flow $\psi_t \in \mathrm{Diff}(M)$, $\psi_t(x(0)) = x(t)$, which consists of symplectic maps (cf. Proposition 1.5). A 1-periodic solution $x(t) = x(t+1)$ is called *non-degenerate* if

$$\det\bigl(I - d\psi_1(x(0))\bigr) \neq 0.$$

Now let x be a nondegenerate and contractible 1-periodic solution of (1.7). Since x is contractible we can find a smooth function $\varphi : D \to M$, where D is the unit disk, such that

$$\varphi(\exp(2\pi i t)) = x(t).$$

Choose a fixed almost complex structure J on M (see Definition 2.5) and define by $g(\cdot,\cdot) = \omega(\cdot, J\cdot)$ a Riemannian metric with the associated Levi-Civita connection ∇. As is shown in [91, Lemma 5.1], for the disk $\varphi : D \to M$ there exists a trivialization

$$D \times \mathbf{R}^{2n} \longrightarrow \varphi^*TM : (z, \zeta) \mapsto \Phi(z)\zeta$$

such that $J\Phi = \Phi J_0$, $\Phi^*\omega = \omega_0$ and $g(\Phi\zeta, \Phi\xi) = \langle \zeta, \xi \rangle$. Here J_0 and ω_0 are the standard complex and symplectic structures of $\mathbf{R}^{2n}$, respectively, and $\langle \zeta, \xi \rangle$ denotes the euclidean scalar product of $\mathbf{R}^{2n}$. All such trivializations are homotopic. (If Ψ is another such trivialization, then $z \mapsto \Psi(z)^{-1}\Phi(z)$ is a smooth map $D \to U(n)$, hence homotopic to the constant map $z \mapsto 1$.) This gives rise to a unitary trivialization of the pull-back x^*TM:

$$\Phi_x(t) := \Phi(\exp(2\pi i t)) : \mathbf{R}^{2n} \to T_{x(t)}M$$

such that $\Phi_x(t+1) = \Phi_x(t)$. Now we assume that the *first Chern class* $c_1(TM)$ vanishes over $\pi_2(M)$. This means that for every smooth map $u : S^2 \to M$ the $\mathbf{C}^n$-bundle u^*TM is trivial. Then the homotopy class of Φ_x is independent of the choice of the extension φ (see [91, Lemma 5.2]).

Define now the path γ_x in $Sp(n, \mathbf{R})$ by

$$\gamma_x(t) = \Phi_x(t)^{-1} d\psi_t(x(0))\, \Phi_x(0), \quad 0 \leq t \leq 1.$$

Because of the non-degeneracy of x the path γ_x is admissible, so that we are able to assign a Maslov index to x by setting:

$$\mu(x) := \mu(\gamma_x).$$

Let now x and y be two non-degenerate contractible 1-periodic solutions of (1.7) and $u : \mathbf{R}^2 \to M$ a smooth function which satisfies $u(s, t+1) = u(s, t)$ and

$$\lim_{s \to -\infty} u(s,t) = x(t), \qquad \lim_{s \to +\infty} u(s,t) = y(t).$$

Here the convergence is understood uniformly in t, $\frac{\partial u}{\partial t}$ has to converge to $\dot{x}$ and $\dot{y}$, respectively, uniformly in t and $\frac{\partial u}{\partial s} \to 0$ as s tends to $\pm\infty$.

Define the Hilbert space $L^2(u)$ as the completion of the space of compactly supported vector fields ξ on u^*TM which are 1-periodic in t with respect to the norm

$$\|\xi\|^2 = \int_{-\infty}^{\infty} \int_0^1 |\xi(s,t)|^2 \, dt \, ds.$$

where of course $|\xi|^2 = g(\xi,\xi)$. The Sobolev space $H^{1,2}(u)$ consists of those vector fields in $L^2(u)$ whose derivatives $\nabla_s\xi$, $\nabla_t\xi$ are in $L^2(u)$. We can define

$$F(u)\xi := \nabla_s\xi + J(u)\nabla_t\xi + \nabla_\xi J(u)\frac{\partial u}{\partial t} - \nabla_\xi \nabla H(t,u)$$

as a linear operator from $H^{1,2}(u)$ to $L^2(u)$. This operator $F(u)$ is the linearization of the operator $F : u \mapsto \frac{\partial u}{\partial s} + J(u)\frac{\partial u}{\partial t} - \nabla H(t,u)$, which will appear in section 5.3. The following result will be used in section 5.4. For its proof see [91, Theorem 5.3].

Theorem 1.15 *The operator $F(u)$ is Fredholm and its index can be expressed in terms of the Maslov indices of the boundary solutions x and y:*

$$\operatorname{ind} F(u) = \mu(x) - \mu(y).$$

2 Darboux' Theorem and Examples of Symplectic Manifolds

The main object of this chapter is first to show that *locally* all finite-dimensional symplectic manifolds look alike. On the other hand, a *global* examination of symplectic structures is usually made difficult by additional geometric properties of the manifold. Therefore we restrict our considerations and illustrating examples to the three most frequently encountered types of symplectic manifolds, namely cotangent bundles, Kähler manifolds and coadjoint orbits.

2.1 The Theorem of Darboux

Remember from the first chapter that a smooth even-dimensional manifold M^{2n} is called symplectic, if and only if there exists a non-degenerate closed 2-form ω on M^{2n}. Our interest in the geometry induced on M^{2n} by ω is restricted in this section to local aspects. More precisely, we concentrate our attention on the most important local result, due to Darboux, which may be formulated in the following way:

For any point y on a symplectic manifold (M^{2n}, ω) of dimension $2n$ there exists an open neighbourhood U of y and a differentiable map $f : (U, \omega) \to (\mathbf{R}^{2n}, \omega_0)$ such that $f^* \omega_0 = \omega|_U$ (where ω_0 denotes the standard symplectic structure on $\mathbf{R}^{2n}$).

In other words: locally, for an appropriate choice of so-called symplectic coordinates $\mathbf{p}$ and $\mathbf{q}$, the symplectic form ω can be written in the standard way: $\omega = d\mathbf{p} \wedge d\mathbf{q} = \sum_{i=1}^n dp_i \wedge dq_i$. Obviously, this result allows us to extend to all symplectic manifolds any assertion of local character which holds for $(\mathbf{R}^{2n}, \omega_0)$. Let us point out that the symplectic situation is totally different from the Riemannian case, where the curvature of the Riemannian metric does not allow such a result.

The current proof of Darboux' theorem uses induction on the dimension and is accessible in different versions in the basic literature (see for example Sternberg [98], Arnold [7]). The proof presented in this lecture, however, is due to J. Moser (consult Guillemin-Sternberg [51]) and is based on standard differential calculus. A useful formula for the proof is given in the following

Lemma 2.1 *Let V and W be differentiable manifolds and $\varphi_t : V \to W$ a smooth one-parameter family of maps. By X_t we denote the tangent field along φ_t, that is. $X_t : V \to TW$ with $X_t(v) = \frac{d}{dt}\varphi_t(v)$, and we let σ_t be a smooth one-parameter family of forms on W. Then the following formula is true*

$$\frac{d}{dt}(\varphi_t^* \sigma_t) = \varphi_t^*\Big(\frac{d\sigma_t}{dt} + i(X_t)d\sigma_t + d(i(X_t)\sigma_t)\Big).$$

Proof (by Moser, Guillemin-Sternberg). **a)** We first proof the Lemma in a special case: Let $V = W = M \times I$ (M a n-dimensional manifold) and, instead of φ_t, take $\psi_t : M \times I \to M \times I$, defined by $\psi_t(x, s) = (x. s + t)$.

The most general differential form on $M \times I$, which depends on $(x, s) \in M \times I$ and a parameter t may be written as

$$\sigma_t = ds \wedge a(x, s, t)dx^k + b(x, s, t)dx^{k+1}$$

(where the k-form $a(x. s, t)dx^k$ is a sum of terms like $a_{i_1,\dots,i_k}(x, s, t)dx_{i_1} \wedge \dots \wedge dx_{i_k}$; we will write $a_i(x, s, t)$ instead of $a_{i_1,\dots,i_k}(x, s, t)$. The analogous notation is used for $b(x, s, t)dx^{k+1}$). Obviously $\psi_t^* \sigma_t = ds \wedge a(x, s + t, t)dx^k + b(x, s + t, t)dx^{k+1}$ and therefore

$$
\begin{aligned}
\frac{d}{dt}(\psi_t^* \sigma_t) &= ds \wedge \frac{\partial a}{\partial s}(x, s + t, t)dx^k + \frac{\partial b}{\partial s}(x, s + t, t)dx^{k+1} \\
&+ ds \wedge \frac{\partial a}{\partial t}(x, s + t, t)dx^k + \frac{\partial b}{\partial t}(x, s + t, t)dx^{k+1}.
\end{aligned}
\tag{2.1}
$$

We also immediately have

$$\psi_t^*\left(\frac{d\sigma_t}{dt}\right) = ds \wedge \frac{\partial a}{\partial t}(x, s + t, t)dx^k + \frac{\partial b}{\partial t}(x, s + t, t)dx^{k+1}.
\tag{2.2}$$

The tangent vector to $\psi_t(x, s)$ is in our case the vector field $\frac{\partial}{\partial s}$ evaluated at $(x, s + t)$. Therefore $i(\frac{\partial}{\partial s})\sigma_t = a(x, s, t)dx^k$, so we find

$$\psi_t^*\left(i\left(\frac{\partial}{\partial s}\right)\sigma_t\right) = a(x, s + t, t)dx^k = \sum_i a_i(x, s + t, t)dx_{i_1} \wedge \dots \wedge dx_{i_k}.$$

Finally, for the exterior derivative one gets the formula

$$
\begin{aligned}
\psi_t^* d\left(i\left(\frac{\partial}{\partial s}\right)\sigma_t\right) &= \sum_i d(a_i(x, s + t, t)) \wedge dx_{i_1} \wedge \dots \wedge dx_{i_k} \\
&= \sum_i \Big(\frac{da_i}{ds}(x, s + t, t)ds \wedge dx_{i_1} \wedge \dots \wedge dx_{i_k} + \\
&\quad + \sum_{j \notin \{i_1,\dots,i_k\}} \frac{da_i}{dx_j}(x, s + t, t)dx_j \wedge dx_{i_1} \wedge \dots \wedge dx_{i_k} \Big)
\end{aligned}
$$

which is written in the abbreviated form

$$d\psi_t^*\left(i\left(\frac{\partial}{\partial s}\right)\sigma_t\right) = \frac{\partial a}{\partial s}(x, s + t, t)ds \wedge dx^k + d_x a(x, s + t, t)dx^{k+1}.
\tag{2.3}$$

Using the same notation, we calculate

$$d\sigma_t = -ds \wedge d_x a(x. s. t)dx^{k+1} + \frac{\partial b}{\partial s}(x, s. t)ds \wedge dx^{k+1} + d_x b(x. s. t)\, dx^{k+2}.$$

Therefore $i(\frac{\partial}{\partial s})d\sigma_t = -d_x a(x,s,t)dx^{k+1} + \frac{\partial b}{\partial s}(x,s.t)dx^{k+1}$ and we conclude

$$\psi_t^* i(\frac{\partial}{\partial s})d\sigma_t = -d_x a(x,s+t,t)dx^{k+1} + \frac{\partial b}{\partial s}(x.s+t,t)\,dx^{k+1}. \tag{2.4}$$

Thus we have proved the formula for the special case, because
$(2.1) = (2.2) + (2.3) + (2.4)$.

b) We now go on with the general case: Let $\varphi : V \times I \to W$ be given by $\varphi(v,s) = \varphi_s(v)$. Under φ a line parallel to I through the point v is taken to the curve $\varphi_s(v)$ in W. That is why $d\varphi(\frac{\partial}{\partial s})_{(v,t)} = X_t(v)$. Obviously $\varphi_t = \varphi \circ \psi_t \circ \imath$, where $\imath : V \to V \times I$ is defined by $\imath(v) = (v,0)$. Thus $\varphi_t^* \sigma_t = \imath^* \psi_t^* \varphi^* \sigma_t$, and

$$\frac{d}{dt}\varphi_t^* \sigma_t = \imath^* \frac{d}{dt}(\psi_t^*(\varphi^* \sigma_t))$$

follows immediately. We now use the result of (2.1) to calculate

$$\frac{d}{dt}(\psi_t^*(\varphi^* \sigma_t)) = \psi_t^* \frac{d(\varphi^* \sigma_t)}{dt} + \psi_t^*(i(\frac{\partial}{\partial s})d(\varphi^* \sigma_t)) + d\psi_t^* i(\frac{\partial}{\partial s})\varphi^* \sigma_t.$$

As $i(\frac{\partial}{\partial s})\varphi^* \sigma_t = \varphi^* i(X_t)\sigma_t$ and $i(\frac{\partial}{\partial s})\varphi^* d\sigma_t = \varphi^* i(X_t)d\sigma_t$, we get

$$\begin{aligned}
\frac{d}{dt}(\varphi_t^* \sigma_t) &= \imath^* \frac{d}{dt}(\psi_t^*(\varphi^* \sigma_t)) \\
&= \imath^* \psi_t^*(\varphi^* \frac{d\sigma_t}{dt} + \varphi^* i(X_t)d\sigma_t + \varphi^* d(i(X_t)\sigma_t)) \\
&= \varphi_t^*(\frac{d\sigma_t}{dt} + i(X_t)d\sigma_t + d(i(X_t)\sigma_t))
\end{aligned}$$

which is the desired formula. $\qquad\qquad\qquad\qquad\qquad\qquad\qquad\qquad \square$

Remark: In the case that $\sigma_t = \sigma$ does not depend on t, the formula of the lemma can be derived from the definition of the Lie derivative.

We are now going on with the main result of this section:

Theorem 2.2 (Darboux) *Let ω_0 and ω_1 be two non-degenerate closed 2-forms on a 2n-dimensional manifold M such that $\omega_0|_y = \omega_1|_y$ for a certain fixed point $y \in M$. Then there exists a neighbourhood U of y and a diffeomorphism $f : U \to f(U) \subseteq M$ with $f(y) = y$ and $f^* \omega_1 = \omega_0$.*

Proof (by Moser, Guillemin-Sternberg). The idea of the proof is to obtain f as the time-1-map of a flow f_t $(0 \leq t \leq 1)$ such that $f_t^* \omega_t = \omega_0$, where $\omega_t = (1-t)\omega_0 + t\omega_1$. The existence of the time-dependent vector field V_t belonging to the flow f_t is proven step by step in the following way:

a) From $\omega_0 = f_t^* \omega_t$ we get the necessary condition

$$0 = \frac{d}{dt}(f_t^* \omega_t) = f_t^*\left(\frac{d}{dt}\omega_t + i(V_t)d\omega_t + d(i(V_t)\omega_t)\right).$$

As $d\omega_t = 0$, this is equivalent to

$$d(i(V_t)\omega_t) = -\frac{d}{dt}\omega_t.$$

b) We have to determine a 1-form $\alpha_t := i(V_t)\omega_t$ such that $d\alpha_t = -\frac{d}{dt}\omega_t$. This is locally possible as the integrability condition is fulfilled:

$$d\left(\frac{d}{dt}\omega_t\right) = \frac{d}{dt}(d\omega_t) = 0.$$

c) Now the equation $\alpha_t = i(V_t)\omega_t$ can be solved, because ω_t proves to be non-degenerate and we get the vector field V_t. Therefore we have established the existence of V_t and thus of f_t.

Let us turn our attention now to some details of the proof:

b) We have to determine a 1-form α_t with $d\alpha_t = -\frac{d}{dt}\omega_t$. Let $\sigma := \omega_0 - \omega_1 = -\frac{d}{dt}((1-t)\omega_0 + t\omega_1) = -\frac{d}{dt}\omega_t$. There exists a neighbourhood U_1 of y and a smooth retraction φ_t of U_1 onto $\{y\}$ (i.e. $\varphi_t : U_1 \to U_1$ with $\varphi_0 : U_1 \to \{y\}$, $\varphi_1 = id$ and $\varphi_t(y) = y$ ($\forall t \in I := [0,1]$)) such that in U_1

$$\sigma - \varphi_0^* \sigma = \int_0^1 \left(\frac{d}{dt}\varphi_t^* \sigma\right)dt = \int_0^1 \varphi_t^*(i(X_t)d\sigma + d(i(X_t)\sigma))dt$$

(where X_t is the vector field induced by φ_t in the neighbourhood of y). Note that $\sigma|_y = 0$. We have $\varphi_0^* \sigma = 0$ and moreover $d\sigma = 0$; consequently

$$\sigma = \int_0^1 \varphi_t^*(d(i(X_t)\sigma))dt = d\int_0^1 \varphi_t^*(i(X_t)\sigma)dt.$$

Hence $\alpha := \int_0^1 \varphi_t^*(i(X_t)\sigma)dt$ is the searched for 1-form:

$$d\alpha = d\int_0^1 \varphi_t^*(i(X_t)\sigma)dt = \sigma = -\frac{d}{dt}\omega_t.$$

Note that $\alpha(y) = 0$.

c) In order to solve the equation $\alpha = i(V_t)\omega_t$ we have to check that ω_t is non-degenerate: Indeed nondegeneracy ($\forall t \in I$) is trivial in the point y, because

$$\omega_t|_y = \omega_0|_y - t\omega_0|_y + t\omega_1|_y = \omega_0|_y = \omega_1|_y$$

and can of course be extended to a neighbourhood $U_0 \subseteq U_1$ of y. Therefore the vector field V_t can be calculated for all $t \in I$ and $x \in U_0$. By integration of V_t we get a one-parameter family of maps f_t, which has to be restricted to a smaller neighbourhood $U \subseteq U_0$ such that $f_t(U) \subseteq U_0$ ($\forall t \in I$). Note the following properties of $\{f_t\}_{t \in I}$:

- $f_t(y) = y$: As $\alpha(y) = 0$, we get from $\alpha = i(V_t)\omega_t$ and the nondegeneracy of ω_t that $V_t(y) = 0$.

- $f_1^*\omega_1 - \omega_0 = 0$:

$$
\begin{aligned}
f_1^*\omega_1 - \omega_0 &= \int_0^1 \frac{d}{dt}(f_t^*\omega_t)dt \\
&= \int_0^1 f_t^*(\frac{d}{dt}\omega_t + i(V_t)d\omega_t + d(i(V_t)\omega_t))dt \\
&= \int_0^1 f_t^*(-\sigma + d(i(V_t)\omega_t))dt = \int_0^1 f_t^*(-\sigma + d\alpha)dt = 0.
\end{aligned}
$$

Thus f_t provides the desired flow and the theorem is proved: $f := f_1$ is the diffeomorphism defined in the neighborhood U. $\qquad\square$

As a corollary we get the more current version of Darboux' result already formulated at the beginning of this section:

Corollary 2.3 *For any point y on a symplectic manifold (M^{2n},ω) there exists an open neighbourhood U of y and a symplectomorphism $f : (U,\omega) \to (f(U) \subseteq \mathbf{R}^{2n},\omega_0)$ such that $f^*\omega_0 = \omega$ (where ω_0 is the standard symplectic structure on $\mathbf{R}^{2n}$).*

Proof. Use the exponential map of some Riemann metric on (M^{2n},ω) to define a local diffeomorphism $\varphi_1 : U_1 \to U$ of some neighbourhood U_1 of the origin in the tangent space $T_y M^{2n} \cong \mathbf{R}^{2n}$ onto some neighbourhood U of y in M^{2n}. In this way we get on U_1 the symplectic form $\omega_1 = \varphi_1^*\omega$ and with a linear transformation we attain without loss of generality that $\omega_1|_0 = \omega_0|_0$. Apply now Darboux' theorem to get a neighbourhood $U_0 \subseteq U_1$ of 0 and a diffeomorphism $\varphi_0 : U_0 \to \varphi_0(U_0) \subseteq U_1$ with $\varphi_0(0) = 0$ and $\varphi_0^*\omega_1 = \omega_0$. Therefore $\varphi_0^*\varphi_1^*\omega = \omega_0$. The symplectomorphism of the corollary is now

$$
f := (\varphi_1 \circ \varphi_0)^{-1} : ((\varphi_1 \circ \varphi_0)(U_0).\omega) \to (U_0,\omega_0)
$$

because $f^*\omega_0 = (\varphi_1 \circ \varphi_0)^{-1^*}\omega_0 = (\varphi_1^{*-1} \circ \varphi_0^{*-1})\omega_0 = \varphi_1^{*-1}\omega_1 = \omega.$ $\qquad\square$

Remark: The coordinates in the corollary are sometimes called *symplectic*, and an atlas of M is said to be symplectic, if the whole manifold is covered with neighbourhoods (and corresponding local coordinates) of the form mentioned in the corollary.

We will now leave local aspects of symplectic geometry and pass, in the last three sections of this chapter to some important examples of symplectic manifolds. Let us just remark that, unlike a Riemannian structure, a symplectic structure does not always exist on a smooth manifold. Beside the restriction of even dimension, there is the limitation that any symplectic manifold has to be orientable. But even an orientable even-dimensional manifold possibly has no symplectic structure: S^{2n} (with $n > 1$) or any compact manifold M such that the cohomology $H^2(M, \mathbf{R}) = 0$, for example, have no symplectic structure (cf. chapter 1).

On the other hand, a smooth orientable closed Riemann surface of genus $g \geq 0$ equipped with the standard two-dimensional Riemann volume form (which is a closed non degenerate exterior 2-form) provides the probably most simple example of a symplectic manifold (besides the Euclidean space $(\mathbf{R}^{2n}, \omega_0)$ with the standard symplectic form ω_0, of course).

Some other important sources of symplectic manifolds are described in the following section.

2.2   The Cotangent Bundle

Starting with an n-dimensional smooth manifold M, our aim is to describe a natural symplectic form on the cotangent bundle T^*M, which is defined by $T^*M := \{\, \text{linear maps } f : T_qM \to \mathbf{R};\ q \in M \}$.

If $q = (q_1, \ldots, q_n)$ is a choice of local coordinates on $U \subseteq M$, then, for a fixed $q \in U$, a 1-form $\sum_{i=1}^n p_i\, dq_i$ on T_qM is determined by the coefficients $p_1, \ldots, p_n$. Local coordinates of an element $l \in T^*M$ are therefore $(\mathbf{p}, \mathbf{q}) = (p_1, \ldots, p_n, q_1, \ldots, q_n)$ (describing the so-called *Liouville form* $\mathbf{l} = \mathbf{p}d\mathbf{q} = \sum_{i=1}^n p_i\, dq_i$).

First, one defines a distinguished 1-form ϑ on T^*M: Let $X \in T_l(T^*M)$ be a vector tangent to the cotangent bundle at the point $l = (\mathbf{p}, \mathbf{q}) \in T^*M$. Under the derivative $\pi_* : T(T^*M) \to TM$ of the natural projection, the tangent vector $X \in T_l(T^*M)$ is mapped to the tangent vector $\pi_* X \in T_qM$. Define the 1-form ϑ on T^*M by the relation $\vartheta(X) = l(\pi_* X)$.

The exterior derivative $\omega := d\vartheta$ turns out to be a symplectic form on T^*M: indeed, it is closed and non-degenerate, as one easily checks using local coordinates. (Write $\vartheta = \sum_{i=1}^n (r_i dp_i + s_i dq_i)$ and $X = \sum_{i=1}^n (\zeta_i \frac{\partial}{\partial p_i} + \eta_i \frac{\partial}{\partial q_i})|_{(\mathbf{p},\mathbf{q})}$; applying $l = \sum_{i=1}^n p_i dq_i = \mathbf{p}d\mathbf{q}$ on $\pi_* X = \sum_{i=1}^n \eta_i \frac{\partial}{\partial q_i}|_{\mathbf{q}}$ and comparing with $\vartheta(X)$, one immediatly sees that $r_i = 0$ and $s_i = p_i\ \forall i \in \{1, \ldots, n\}$; so $\vartheta = \mathbf{p}d\mathbf{q}$ in local coordinates. Consequently $\omega = d\vartheta = d\mathbf{p} \wedge d\mathbf{q}$ and ω is clearly non-degenerate). The result is thus:

The cotangent bundle T^*M of an n-dimensional smooth manifold M has a natural symplectic structure ω, which, in the local coordinates $(\mathbf{p}, \mathbf{q})$ described above, takes the form $\omega = d\mathbf{p} \wedge d\mathbf{q}$.

Consequence: Every diffeomorphism $f : M \to M$ has a natural extension to T^*M, namely $\hat{f} := (f^{-1})^* = f^{*-1}$. The definition of $\hat{f} : T^*M \to T^*M$ does not depend on a choice of coordinates and $\hat{f}$ is symplectic. i.e. $\hat{f}^*\omega = \omega$. (Of course, if $l = (\mathbf{p}, \mathbf{q}) \in T^*M$, then $\hat{f}(l)$ is understood as the 1-form $(f^{-1})^*$ $(\mathbf{p}d\mathbf{q})$ on $T^*_{f(q)}M$.)

Proof. It is enough to show that $\hat{f}^*\vartheta = \vartheta$ (because $\hat{f}^*\omega = \hat{f}^*d\vartheta = d(\hat{f}^*\vartheta) = d\vartheta = \omega$). We have $\pi \circ \hat{f} = f \circ \pi$, and so $\pi_* \circ \hat{f}_* = f_* \circ \pi_*$. Then

$$
\begin{aligned}
(\hat{f}^*\vartheta)_l(X) &= \vartheta_{\hat{f}(l)}(\hat{f}_*X) = \hat{f}(l)(\pi_*(\hat{f}_*X)) = (f^{*-1}l)(\pi_*(\hat{f}_*X)) = \\
&= (f^{*-1}l)(f_*(\pi_*X)) = [f^*(f^{*-1}l)](\pi_*X) = l(\pi_*X) = \vartheta_l(X).
\end{aligned}
$$

$\square$

Note that not every noncompact symplectic manifold can be constructed as a cotangent bundle, because the symplectic form $\omega = d\mathbf{p} \wedge d\mathbf{q}$ is exact (i.e. of the form $\omega = d\vartheta$ for a certain 1-form ϑ).

2.3 Kähler Manifolds

As a part of the theory of Riemannian manifolds as well as a proper subject of great interest, Kähler manifolds have been investigated extensively and many important results have been obtained (cf. Weil [105], for instance). By definition, Kähler manifolds are symplectic and therefore provide a rich source of examples for symplectic manifolds.

Definition 2.4 *A complex structure on a $2n$-dimensional real vector space V is a real linear transformation $J : V \to V$ with $J^2 = -I$ (where $I : V \to V$ is the identity transformation).*

Note that V, endowed with a complex structure, may be understood as an n-dimensional complex vector space: it suffices to define the scalar multiplication by a complex number by means of $(x + iy)v := xv + yJv$ with $x, y \in \mathbf{R}$ and $v \in V$.

Conversely, an n-dimensional complex vector space can be interpreted as a $2n$-dimensional real vector space with a complex structure J: simply define $Jv = iv$ $(v \in V)$.

Remember that a n-dimensional *complex manifold* is a manifold modelled on $\mathbf{C}^n$ for which all coordinate transformations are biholomorphic. As a useful example we mention that on every tangent space T_pM of a n-dimensional complex manifold M, a complex structure J_p is induced in a natural way: if $\mathbf{z} = \mathbf{x} + i\mathbf{y} : U \to \mathbf{C}^n$ (with $\mathbf{z} = (z_1, \ldots, z_n) \in \mathbf{C}^n$ and $\mathbf{x} = (x_1, \ldots, x_n), \mathbf{y} = (y_1, \ldots, y_n) \in \mathbf{R}^n$) are local coordinates in a neighbourhood U of the point $p \in M$, then the linear transformation $J_p : T_pM \to T_pM$ (brought about by the multiplication by i) is given by

$$
J_p(\frac{\partial}{\partial x_j}|_p) := (\frac{\partial}{\partial y_j}|_p) \text{ and } J_p(\frac{\partial}{\partial y_j}|_p) := -(\frac{\partial}{\partial x_j}|_p) \quad (\text{ for } j = 1, \ldots, n).
$$

Moreover, the induced map $J : p \mapsto J_p$ defined on the complex manifold satisfies the *integrability condition*

$$h_* \circ J = J \circ h_*$$

for every coordinate transformation h. An equivalent condition is that h is holomorphic.

The just described complex structure J on a complex manifold is a special case of the important notion of almost complex structure explained in the following

Definition 2.5 *An almost complex structure J on a real differentiable manifold M is a C^∞-tensor field of type $(1,1)$ (i.e. an element in $\Gamma(T^{(1,1)}M)$) such that J_p, interpreted as a linear map : $T_pM \to T_pM$, has the property $J_p^2 = -I$ ($\forall p \in M$). Moreover, if the integrability condition is fulfilled, J is called a complex structure on M.*

A **Theorem of Newland and Nirenberg** states that an almost complex structure is a complex structure if and only if it has no torsion, i.e. $N(X,Y) := [JX, JY] - [X,Y] - J[X,JY] - J[JX,Y] = 0$ for vector fields X and Y (cf. [62] vol 2, p. 124). Let us come back to our main purpose:

Definition 2.6 *A Kähler vector space (V, ω, J) is a symplectic vector space (V, ω) together with a complex structure $J \in Sp(V)$ such that $\omega(v. Jv) > 0 \, \forall v \in V, v \neq 0$.*

Note that some authors (like Guillemin-Sternberg [51]) do not require $\omega(v, Jv)$ to be positive definite.

Definition 2.7 *A complex manifold M with symplectic structure ω is called a Kähler manifold, if at every point $p \in M$ the vector space (T_pM, ω_p, J_p) is Kählerian. (J_p as described above).*

We should mention another often encountered equivalent definition of Kähler manifolds:

Definition 2.8 *Let M be a complex manifold provided with a Hermitian metric h. M is a Kähler manifold, if $\omega(\cdot, \cdot) := h(J\cdot, \cdot)$ is a closed differential form on M.*

Remarks: 1. A Hermitian metric h is a Riemannian metric such that for every point $p \in M$, h_p is a J_p-invariant inner product on the $2n$-dimensional *real* vector space T_pM: $h_p(J_pv, J_pw) = h_p(v, w)$ ($\forall u, v \in T_pM$). It is worthwhile to note that h_p is the real part of a Hermitian scalar product (in the usual sense) on the n-dimensional *complex* vector space T_pM, namely $\langle \cdot . \cdot \rangle_h := h_p(\cdot, \cdot) + ih_p(J\cdot, \cdot)$.

2. A Hermitian metric h, which satisfies the conditions of the definition (i.e. $\omega(\cdot,\cdot) := h(J\cdot,\cdot)$ closed) is called a *Kähler metric*.

The equivalence of the two definitions of Kähler manifolds is a good occasion to discuss some more details about Kähler structures: On a Kähler vector space (V,ω,J) a bilinear form b is induced by

$$b(v,w) := \omega(v, Jw).$$

It is easy to check the properties of b listed below

- b is symmetric, i.e. $b(v,w) = b(w,v)$

- b is positive definite, i.e. $b(v,v) > 0 \quad \forall v \neq 0$ in V

- b is J-invariant, i.e. $b(Jv, Jw) = b(v,w)$

- b is non-degenerate, i.e. $b(v,w) = 0 \quad \forall v \in V \Rightarrow w = 0$

- and finally $\omega(v,w) = b(Jv, w)$.

On the other hand, we can start with a complex structure J on V and a non-singular, symmetric, positive definite and J-invariant bilinear form b. Then a symplectic form ω is induced by

$$\omega(v,w) = b(Jv, w)$$

which makes (V,ω,J) a Kähler vector space.

What are the conclusions? Starting with a complex structure $J \in Sp(V)$ on a vector space V, we can proceed in two different ways in order to have the same effect, namely:

- either we endow (V, J) with a symplectic structure ω, which turns (V,ω,J) into a Kähler vector space, provided that $\omega(v, Jv) > 0 \ \forall v \neq 0$ in V and $\omega(Ju, Jv) = \omega(u,v) \quad \forall u, v \in V$. As a consequence we get a Hermitian scalar product h by setting $h(v,w) := \omega(v, Jw)$.

- or we equip (V, J) with a Hermitian scalar product h. The consequences: by definition h has all the properties needed to define a symplectic form $\omega(v,w) := h(Jv, w)$ on (V, J); so (V,ω,J) is again a Kähler vector space.

If we now transfer these concepts to the tangent spaces of a complex manifold, we get exactly the two equivalent definitions of a Kähler manifold, with the exception of one difficulty occurring in the second case: starting with a Hermitian metric on M it is clear that the resulting ω is a non-degenerate differential form, which, however, is not necessarily closed. So the requirement of $d\omega = 0$ enters into the definition.

In practice, using the second definition, the question of course is, whether ω is closed or not. Let us cite a useful criterion due to Mumford [81] : G is supposed to be a group of diffeomorphisms of M preserving the complex structure and the Hermitian metric h. Moreover, let G_p denote the isotropy subgroup of $p \in M$, that is $G_p := \{\gamma \in G : \gamma(p) = p\}$. Note that $\rho_p(\gamma) := d\gamma_p$ defines a representation $\rho_p : G_p \to Aut_{\mathbf{C}}(T_pM)$ of G_p.

Criterion (Mumford) If $J_p \in \rho_p(G_p)$ for all $p \in M$, then $d\omega = 0$.

Proof. By assumption G preserves J and h. Therefore, ω and $d\omega$ are preserved too, that is to say

$$d\omega_p(\rho_p(\gamma)u, \rho_p(\gamma)v, \rho_p(\gamma)w) = d\omega_p(u, v, w)$$

for any $\gamma \in G_p$ and $u, v, w \in T_pM$. Take $\rho_p(\gamma) = J_p$ and apply this equation twice to get

$$\begin{aligned}
d\omega_p(u, v, w) &= d\omega_p(J_pu, J_pv, J_pw) = d\omega_p(J_p^2u, J_p^2v, J_p^2w) \\
&= d\omega_p(-u, -v, -w) = -d\omega_p(u, v, w) = 0.
\end{aligned}$$

Therefore $d\omega = 0$. $\square$

This criterion will be used in section 2.5, where we shall deal with the complex projective space $\mathbf{C}P^n$ as an example of a Kähler manifold.

Another useful framework, giving an equivalent condition for $d\omega = 0$, is the notion of *Riemannian connection*. Given a Riemannian manifold (M, g), let $\Gamma(TM)$ denote the set of all smooth vector fields on M. A Riemannian connection is a map which assigns to each $X \in \Gamma(TM)$ a linear mapping $\nabla_X : \Gamma(TM) \to \Gamma(TM)$ satisfying the following conditions

- $\nabla_{fX+gY} = f\nabla_X + g\nabla_Y \qquad$ for $f, g \in C^\infty(M)$

- $\nabla_X(fY) = f\nabla_X(Y) + (Xf)Y \qquad$ for $f \in C^\infty(M)$

- $\nabla_X Y - \nabla_Y X = [X, Y]$

- $Zg(X, Y) = g(\nabla_Z X, Y) + g(X, \nabla_Z Y)$.

For each $X \in \Gamma(TM)$ the map ∇_X can be uniquely extended to the algebra of all smooth tensor fields on M as a derivation which commutes with contractions. If T is a tensor field of type (r, s), then $\nabla_X T$ is again a (r, s) tensor field given by

$$\begin{aligned}
(\nabla_X T)(X_1, \ldots, X_s) &= \nabla_X(T(X_1, \ldots, X_s)) \\
&\quad - \sum_{i=1}^{s} T(X_1, \ldots, \nabla_X X_i, \ldots, X_s)
\end{aligned}$$

(see [62] or [53]).

Criterion The condition $d\omega = 0$ is equivalent to

$$\nabla_X J = 0 \qquad \text{for all } X \in \Gamma(TM),$$

where ∇ denotes the Riemannian connection with respect to the Hermitian metric h.

We give some indications about the proof. Two important formulas are necessary, namely:

$$3d\omega(X,Y,Z) = \nabla_X\omega(Y,Z) + \nabla_Y\omega(Z,X) + \nabla_Z\omega(X,Y) \qquad (2.5)$$
$$\nabla_Z\omega(X,Y) = h((\nabla_Z J)X,Y). \qquad (2.6)$$

For the first one consult [62, vol. 1, p. 149]. It is a special case of the well known fact that $d\omega$ is the alternation of $\nabla\omega$, if the torsion T^∇ vanishes (i.e. $T^\nabla(X,Y) := \nabla_X Y - \nabla_Y X - [X,Y] = 0$). For the second formula verify the following equalities:

$$
\begin{aligned}
\nabla_Z\omega(X,Y) &= \nabla_Z[\omega(X,Y)] - \omega(\nabla_Z X,Y) - \omega(X,\nabla_Z Y) \\
&= \nabla_Z[h(JX,Y)] - h(J\nabla_Z X,Y) - h(JX,\nabla_Z Y) \\
&= (\nabla_Z h)(JX,Y) + h(\nabla_Z JX,Y) - h(J\nabla_Z X,Y) \\
&= h(\nabla_Z JX - J\nabla_Z X,Y) = h((\nabla_Z J)X,Y)
\end{aligned}
$$

where $\nabla_Z h = 0$ (equivalent to the last property of a Riemannian connection) has been used. Now, what about the equivalence $d\omega = 0 \Leftrightarrow \nabla_X J = 0$?

$\Rightarrow$) We state that

$$2h((\nabla_X J)Y,Z) = 3d\omega(X,Y,Z) - 3d\omega(X,JY,JZ).$$

Then $d\omega = 0$ will immediatly imply $\nabla_X J = 0$. Start with the simple relation

$$
\begin{aligned}
h((\nabla_X J)Y,Z) &= h(\nabla_X(JY),Z) - h(J(\nabla_X Y),Z) \\
&= h(\nabla_X(JY),Z) + h(\nabla_X Y,JZ).
\end{aligned}
$$

Now apply to the two terms on the right the formula

$$
\begin{aligned}
2h(\nabla_X Y,Z) = {} & X(h(Y,Z)) + Y(h(X,Z)) - Z(h(X,Y)) \\
& + h([X,Y],Z) + h([Z,X],Y) + h(X,[Z,Y])
\end{aligned}
$$

(which is easily verified using the last two properties of Riemannian connections). Moreover, by means of formula (2.5) and Riemannian connection properties one computes:

$$
\begin{aligned}
3d\omega(X,Y,Z) = {} & X(\omega(Y,Z)) + Y(\omega(Z,X)) + Z(\omega(X,Y)) \\
& - \omega([X,Y],Z) - \omega([Z,X],Y) - \omega([Y,Z],X).
\end{aligned}
$$

Inserting all this into the stated formula we get a difference amounting to $h(N(Y,Z), JX)$. But the torsion N of a complex structure J is zero by the theorem of Newland and Nirenberg.

$\Leftarrow$) Conversely, we assume $\nabla_X J = 0$. Then $d\omega = 0$ follows immediately by the formulas (2.5) and (2.6). $\qquad\qquad\qquad\qquad\qquad\qquad\qquad\qquad\qquad\qquad\qquad$ $\square$

Remark: There are symplectic manifolds, which have no Kähler structure. Consult for instance [5].

As an illustrating example we consider now the unit ball in $\mathbf{C}^n$ equipped with the Bergman metric.

<u>Example</u> On the unit ball $B^n = \{|z| < 1\} \subset \mathbf{C}^n$ the Bergman kernel function is defined by

$$K_n(z.\zeta) := \frac{n!}{\pi^n} \frac{1}{(1 - z \cdot \bar{\zeta})^{n+1}} \qquad (z, \zeta \in B^n).$$

The Bergman metric derived from K and defined by

$$g \quad := \quad \sum_{i,j=1}^{n} g_{ij} dz_i d\bar{z}_j, \quad \text{with}$$

$$g_{ij}(z) \quad := \quad \frac{\partial^2}{\partial z_i \partial \bar{z}_j} \log K_n(z, z) \quad (i, j = 1, 2, \ldots, n)$$

has the important property that it is invariant under biholomorphic transformations. Indeed, the Bergman metric can be considered as a generalisation of the Poincaré metric. For $n = 1$ we have

$$g \quad = \quad g_{11} dz d\bar{z} = \frac{\partial^2}{\partial z \partial \bar{z}} (\log \frac{1}{\pi(1 - |z|^2)^2}) |dz|^2$$

$$= \quad \frac{\partial}{\partial z}(\frac{2z}{1 - |z|^2}) |dz|^2 = \frac{2 |dz|^2}{(1 - |z|^2)^2}.$$

We shall show that $h := \mathrm{Re}\, g$ is a Kähler metric. First calculate g_{ii} and g_{ij} from the definition:

$$g_{ii} \quad = \quad \frac{n + 1}{(1 - |z|^2)^2}(1 - |z|^2 + z_i \bar{z}_i)$$

$$g_{ij} \quad = \quad \frac{n + 1}{(1 - |z|^2)^2} \bar{z}_i z_j.$$

As $g_{ji} = \overline{g_{ij}}$, the bilinear form g proves to be a Hermitian scalar product on $T_z B^n$ for every $z \in B^n \subset \mathbf{C}^n$, if one can show that g is positive definite. Indeed, the matrix $(g_{ij})_{1 \le i,j \le n}$ is of the form

$$\frac{n + 1}{(1 - |z|^2)^2}(cI + A)$$

with $c = 1 - |z|^2 > 0$ and A the singular matrix

$$\begin{pmatrix} z_1\bar{z}_1 & z_1\bar{z}_2 & \cdots & z_1\bar{z}_n \\ z_2\bar{z}_1 & z_2\bar{z}_2 & \cdots & z_2\bar{z}_n \\ \cdots & \cdots & \cdots & \cdots \\ \cdots & \cdots & \cdots & \cdots \\ z_n\bar{z}_1 & z_n\bar{z}_2 & \cdots & z_n\bar{z}_n \end{pmatrix}$$

of rank 1. Calculating the corresponding characteristic polynomials it is easy to show that all the principal minors of $\frac{n+1}{(1-|z|^2)^2}(cI + A)$ are positive and we conclude that g is positive definite.

As we know, the real part h of g is a Hermitian metric (i.e., a J-invariant Riemannian metric) on $M := B^n$, now interpreted as a $2n$-dimensional real manifold with complex structure J (induced by multiplication with i). The last task is to check that $\omega(\cdot, \cdot) := h(J\cdot, \cdot)$ is a closed differential form or, by the second criterion, that $\nabla_X J = 0$ for every $X \in \Gamma(TM)$. Following [62], we extend our definition

$$h(X, Y) := \frac{1}{2}[g(X, Y) + g(Y, X)]$$

to complex vector fields X, Y, i.e. X and Y belong to the complexification $TM^{\mathbf{C}} = TM^{1,0} + TM^{0,1}$ of the tangent space and can be written in the form

$$X = \sum_k \xi_{1k}\frac{\partial}{\partial z_k} + \sum_k \xi_{2k}\frac{\partial}{\partial \bar{z}_k},$$

$$Y = \sum_k \eta_{1k}\frac{\partial}{\partial z_k} + \sum_k \eta_{2k}\frac{\partial}{\partial \bar{z}_k}.$$

Remember that

$$TM^{\mathbf{C}} = \{(a + ib)X;\ X \in TM \text{ and } a, b \in \mathbf{R}\}$$

and that $TM^{\mathbf{C}} = TM^{1,0} + TM^{0,1}$, where

$$TM^{1,0} = \{Z \in TM^{\mathbf{C}}; JZ = iZ\}$$
$$TM^{0,1} = \{Z \in TM^{\mathbf{C}}; JZ = -iZ\}.$$

For instance $\frac{\partial}{\partial z} \in TM^{1,0}$ and $\frac{\partial}{\partial \bar{z}} \in TM^{0,1}$. Obviously

$$h_{ij} := h\left(\frac{\partial}{\partial z_i}, \frac{\partial}{\partial z_j}\right) = 0$$

$$h_{i^*j^*} := h\left(\frac{\partial}{\partial \bar{z}_i}, \frac{\partial}{\partial \bar{z}_j}\right) = 0 \quad \text{and}$$

$$h_{ij^*} := h\left(\frac{\partial}{\partial z_i}, \frac{\partial}{\partial \bar{z}_j}\right) = h_{j^*i} = g_{ij}.$$

For simplicity use the notation $Z_i := \frac{\partial}{\partial z_i}$ and $Z_{i^*} := \frac{\partial}{\partial \bar{z}_i}$. Then we define the functions Γ_{ij}^k, $\Gamma_{ij}^{k^*}$ by

$$\nabla_{Z_i}(Z_j) = \sum_k \Gamma_{ij}^k Z_k + \sum_k \Gamma_{ij}^{k^*} Z_{k^*}.$$

Note that $\nabla_{Z_i}(Z_j) = \nabla_{Z_j}(Z_i)$ and so $\Gamma_{ij}^k = \Gamma_{ji}^k$. In a similar way we define $\Gamma_{i^*j}^k, \Gamma_{i^*j}^{k^*}, \ldots, \Gamma_{i^*j^*}^k. \Gamma_{i^*j^*}^{k^*}$.

Setting $X = Z_j. Y = Z_\gamma$ and $Z = Z_\alpha$ (where α, β, γ and, for later use, also δ are any indices. starred or not) in the standard formula

$$\begin{aligned}
2h(\nabla_X Y, Z) &= X(h(Y, Z)) + Y(h(X, Z)) - Z(h(X.Y)) \\
&+ h([X, Y], Z) + h([Z, X], Y) + h(X. [Z, Y])
\end{aligned}$$

and using the above notations we get

$$2 \sum_\delta h_{\alpha\delta} \Gamma_{\beta\gamma}^\delta = Z_\beta h_{\alpha\gamma} + Z_\gamma h_{\alpha\beta} - Z_\alpha h_{\gamma\beta}.$$

From this and the non-degeneracy of h we derive

$$\Gamma_{ij}^{k^*} = \Gamma_{i^*j}^k = \Gamma_{ij^*}^{k^*} = \Gamma_{i^*j^*}^k = 0.$$

In general, $\nabla_{Z_i}(JZ_j) = J\nabla_{Z_i}(Z_j) + (\nabla_{Z_i}J)Z_i$. However, $\Gamma_{ij}^{k^*} = 0$ implies that $J\nabla_{Z_i}(Z_j) = i\nabla_{Z_i}(Z_j) = \nabla_{Z_i}(JZ_j)$ (because $\nabla_{Z_i}(Z_j) \in TM^{1,0}$) and therefore $\nabla_{Z_i}J = 0$.

Using the other relations in an analogous way, we get $\nabla_X J = 0$ for any tangent vector X. Therefore $d\omega = 0$. $\qquad\qquad\qquad\qquad\square$

We close this section with an appendix describing

Some results about submanifolds

We intend to give a summary of some definitions and well-known facts, which are used later, especially in Chapters 6 and 7. For details and proofs we recommend Lawson's book on minimal submanifolds [66].

By a submanifold of a Riemannian manifold M we mean an immersion $\psi : N \to M$ where N is some differentiable manifold. N usually carries the induced Riemannian metric. With regard to minimal properties of submanifolds, we introduce the mean curvature vector field: Note first that for any p in a submanifold N of M we have an orthogonal splitting

$$T_p M = T_p N \oplus S_p N$$

into the tangent space $T_p N$ and the normal space $S_p N$ of N at p. Any tangent vector $X \in T_p M$ may be written as $X = (X)^T + (X)^S$.

If ∇ denotes the Riemannian connection on M, then the unique Riemannian connection ∇^N on N is defined in the following way: For tangent vector fields X, Y of N, both defined in a neighbourhood of a point $p \in N$, let

$$\nabla^N_X Y := (\nabla_X Y)^T$$

In an analogous manner one defines locally a normal vector field in a neighbourhood of p, namely

$$B_{X,Y} := (\nabla_X Y)^S$$

It is easy to see that $B_{X,Y} = B_{Y,X}$ depends only on the vectors X_p and Y_p, not on the choice of local fields X, Y. As a C^∞-section of the bundle $T^*N \otimes T^*N \otimes SN$, B is called the *second fundamental form* of the submanifold N. B_p is a symmetric bilinear map of T_pN into S_pN, which enables us to define

$$K_p := \text{trace } (B_p)$$

for each p. K is a smooth field of normal vectors on N and is called the *mean curvature vector field*.

As a rough interpretation, the mean curvature vector field may be considered as the gradient of the area function on the space of immersions $\psi : N \to M$. Indeed, more information on this fact would be given by an investigation of smooth variations of submanifolds. Nevertheless, we have the idea of a motivation for the following definition: An immersion $\psi : N \to M$ is called a *minimal submanifold* if $K = 0$.

Definition 2.9 *By a* complex submanifold *of a complex manifold M we mean an immersion $\psi : (N, J^N) \to (M, J)$ of a complex manifold (N, J^N) satisfying the condition*

$$J \circ d\psi_p = d\psi_p \circ J^N \qquad \text{for all } p \in N.$$

Proposition 2.10 *Every complex submanifold of a Kähler manifold is Kählerian in the induced metric and minimal.*

For the proof consult [66]. Of course with this proposition it is easy to write down a great number of Kähler manifolds or, in other words, to give some more examples of symplectic manifolds. In addition, with regard to an important property of holomorphic curves used in Chapters 6 and 7 we formulate the following

Proposition 2.11 (Wirtinger's Inequality) *Let (M, ω, μ) be a Kähler manifold with symplectic structure ω and hermitian metric μ, $\psi : N \to M$ any $2n$-dimensional real submanifold. At any point $p \in N$ we denote the volume form of the induced metric on N by $(\sigma_{\psi \cdot \mu})_p$. Then the pull back $\psi^*\omega^n$ of $\omega^n = \omega \wedge \ldots \wedge \omega$ satisfies*

$$\frac{(\psi^*\omega^n)_p}{n!} \leq (\sigma_{\psi \cdot \mu})_p$$

and equality holds if and only if T_pN is a complex subspace of T_pM.

For the special case that N is compact (possibly with boundary) we get as an immediate consequence that every complex submanifold N with $\dim_{\mathbf{R}} N = 2n$ minimizes the volume (with respect to the induced metric) among all submanifolds of dimension $2n$ homologous to N in M. Note that a submanifold N' will have volume equal to N if and only if $\psi' : N' \to M$ is also a complex submanifold.

Remark: In Chapters 6 and 7 we will see that pseudoholomorphic curves are area minimizing.

2.4   Coadjoint Orbits

Our third class of symplectic manifolds are coadjoint orbits. A good reference, especially for people interested also in physical aspects of coadjoint orbits, is the book of Guillemin-Sternberg [51]. Again, the intention of this section can only be to give a short introduction into the constructions around coadjoint orbits and to present a central theorem due to Kostant-Souriau. In order to show the aim of our following investigation, we formulate here the result of Kostant-Souriau:

Given a Lie group G with corresponding Lie algebra g. If $H^1(g) = H^2(g) = \{0\}$ (where $H^k(g)$ denotes the kth cohomology of g), then there is (up to coverings) a one-one relation between homogeneous symplectic manifolds for G (that means homogeneous manifolds of the form G/H, which are symplectic) and G-orbits in the dual space g^* of g.

With this introducing statement, let us pass to some

Definitions and Notions

In this section, G always denotes a Lie group, and its Lie algebra is $g := T_e G$. For any $a \in G$ the left multiplication $l_a : G \to G$ is given by $l_a(c) := ac$. The right multiplication is $r_a : G \to G$ with $r_a(c) := ca$.

A vector field X on G is called *left-invariant*, if $l_{a*}X = X \circ l_a$ $(\forall a \in G)$ and the vector space of left invariant vector fields is isomorphic to g. Similarly, a differential form ω on G is called *left invariant*, if $l_a^*\omega = \omega$ $(\forall a \in G)$ and the vector space of left invariant differential q-forms is isomorphic to $\bigwedge^q(g^*)$, the space of alternating q-forms on g. We may thus identify the space of left invariant q-forms with $\bigwedge^q(g^*)$. The exterior derivative then induces a linear map δ from $\bigwedge^q(g^*)$ to $\bigwedge^{q+1}(g^*)$.

The Lie derivative with respect to the vector field X with local flow φ_t of a differential form ω and a vector field Y, respectively, is defined by

$$\mathcal{L}_X\omega := \frac{d}{dt}(\varphi_t^*\omega)\big|_{t=0} \quad \text{and} \quad (\mathcal{L}_X Y)_a := \frac{d}{dt}(\varphi_{-t*}Y_{\varphi_t(a)})\big|_{t=0}$$

and has the following properties:

- $\mathcal{L}_X = i(X) \circ d + d \circ i(X)$ where i denotes interior multiplication and d the exterior derivative.

- $\mathcal{L}_X \circ d = d \circ \mathcal{L}_X$

- $\mathcal{L}_X(\omega(Y_1, Y_2)) = (\mathcal{L}_X\omega)(Y_1, Y_2) + \omega(\mathcal{L}_X Y_1, Y_2) + \omega(Y_1, \mathcal{L}_X Y_2)$

- for $f \in C^\infty(G)$ we have $\mathcal{L}_X f = X(f)$

- $\mathcal{L}_X Y = [X, Y]$

The kth cohomology of g is defined by

$$H^k(g) := Z^k(g)/B^k(g)$$

where $B^k(g) = \delta(\bigwedge^{k-1}(g^*))$ and $Z^k(g) = \ker \delta \subseteq \bigwedge^k(g^*)$.

We come now to the definition of the coadjoint representation: Let $A(a)$ be the inner automorphism given by $A(a) := (l_{a^{-1}} \circ r_a) : G \to G$. By setting $Ad_a := (A(a))_{*e} : g \to g$, we introduce the usual *adjoint representation*

$$Ad : G \to Aut(g).$$

In a similar way one gets the *coadjoint representation*

$$Ad^{\#} : G \to Aut(\textstyle\bigwedge^q(g^*))$$

defined by $Ad^{\#}\omega := (l_{a^{-1}} \circ r_a)^* \omega = (r_a^* \circ l_{a^{-1}}^*)\omega = r_a^*\omega$ (the last equation holds because $\omega \in \bigwedge^q(g^*)$ is left-invariant). Our next purpose is to speak about

Orbits in $\bigwedge^2(g^*)$

Let (M, Ω) be a symplectic manifold, on which we have a left action of the Lie group G, that is a differentiable map: $G \times M \to M$. From this we derive two different maps:

$$
\begin{aligned}
\varphi_a : M &\to M \quad \text{given by } \varphi_a(m) := am \quad (\text{ for any } a \in G) \\
\psi_m : G &\to M \quad \text{given by } \psi_m(a) := am \quad (\text{ for any } m \in M).
\end{aligned}
$$

If Ω is invariant with respect to φ_a^*, i.e.

$$\varphi_a^*\Omega = \Omega \qquad \forall a \in G,$$

then $(a, m) \mapsto am$ is called a *symplectic action*. The following relations are easy to verify:

$$
\begin{aligned}
\varphi_b \circ \psi_m &= \psi_m \circ l_b \\
\psi_{am} &= \psi_m \circ r_a.
\end{aligned}
$$

Theorem 2.12 *(a)   A symplectic action $G \times (M,\Omega) \to (M.\Omega)$ defines a G-morphism $\Psi : M \to Z^2(g)$ given by*

$$\Psi(m) := \psi_m^* \Omega.$$

That Ψ is a G-morphism means that $\Psi(am) = Ad_a^{\#} \Psi(m)$.

(b) $\Psi(M)$ is a union of G-orbits in $Z^2(g^*)$. If the action is transitive, then the image of Ψ consists of a single orbit.

Proof. (a) $\psi_m^* \Omega$ is left-invariant:

$$
\begin{aligned}
l_b^* \iota_m^* \Omega &= (\psi_m \circ l_b)^* \Omega = (\varphi_b \circ \psi_m)^* \Omega \\
&= \psi_m^* \circ \varphi_b^* \Omega = \psi_m^* \Omega.
\end{aligned}
$$

$\Psi(am) = Ad_a^{\#} \Psi(m)$ because

$$\psi_{am}^* \Omega = (\iota_m \circ r_a)^* \Omega = r_a^* \psi_m^* \Omega = r_a^* \Psi(m) = Ad_a^{\#} \Psi(m).$$

$\Psi(m)$ is closed, as $d(\Psi(m)) = d(\psi_m^* \Omega) = \psi_m^*(d\Omega) = 0$.

(b)   The orbit through the point $\Psi(m)$ is given by

$$\Gamma(m) = \{\Psi(am) = Ad_a^{\#} \Psi(m); a \in G\}.$$

We have $\Psi(M) = \bigcup_{m \in M} \Gamma(m)$. Obviously a transitive action of G gives a single orbit and $\Psi(M) = \Gamma(m)$. $\square$

A natural question in the situation at hand is, of course, the following: Given a G-orbit in $Z^2(g)$, say $\Gamma(\omega) = \{Ad_a^{\#}\omega; a \in G\}$. does there exist a symplectic manifold M and a map $\Psi : M \to Z^2(g)$ (as above) such that $\Gamma(\omega) = \Psi(M)$?

An affirmative answer will be given in the course of the following constructions. Before, however. let us remember some elementary

Notions and Results from Lie Group Theory

Definitions

1. *Let M be a n-dimensional manifold. A c-dimensional distribution Δ $(1 \le c \le n)$ on M is given by a map which attributes to each $m \in M$ a c-dimensional subspace $\Delta(m)$ of the tangent space $T_m M$.*

2. *Δ is called a* smooth distribution *if for every $m \in M$ there exists a neighbourhood $U(m)$ and C^{∞} vector fields $Z_1, \ldots, Z_c$ on $U(m)$ such that for each point p in $U(m)$ the vectors $Z_{1p}, \ldots, Z_{cp}$ are a basis for $\Delta(p)$.*

3. Δ *is called* involutive *if it is smooth and if* $[X,Y] \in \Delta$ *for all* $X, Y \in \Delta$. *($Z \in \Delta :\Leftrightarrow Z_m \in \Delta(m)$ $\forall m \in M$).*

4. *A submanifold* $(N, \imath)$ *of* M *is called an* integral manifold *of* Δ *on* M *if* $\imath_*(T_q N) = \Delta(\imath(q))$ *($\forall q \in N$).*

Theorem 2.13 (Frobenius) *Let* Δ *be a c-dimensional smooth distribution on* M.

1. Δ *is involutive if and only if every point of* M *lies in an integral manifold of* Δ.

2. *Locally there exist coordinates* $x_1, \ldots, x_n$ *such that an integral manifold is of the form* $\{x; \; x_i = \mathrm{const}_i \; for \; i = c+1, \ldots, n\}$.

Theorem 2.14 *Given a Lie group* G *with Lie algebra* g *and a subalgebra* $\tilde{h} \subset g$. *Then there exists (up to a Lie group isomorphism) exactly one connected Lie subgroup* $(H, \imath)$ *of* G *such that* $\imath_*(h) = \tilde{h}$, *where* h *is the Lie algebra of* H.

Coming back to our question we shall now. starting with a fixed orbit $\Gamma(\omega)$, give the

Construction of a Symplectic Manifold M, corresponding to $\Gamma(\omega)$

The idea of the construction consists in realizing M as a homogeneous space of the form G/H, where H is a closed Lie subgroup of G. The problem will be the determination of H in such a way that a symplectic form $\bar{\omega}$ on G/H exists, which, under the projection $\rho : G \to G/H$, is pulled back to $\rho^*\bar{\omega} = \omega$, the given form in $Z^2(g^*)$.

Given the closed 2-form $\omega \in Z^2(g)$ on G, we set $h_\omega := \{X \in g; i(X)\omega = 0\}$. Note that if ω is non-degenerate, then $h_\omega = \{0\}$.

Lemma 2.15 h_ω *is a subalgebra.*

Proof. Let $X, Y \in h_\omega$. Since ω is closed, we have

$$\mathcal{L}_Y \omega = d(i(Y)\omega) + i(Y)d\omega = 0.$$

Hence the computation

$$
\begin{aligned}
0 = \mathcal{L}_Y(\omega(X,Z)) &= (\mathcal{L}_Y\omega)(X,Z) + \omega(\mathcal{L}_Y X, Z) + \omega(X, \mathcal{L}_Y Z) \\
&= \omega([Y,X], Z)
\end{aligned}
$$

shows that $[Y, X] \in h_\omega$. $\qquad\square$

The next step is to define a distribution Δ: At the origin we set $\Delta(e) := h_\omega$ and for any other point $m \in M$ we define, using left translation.

$$\Delta(m) := l_{m*}(h_\omega).$$

The point is now that Δ is involutive. This statement follows immediately from the lemma: If X, Y are two vector fields in Δ, then $X_m = l_{m*}X_e$ and $Y_m = l_{m*}Y_e$; so

$$\begin{aligned}
[X, Y]\,|_m &= [l_{m*}X_e, l_{m*}Y_e] = l_{m*}[X_e, Y_e] \\
&\in l_{m*}(h_\omega) = \Delta(m).
\end{aligned}$$

Using now the one-one relation between Lie subalgebras and connected Lie subgroups described in Theorem 2.14 we get the existence of exactly one connected Lie subgroup (H_ω, i) of G such that $i_*(T_e(H_\omega)) = h_\omega$. At the same time, we know from Frobenius' theorem 2.13 that through each point of M there exists an integral manifold, as Δ is involutive. We know even more: The integral manifold through the point $a \in G$ is (aiH_ω, id_G). Moreover, the right cosets aiH_ω of iH_ω are the leaves of a foliation of G.

The verification is easy: (H_ω, i) is the integral manifold through the origin e by construction: $i_*(T_e(H_\omega)) = h_\omega = \Delta(e)$. But we also have $T_a(aiH_\omega) = l_{a*}T_e(iH_\omega) = l_{a*}h_\omega = \Delta(a)$, and so (aiH_ω, id_G) is the integral manifold through the point $a \in G$. For simplicity we write aH_ω instead of (aiH_ω, id_G). Note that the cosets aH_ω are, of course, connected, as H_ω is connected, and that we have local coordinates $x_1, \ldots, x_n$ such that in an appropriate neighbourhood aH_ω is of the form $\{x;\ x_i = \mathrm{const}_i \text{ for } i = 1, \ldots, k\}$ (where $k = n - \dim h_\omega$).

We extend our results by the following important

Theorem 2.16 *If H_ω is closed, then there exists exactly one symplectic form $\bar\omega$ on $M_\omega := G/H_\omega$ with $\omega = \rho^*\bar\omega$, where $\rho : G \to M_\omega$ is the canonical projection.*

Remarks: That H_ω to be closed is necessary to ensure that G/H_ω is a differentiable manifold. The fibres of the projection ρ are just the leaves of the foliation and at the same time the integral manifolds.

Proof (Sketch). (For details consult [51].) We begin with the local description of the integral manifold, following the theorem of Frobenius: There exist local coordinates $x_1, \ldots, x_n$ such that the integral manifolds are (in an appropriate neighbourhood) of the form

$$\{x;\ x_i = \mathrm{const}_i \text{ for } i = 1, \ldots, k\} \quad (k = n - \dim h_\omega).$$

For the tangent space we have the basis $\frac{\partial}{\partial x_{k+1}}, \ldots, \frac{\partial}{\partial x_n}$. and the symplectic form on G may be written as

$$\omega = \sum_{i<j}^{n} a_{ij}(x)dx_i \wedge dx_j.$$

Step 1: ω does not depend on $x_{k+1}, \ldots, x_n$. Indeed, $a_{ij} = 0$ if $i > k$ or $j > k$ and a_{ij} does not depend on $x_{k+1}, \ldots, x_n$ for $1 \leq i < j \leq k$, because $d\omega = 0$. Therefore we have

$$\omega = \sum_{i<j}^{k} a_{ij}(x_1, \ldots, x_k)dx_i \wedge dx_j.$$

Step 2: As the integral manifolds correspond to the fibres of $\rho : G \to M_\omega$, we can use $x_1, \ldots, x_k$ as local coordinates for M_ω. $\rho(x_1, \ldots, x_n) = (x_1, \ldots, x_k)$. Define

$$\bar{\omega} = \sum_{i<j}^{k} a_{ij}(x_1, \ldots, x_k)dx_i \wedge dx_j.$$

It is easy to check the following properties: $\rho^*\bar{\omega} = \omega$, $\bar{\omega}$ is closed and $\bar{\omega}$ is non-degenerate.

Step 3: $\bar{\omega}$ is well defined and uniquely determind on the whole manifold M_ω. This follows from the fact, that $\rho^{-1}(m)$ is a leaf and therefore connected. So we see that any two points in the same preimage $\rho^{-1}(m)$ will give the same form $\bar{\omega}_m$.

$$\square$$

What have we won in regard to our question? Is M_ω really the manifold corresponding to the given orbit $\Gamma(\omega)$? By setting $m := H_\omega$ we have the relations

$$\rho(a) = aH_\omega = am = \psi_m(a) \qquad (\text{so } \rho = \psi_m).$$

As in Theorem 2.12 we define

$$\Psi(m) := \psi_m^*\bar{\omega}.$$

Therefore $\Psi(m) = \psi_m^*\bar{\omega} = \rho^*\bar{\omega} = \omega$. Of course, Ψ is a G-morphism: $\Psi(am) = Ad_a^{\#}\Psi(m) = Ad_a^{\#}\omega$, and so

$$\Psi(M_\omega) = \Gamma(\omega)$$

because G acts transitively on M_ω. It is important to point out that the construction *only works if H_ω is closed* !

Let us turn now to the question of uniqueness. In other words: we ask whether there are homogeneous

Symplectic Manifolds with the same Orbit

We start with the homogeneous manifold $M = G/H$, where H is a closed Lie subgroup (which may be non-connected!) of the Lie group G. The Lie algebra of H is denoted by h. On the manifold M we give a symplectic form Ω and require a transitive symplectic action $G \times M \to M$ with the induced map $\psi_{bH} : G \to M$ given by $\psi_{bH}(a) = abH$ $(\forall b \in G)$. Theorem 2.12 provides the mapping $\Psi : M \to Z^2(g)$, and we define $\omega := \Psi(H) = \iota_H^* \Omega$. Then $\Psi(M) = \{Ad_a^\# \omega; a \in G\}$ consists of only one orbit and we set $\Gamma(\omega) := \Psi(M)$.

The question is now the following: If we go back from $\Gamma(\omega)$ to M_ω, exactly as we have done above, do we get $M_\omega = M$? Following the construction above let us go back step by step and first calculate

$$\begin{aligned} h_\omega &= \{X \in g; i(X)\omega = 0\} \\ &= \{X \in g; \psi_H^* \Omega(X, \cdot) = 0\} \\ &= \{X \in g; \psi_{H*}X = 0\}. \end{aligned}$$

We recognize that $h = h_\omega$, because

$$\psi_{H*}|_g : g = T_e G \to T_H M = g/h, \qquad X \mapsto X + h$$

and so $\psi_{H*}X = 0 \Leftrightarrow X \in h$. Now H_ω is the connected component of H and so $M_\omega = G/H_\omega$ is a covering of $M = G/H$. So the result is:

Proposition 2.17 *Homogeneous symplectic manifolds G/H are, up to coverings, parametrized by G-orbits in $Z^2(g)$, provided that ω produces a closed H_ω in the above construction.*

The Theorem of Kostant-Souriau

After all these preparations there is only a small step from the above result to the

Theorem 2.18 (Kostant-Souriau) *If $H^1(g) = H^2(g) = \{0\}$, then there is, up to coverings, a one-one relation between homgeneous symplectic manifolds for G and G-orbits in g^*.*

Proof. Since $H^2(g) = \{0\}$, for every $\omega \in Z^2(g)$ there exists a β in g^* with $d\beta = \omega$. Moreover, $H^1(g) = \{0\}$ is equivalent to $Z^1(g) = B^1(g) = \delta(\bigwedge^0(g^*)) = \{0\}$. So if $\beta_1, \beta_2 \in g^*$ are such that $d\beta_1 = d\beta_2$, it follows immediately that $\beta_1 = \beta_2$. So far we have proved that to any $\omega \in Z^2(g)$ there exists one and only one $\beta \in g^*$ such that $\omega = d\beta$.

We are able to establish now a one-one correspondence between G-orbits in $Z^2(g)$ and G-orbits in g^*, that means between

$$\begin{aligned} \mathrm{orb}_G(\omega) &:= \{Ad_a^\# \omega; a \in G\} \quad \text{and} \\ \mathrm{orb}_G(\beta) &:= \{Ad_a^\# \beta; a \in G\}. \end{aligned}$$

Obviously the mapping $Ad_a^\# \beta \mapsto Ad_a^\# \omega$ $(\forall a \in G)$ is bijective, because

$$d(Ad_a^\# \beta) = dr_a^* \beta = r_a^* \omega = Ad_a^\# \omega.$$

As the last step, it is enough to prove now the following statement: If we have the unique relation between $\omega \in Z^2(g)$ and $\beta \in g^*$ such that $\omega = d\beta$, then H_ω is the connected component of the (closed) isotropy group

$$G_\beta := \{a \in G : Ad_a^\# \beta = \beta\}$$

(and therefore H_ω is closed itself). The confirmation of this statement is not too hard: The Lie subalgebra corresponding to the isotropy group G_β is

$$g_\beta = \{X \in g : \mathcal{L}_X \beta = 0\}.$$

$Ad_a^\# \beta = \beta \Leftrightarrow r_a^* \beta = \beta$ with $a = \exp(tX)$. Differentiation at time zero gives $\mathcal{L}_X \beta = 0$ (cf. [2] p. 268, for instance). If $X \in g$ then $\mathcal{L}_X \beta = i(X)d\beta + d(i(X)\beta) = i(X)d\beta = i(X)\omega$ (note that $i(X)\beta$ is locally constant, because β and X are both left-invariant). So $X \in h_\omega \Leftrightarrow \mathcal{L}_X \beta = 0$. Therefore $h_\omega = g_\beta$. Clearly H_ω is the connected component of G_β and consequently closed. $\square$

Remark: There are other criteria whether H_ω is closed or not. We just mention a result of Chu: If G is simply connected, then H_ω is closed.

At the end of this chapter just one additional result, which provides a greater ability in practice to handle the symplectic form on a coadjoint orbit:

Theorem 2.19 *Given a Lie group G such that $H^1(g) = H^2(g) = \{0\}$. Let* $\mathrm{orb}_G(\beta)$ *denote the symplectic manifold and let* $\omega := d\beta$. *Then the symplectic form $\bar{\omega}$ on* $\mathrm{orb}_G(\beta) \cong G/H_\omega$ *can be calculated by*

$$\bar{\omega}(X_\beta, Y_\beta) = -\beta([X, Y])$$

where $X_\beta = \rho_* X$ and $Y_\beta = \rho_* Y$ with $X, Y \in g$ and $\rho : G \to G/H_\omega$.

Proof. $\bar{\omega}$ is the symplectic form on G/H_ω with $\rho^* \bar{\omega} = \omega$, uniquely determined by Theorem 2.16. We have

$$\bar{\omega}(X_\beta, Y_\beta) = \bar{\omega}(\rho_* X, \rho_* Y) = \rho^* \bar{\omega}(X, Y) = \omega(X, Y).$$

Further $-\beta([X, Y]) = d\beta(X, Y) = \omega(X, Y)$, because

$$
\begin{aligned}
0 = \mathcal{L}_X(\beta(Y)) &= (\mathcal{L}_X \beta)(Y) + \beta(\mathcal{L}_X Y) \\
&= (i(X)d\beta)(Y) + (di(X)\beta)(Y) + \beta([X, Y]) \\
&= d\beta(X, Y) + \beta([X, Y]).
\end{aligned}
$$

$\square$

An illustrating example for a symplectic manifold realized by a coadjoint orbit will be given in the next section.

2.5   An Example: CP^n as a Symplectic Manifold

One classical example of a symplectic manifold is the complex projective space. We close this chapter with an illustration of two different ways of construction of a symplectic structure on CP^n: We first show that the complex projective space supports a Kähler structure. Later, we shall see how a realization of CP^n as a symplectic manifold is possible by means of the coadjoint orbit theory.

Let us first define the Hermitian scalar product on CP^n, which of course is induced by the usual Hermitian inner product $\langle \cdot, \cdot \rangle$ on $\mathbf{C}^{n+1}$ via the projection $\pi : \mathbf{C}^{n+1} \setminus \{0\} \to CP^n$ sending $z \in \mathbf{C}^{n+1} \setminus \{0\}$ into the line through z: If $\eta_1, \eta_2 \in T_{\pi(z)} CP^n$, then the Hermitian scalar product is given by

$$\ll \eta_1 \cdot \eta_2 \gg := \frac{\langle \xi_1, \xi_2 \rangle \langle z, z \rangle - \langle \xi_1, z \rangle \langle z, \xi_2 \rangle}{\langle z, z \rangle}$$

where ξ_1, ξ_2 are any vectors in $T_z \mathbf{C}^{n+1}$ satisfying $\pi_* \xi_k = \eta_k \in T_{\pi(z)} CP^n$.

For better understanding we give some details about π_* and the tangent space $T_{\pi(z)} CP^n$ (with $z \in \mathbf{C}^{n+1}$): Let W be the orthogonal complement to z in $\mathbf{C}^{n+1}$ (with respect to the Hermitian scalar product $\langle \, , \, \rangle$). Identifying $\mathbf{C}^{n+1}$ with $T_z \mathbf{C}^{n+1}$, W may be considered as a subspace of $T_z \mathbf{C}^{n+1}$. Therefore an element $\xi \in T_z \mathbf{C}^{n+1}$ has the form $\xi = cz + \eta \in T_z \mathbf{C}^{n+1}$ for a unique $c \in \mathbf{C}$ and a unique vector $\eta \in W$. Clearly $\pi_*(cz + \eta) = \pi_*(\eta)$ is a point in $T_{\pi(z)} CP^n$, independent of $c \in \mathbf{C}$. Consequently π_* maps W isomorphically onto $T_{\pi(z)} CP^n$, as expected. The constant c is calculated to be $\langle \xi, z \rangle \cdot \langle z, z \rangle^{-1}$, which implies

$$\eta = \frac{\langle z, z \rangle \xi - \langle \xi, z \rangle z}{\langle z, z \rangle}.$$

This motivates the definition of $\ll \cdot, \cdot \gg$. Now $h(\cdot, \cdot) := \mathrm{Re} \ll \cdot, \cdot \gg$ is a Hermitian metric in the sense of the definition in section 2.3. namely a J-invariant Riemannian metric.

With regard to Mumford's criterion presented in section 2.3. we are interested in a group G of diffeomorphisms of CP^n preserving the complex structure as well as the Hermitian metric h. Evidently, the classical diffeomorphism $CP^n \cong SU(n+1)/U(n)$ in mind, we choose G to be the group $SU(n+1)$, which acts in the right way on CP^n, that is to say as a group of holomorphic diffeomorphisms preserving the metric h. The isotropy subgroup of $\pi(z) \in CP^n$ is

$$(SU(n+1))_{\pi(z)} \cong U(W).$$

As above W is the orthogonal complement of z in $\mathbf{C}^{n+1}$, whereas $U(W)$ denotes the unitary group of W. The representation $\rho_{\pi(z)}$ used in Mumford's criterion maps $(SU(n+1))_{\pi(z)}$ to $U(W)$ and, as $J_{\pi(z)}$ on $T_{\pi(z)} CP^n$ is nothing else than multiplication by i on W. the hypotheses of the criterion are fulfilled. Therefore $\omega(\cdot, \cdot) := h(J\cdot, \cdot)$ is closed. We conclude that (CP^n, h) is a Kähler manifold.

What about the realization of $\mathbf{C}P^n$ as a coadjoint orbit? The Lie group certainly is

$$G := SU(n+1) = \{S \in SL(n+1,\mathbf{C}):\ S\bar{S}^T = E\}$$

with Lie algebra given by

$$g := su(n+1) = \{A \in gl(n+1,\mathbf{C}):\ A = -\bar{A}^T \text{ and } trace(A) = 0\}.$$

Introducing an appropriate real scalar product $(\cdot,\cdot)$ on g

$$(A,B) := Re[trace(A\bar{B}^T)]$$

we get a one-one correspondence between g and its dual space g^*, given by the condition

$$c^A(B) := (A,B) \qquad (\text{ with } A,B \in g \text{ and } c^A \in g^*).$$

From now on we identify $A \in su(n+1)$ with $c^A \in g^*$.

Consider the orbit $\mathrm{orb}_G(B) := \{Ad_a^\# B;\quad a \in G\}$ for a special $B \in su(n+1)$, namely $B := i \cdot E_-$ with

$$E_- := \begin{pmatrix} E_n & 0 \\ 0 & -1 \end{pmatrix} \qquad (E_n = \frac{1}{n} \text{ times the identity}).$$

$\mathrm{orb}_G(B) = \{a^{-1}Ba;\ a \in G\} = \{i \cdot a^{-1}E_- a:\ a \in G\}$ is an orbit in g^*, and we claim that it represents $\mathbf{C}P^n$.

First recognize that $a^{-1}E_- a = E_-$ if and only if a is of the form

$$\begin{pmatrix} T & 0 \\ 0 & (detT)^{-1} \end{pmatrix} \qquad \text{with } T \in U(n).$$

Therefore $\mathrm{orb}_G(B)$ is isomorphic to $SU(n+1)/U(n) \cong \mathbf{C}P^n$ and we have realized $\mathbf{C}P^n$ as an orbit in g^*. Using Theorem 2.19, a symplectic form on $\mathbf{C}P^n$ is given by

$$\omega(X,Y) := (-B,[A^X,A^Y]),$$

where $[\cdot,\cdot]$ denotes the usual commutator and $A^X, A^Y \in g$ are any tangent vectors with $\rho_* A^X = X$ and $\rho_* A^Y = Y$ for the projection $\rho : SU(n+1) \to SU(n+1)/U(n)$.

3 Generating Functions

If S is a real valued function defined on the manifold N, then its differential dS can be considered as a mapping $dS : N \to T^*N$ from N into its cotangent space. The image L_S of this mapping is then an exact Lagrangian submanifold of T^*N:

$$(dS)^*(pdq) = dS$$

and consequently $(dS)^*(dp \wedge dq) = 0$. The function S is called a naive generating function for L_S.

Given a symplectic mapping $\varphi : \mathbf{R}^{2n} \to \mathbf{R}^{2n}$, $\varphi(x,y) = (X,Y)$, its graph Γ_φ is a Lagrangian submanifold of $(\mathbf{R}^{2n} \times \mathbf{R}^{2n}, -\omega_0 \ominus \omega_0)$. Make the symplectic variable transform $h : \mathbf{R}^{4n} \to \mathbf{R}^{4n}$

$$(q, Q, p, P) = (\frac{x+X}{2}, \frac{y+Y}{2}, y - Y, X - x).$$

Then $\tilde{\Gamma}_\varphi = h(\Gamma_\varphi)$ can be considered as a Lagrangian submanifold of $T^*\mathbf{R}^{2n} \cong \mathbf{R}^{4n}$. (If $\varphi = id$, then $\tilde{\Gamma}_\varphi$ is the zero section.) The question arises whether $\tilde{\Gamma}_\varphi$ can be described by a generating function. If there exists a naive generating function S for $\tilde{\Gamma}_\varphi$, then the symplectic mapping φ is determined by this function.

In general, naive generating functions can only be constructed locally. If the notion of generating function is taken in the sense defined in section 3.3, then it turns out that generating functions (quadratic at infinity) exist at least for symplectic mappings isotopic to the identity (see section 3.4 for the precise statement and proof). In this way, symplectomorphisms are replaced by functions. Morse theory then provides an efficient tool for analysing the topology of the level sets of the generating functions. The critical points of the generating function for $\tilde{\Gamma}_\varphi$ are in bijective correspondence with the fixed points of the symplectomorphism φ so that the topology also provides information about the fixed points of φ (see e.g. Hofer's theorem 3.16). Furthermore certain invariants for the symplectomorphisms can be defined via the topology of the level sets of the generating function. These invariants are due to Viterbo (section 3.7). They lead to a notion of capacity for subsets of $\mathbf{R}^{2n}$ (see section 3.8).

3.1 Minimax Principle and Lusternik-Schnirelman Theory

In this section we consider a smooth function $f : M \to \mathbf{R}$, where M is a smooth compact manifold or, more generally, M is a finite dimensional vector bundle over a compact base manifold B and f is *quadratic at infinity*, i.e. outside a compact subset of M, the restriction of f to each fibre is a nondegenerate quadratic form. By $\mathrm{Crit}(f)$ we denote the closed (hence compact) set of critical points of f. Lusternik-Schnirelman theory provides a lower bound on the number of critical points of f. It is an application of the minimax principle which itself has many other useful applications. Our exposition will follow quite closely the elementary parts of [84].

The minimax principle will be proved with the aid of a deformation theorem. For this we need some notation and a lemma. Let

$$f^c = \{\, x \in M : f(x) \leq c \,\}$$

denote the sublevel set and φ_t the flow generated by $-\nabla f$. the gradient being taken with respect to any fixed Riemannian metric on M which is constant on each fibre.

Lemma 3.1 *Let $x_0 \in M$ be a regular point of f and set $f(x_0) = c$. Then there exists an $\varepsilon > 0$ and a neighbourhood U of x_0 such that $\varphi_1(U) \subseteq f^{c-\varepsilon}$.*

Proof. Observe

$$\frac{d}{dt} f(\varphi_t(x_0)) = df(-\nabla f_{\varphi_t(x_0)}) = -\|\nabla f_{\varphi_t(x_0)}\|^2 \leq 0.$$

At $t = 0$ the inequality is strict, hence $f(\varphi_1(x_0)) < f(x_0) = c$. Thus for some $\varepsilon > 0$ and some neighbourhood U of x_0, $f(\varphi_1(x)) < c - \varepsilon$ for $x \in U$. $\qquad\square$

Theorem 3.2 (Deformation Theorem). *Given $c \in \mathbf{R}$ and a neighbourhood U of $\mathrm{Crit}(f) \cap f^{-1}(c)$ in M, there exists $\varepsilon > 0$ such that $\varphi_1(f^{c+\varepsilon} \setminus U) \subseteq f^{c-\varepsilon}$. In particular, if c is a regular value of f there exists $\varepsilon > 0$ so that $\varphi_1(f^{c+\varepsilon}) \subseteq f^{c-\varepsilon}$.*

Proof. Let $K \supset U$ be a compact set outside which f is quadratic. Since B is compact and the metric is constant on the fibres, we can replace K by a larger compact set so that $\|\nabla f\|$ is bounded away from zero in $M \setminus K$. Therefore there exists $\varepsilon > 0$ such that $\varphi_1(f^{c+\varepsilon} \setminus K) \subseteq f^{c-\varepsilon}$. Thus from now on we may assume M is compact.

For each x in the compact set $X = f^{-1}(c) \setminus U$ choose a neighbourhood V_x of x in M and a number $\delta_x > 0$ such that $\varphi_1(V_x) \subseteq f^{c-\delta_x}$. Set $\delta = \min(\delta_{x_1}, \ldots, \delta_{x_m})$, where $(V_{x_1}, \ldots, V_{x_m})$ is a finite cover of X. Then, $\varphi_1(V_{x_1} \cup \ldots \cup V_{x_m}) \subseteq f^{c-\delta}$. Since M is compact and $W = U \cup V_{x_1} \cup \ldots \cup V_{x_m}$ is a neighbourhood of $f^{-1}(c)$, there exists $\varepsilon \in (0, \delta)$ such that $f^{-1}([c-\varepsilon, c+\varepsilon]) \subseteq W$. Hence, $\varphi_1(f^{-1}([c-\varepsilon, c+\varepsilon]) \setminus U) \subseteq \varphi_1(V_{x_1} \cup \ldots \cup V_{x_m}) \subseteq f^{c-\varepsilon}$. The result now follows from $f^{c+\varepsilon} \setminus U \subseteq f^{c-\varepsilon} \cup f^{-1}([c-\varepsilon, c+\varepsilon]) \setminus U$. $\qquad\square$

We can now formulate the minimax principle. For a family $\mathcal{F}$ of subsets of M define the *minimax of f over $\mathcal{F}$* by

$$\mathrm{minimax}(f, \mathcal{F}) = \inf_{F \in \mathcal{F}} \sup\{f(x) : x \in F\}.$$

Equivalently,

$$\text{minimax}\,(f, \mathcal{F}) = \inf \left\{\, c \in \mathbf{R} : \exists\, F \in \mathcal{F} \text{ with } F \subseteq f^c \,\right\}.$$

The family $\mathcal{F}$ is called *ambient isotopy invariant* if for every isotopy g_t of M (i.e. C^∞ map $g : [0, 1] \times M \to M$ with each g_t a diffeomorphism of M and g_0 the identity), $F \in \mathcal{F}$ implies $g_1(F) \in \mathcal{F}$.

Theorem 3.3 (Minimax principle) *If the family $\mathcal{F}$ is ambient isotopy invariant, then* $\text{minimax}\,(f, \mathcal{F})$ *is a critical value of f.*

Proof. Suppose $c = \text{minimax}\,(f, \mathcal{F})$ is a regular value. For ε as in the deformation theorem there exists $F \in \mathcal{F}$ so that $F \subseteq f^{c+\varepsilon}$. Since $\{\varphi_t\}_{0 \leq t \leq 1}$ is an isotopy of M, it follows that $\varphi_1(F) \in \mathcal{F}$. Now, $\varphi_1(F) \subseteq \varphi_1(f^{c+\varepsilon}) \subseteq f^{c-\varepsilon}$ implies $\text{minimax}\,(f, \mathcal{F}) \leq c - \varepsilon$, a contradiction. $\qquad\square$

Examples and applications. (1) Let X be any space and let $[X, M]$ denote the set of homotopy classes of maps $X \to M$. For $\gamma \in [X, M]$, set $\mathcal{F}(\gamma) = \{g(X) : g \in \gamma\}$. Then $\text{minimax}\,(f, \mathcal{F}(\gamma))$ is a critical value of f. In particular (for $X = S^k$) this associates critical values of f to each element of $\pi_k(M)$.

(2) Let H_k be the k-dimensional homology functor (with arbitrary coefficients). Given $\gamma \in H_k(M) \setminus \{0\}$, let $\mathcal{F}$ denote the set of subsets $F \subset M$ such that γ is in the image of $(i_F)_* : H_k(F) \to H_k(M)$, where $i_F : F \hookrightarrow M$ is the inclusion map. This gives a critical value of f for every nonvanishing homology class of M.

(3) A similar construction works for cohomology. For $\gamma \in H^k(M)$, $\gamma \neq 0$, choose $\mathcal{F}$ as the family of subsets F of M such that γ is not annihilated by the restriction map $i_F^* : H^k(M) \to H^k(F)$. Thus we get a critical value of f for every nontrivial cohomology class of M.

Let us now come to Lusternik-Schnirelman theory.

Definition 3.4 *Let M be a topological space, A a subset of M. The* Lusternik-Schnirelman *category of A in M is defined by*

$$\text{cat}\,(A; M) \quad = \quad \min \left\{\, k : A \text{ can be covered by } k \text{ closed subsets of } M \right.$$
$$\left. \text{which are contractible to a point in } M \,\right\}$$

and we write $\text{cat}\,(M)$ *for* $\text{cat}\,(M; M)$. *Explicitly.* $U \subseteq M$ *is contractible to a point in M if there is a homotopy $g_t : U \to M$ such that g_0 is the inclusion $U \hookrightarrow M$ and g_1 is a constant map.*

Some obvious properties of cat are:

(a) $\mathrm{cat}(A; M) = 0 \iff A = \emptyset$.

(b) $\mathrm{cat}(A; M) = 1 \iff \bar{A}$ is contractible in M.

(c) $\mathrm{cat}(A; M) = \mathrm{cat}(\bar{A}; M)$.

(d) $A \subseteq B \subseteq M \implies \mathrm{cat}(A; M) \leq \mathrm{cat}(B; M)$ (monotonicity).

(e) $\mathrm{cat}(A \cup B; M) \leq \mathrm{cat}(A; M) + \mathrm{cat}(B; M)$ (subadditivity).

(f)   If A is closed and *deformable through M into B* (i.e. the inclusion $A \hookrightarrow M$ is homotopic to a map into B), then $\mathrm{cat}(A; M) \leq \mathrm{cat}(B; M)$.

(g) $\mathrm{cat}(h(A); M) = \mathrm{cat}(A; M)$ for every homeomorphism $h : M \to M$.

Proof of (f). Let $h_t : A \to M$ be the homotopy. If $B \subseteq F_1 \cup \ldots \cup F_m$ with each F_j closed and contractible in M, then $A = G_1 \cup \ldots \cup G_m$, where $G_j = h_1^{-1}(F_j)$ is closed. Since $h_t|_{G_j}$ is a homotopy of the inclusion of G_j with a map into F_j, G_j is contractible in M. $\square$

By property (g),

$$\mathcal{F}_m = \{\, F \subseteq M : \mathrm{cat}(F; M) \geq m \,\}$$

is ambient isotopy invariant for $1 \leq m \leq \mathrm{cat}(M)$. Hence we get a critical value

$$
\begin{aligned}
c_m(f) \;&=\; \mathrm{minimax}\,(f, \mathcal{F}_m) \;=\; \inf_{\mathrm{cat}(A; M) \geq m} \sup\{f(x) : x \in A\} \\
&=\; \inf\{c \in \mathbf{R} : \exists\, A \subseteq f^c \text{ with } \mathrm{cat}(A; M) \geq m\} \\
&=\; \inf\{c \in \mathbf{R} : \mathrm{cat}(f^c; M) \geq m\},
\end{aligned}
$$

where the last equality follows from monotonicity. Clearly, $c_{m+1}(f) \geq c_m(f)$. Here, equality can occur. For example, if f is constant, all $c_m(f)$ are equal. If equality occurs, however, it is made up for by there being several critical points on that level.

Theorem 3.5 (Lusternik-Schnirelman multiplicity theorem) *If*

$$c = c_{j+1}(f) = c_{j+2}(f) = \ldots = c_{j+k}(f),$$

then f has at least k critical points on the level c. Hence for $1 \leq m \leq \mathrm{cat}(M)$, f has at least m critical points in f^{c_m}. In particular, f has at least $\mathrm{cat}(M)$ critical points.

Proof. We may assume there are only finitely many critical points in $f^{-1}(c)$, say $x_1, \ldots, x_r$ and we must show $r \geq k$. Choose open neighbourhoods V_i of x_i whose closures are disjoint and contractible. Then for $U = V_1 \cup \ldots \cup V_r$,

$\text{cat}(U; M) \leq r$. By the deformation theorem 3.2, for some $\varepsilon > 0$, $f^{c+\varepsilon} \setminus U$ can be deformed into $f^{c-\varepsilon}$. Since

$$c - \varepsilon < c_{j+1}(f) = \inf \{a \in \mathbf{R} : \text{cat}(f^a : M) \geq j + 1\}.$$

it follows from property (f) that $\text{cat}(f^{c+\varepsilon} \setminus U; M) \leq j$. By monotonicity and subadditivity. $\text{cat}(f^{c+\varepsilon}; M) \leq \text{cat}((f^{c+\varepsilon} \setminus U) \cup U; M) \leq j + r$, hence

$$c_{j+k}(f) < c + \varepsilon \leq \inf \{a \in \mathbf{R} : \text{cat}(f^a; M) \geq j + r + 1\} = c_{j+r+1}(f).$$

Thus. $j + r + 1 > j + k$, i.e. $r \geq k$. $\qquad\qquad\qquad\qquad\qquad\qquad\qquad\quad\square$

Instead of the Lusternik-Schnirelman category. one often uses the cup-length, which is easier to handle.

Definition 3.6 *The* cup-length cl(M) *of a space* M *is the largest integer* k *such that there exists a ring* R *and cohomology classes* $\alpha_1. \ldots, \alpha_{k-1} \in H^*(M; R)$ *with positive dimensions such that their cup product does not vanish.*

Proposition 3.7 *If* M *is connected, then* $\text{cat}(M) \geq \text{cl}(M)$.

Proof. Set $\text{cat}(M) = m$. Then, $M = \cup_{i=1}^{m} M_i$, for some closed and contractible M_i. Let $\alpha_1. \ldots, \alpha_m \in H^*(M)$ be cohomology classes of positive dimension (for arbitrary coefficients). Since M_i is contractible, $H^k(M_i) = 0$ for $k \neq 0$. Hence, by the exactness axiom of cohomology $(H^k(M, M_i) \to H^k(M) \to H^k(M_i) = 0)$, α_i lies in the image of $H^*(M, M_i)$. It follows (see [78. App. A, p. 265]) that $\alpha_1 \cup \ldots \cup \alpha_m$ lies in the image of $H^*(M, \cup_{i=1}^{m} M_i) = 0$. Hence, cl$(M) \leq m$. $\quad\square$

Remark. The same can be proved for a cuplength defined with de Rham cohomology, where the cup product is induced by the wedge product of forms.

Corollary 3.8 *A smooth function* f *on a vector bundle* M *(possibly of rank 0) over a compact connected base manifold with* f *quadratic at* ∞ *(if* rank $M > 0$) *has at least* cl(M) *critical points.*

Remark. The minimax principle and the lower bound $\text{cat}(M)$ for the number of critical points remain valid in infinite dimensions under the following conditions: M is a paracompact C^1 Banach manifold with a Finsler structure on TM, f is bounded from below, f^c is complete in the Finsler metric and satisfies the Palais-Smale condition (see [84, Theorems 4.7 and 4.11]). However, in chapter 5 we will have to consider a variational problem with a function unbounded from above and below.

3.2   Lagrange submanifolds

A submanifold L of a symplectic manifold (M, ω) is *Lagrangian* if at every point $q \in L$ the tangent space $T_q L$ is a Lagrangian subspace of $T_q M$. We recall that

$$(T_q L)^{\perp} = T_q L$$

or equivalently, $2 \dim L = \dim M$ and $\omega|_{TL} = 0$. An embedding (or an immersion) $f : N \to M$ is Lagrangian if $2 \dim N = \dim M$ and $f^* \omega = 0$. The embedded manifold is then a Lagrangian submanifold of M. In case of an immersion we say that $f(N)$ is an *immersed Lagrangian submanifold*.

Symplectomorphisms $f : M_1 \to M_2$ between symplectic manifolds (M_1, ω_1) and (M_2, ω_2) give rise to Lagrangian subspaces in the product manifold $(M_1, \omega_1) \times (M_2, -\omega_2)$, i.e. the symplectic manifold $M_1 \times M_2$ with symplectic form $\omega = \pi_1^* \omega_1 + \pi_2^*(-\omega_2)$, where $\pi_i : M_1 \times M_2 \to M_i$ is the projection $(i = 1, 2)$. In fact, the graph $\Gamma_f = \{(x, f(x)) \in M_1 \times M_2\}$ of a diffeomorphism $f : M_1 \to M_2$ is Lagrangian in $(M_1, \omega_1) \times (M_2, -\omega_2)$ if and only if f is a symplectomorphism. Just consider the embedding

$$g : M_1 \to M_1 \times M_2, \quad g(x) = (x, f(x))$$

and verify that

$$\begin{aligned} g^* \omega &= g^* \pi_1^* \omega_1 + g^* \pi_2^*(-\omega_2) \\ &= \omega_1 - f^* \omega_2 = 0. \end{aligned}$$

The Lagrange intersection theory is motivated by the simple observation that the intersection points of the graph Γ_f of a mapping $f : M \to M$ with the diagonal $\Delta = \Gamma_{id}$ are the fixed points of f. For symplectomorphisms C^1-close to the identity this leads to the following result [107, p. 29].

Theorem 3.9 (Weinstein) *Assume that (M, ω) is a compact symplectic manifold with $H^1(M, \mathbf{R}) = 0$. Then a symplectomorphism $f : M \to M$ which is sufficiently C^1-close to the identity has at least two fixed points.*

The argument in [107] shortly reads as follows: There is a symplectomorphism h defined in a neighbourhood U of Δ, mapping Δ onto the zero section 0_M in T^*M. If f is sufficiently C^1-close to the identity, then $\Gamma_f \subset U$ and the projection of $h\,\Gamma_f$ onto 0_M is a diffeomorphism. So $h\,\Gamma_f$ is the graph of a 1-form α on M which is closed because $h\,\Gamma_f$ is Lagrangian. Since $H^1(M, \mathbf{R}) = 0$ there exists a function S on M with $\alpha = dS$. The points $x \in M$ where S attains its extremal values will be fixed points for f, since

$$dS(x) = \alpha(x) = 0.$$

An existence theorem of a type similar to Weinstein's theorem will be proved in section 3.5 (Theorem of Hofer). It applies to symplectomorphisms which arise from Hamiltonian flows.

3.3 Generating functions

Let $\pi : E \to B$ be a vector bundle over the manifold B and $S : E \to \mathbf{R}$ a C^2-function. The fiber derivative at $e \in E$ will be denoted by $\frac{\partial S}{\partial \xi}(e)$. The function S is said to be *nondegenerate* if the map $\frac{\partial S}{\partial \xi} : E \to E^*$ is transverse to the zero section in the dual bundle E^*. If $\frac{\partial S}{\partial \xi}(e) = 0$, the tangent space $T_{(\pi(e),0)}E^*$ is thus spanned by the tangent space to the zero section $0_{E^*} \cong B$ and the image of $T_e E$ under the tangent mapping of $\frac{\partial S}{\partial \xi}$. Let us introduce coordinates

$$(q.\xi) : \pi^{-1}U \to \mathbf{R}^n \times \mathbf{R}^k, \quad (q,\xi) = (q_1.\ldots.q_n.\xi_1,\ldots,\xi_k),$$

describing a local trivialization of the k-dimensional bundle E over a neighbourhood $U \subset B$. The tangent mapping to the mapping

$$(q,\xi) \mapsto (q, \frac{\partial S}{\partial \xi}) = (q_1,\ldots,q_n,\frac{\partial S}{\partial \xi_1}.\ldots.\frac{\partial S}{\partial \xi_k})$$

is described by the matrix

$$\begin{pmatrix} I & 0 \\ \frac{\partial^2 S}{\partial q \partial \xi} & \frac{\partial^2 S}{\partial \xi^2} \end{pmatrix}$$

and the transversality requirement amounts to the condition

$$\mathrm{rank}\begin{pmatrix} \frac{\partial^2 S}{\partial q \partial \xi} & \frac{\partial^2 S}{\partial \xi^2} \end{pmatrix} = k$$

at every point e of

$$\Sigma_S = \{e \in E : \frac{\partial S}{\partial \xi}(e) = 0\}.$$

The implicit function theorem then implies that Σ_S is a C^1-submanifold. It is called the *critical manifold* of the nondegenerate function S. The mapping

$$\begin{aligned} i_S : \Sigma_S &\to T^*B \\ e &\mapsto (\pi(e), dS(e)) \end{aligned}$$

defines an immersion of Σ_S into T^*B. Observe that since $\frac{\partial S}{\partial \xi}(e) = 0$ for $e \in \Sigma_S$, we can identify $(\pi(e), dS(e))$ with an element in $T^*_{\pi(e)}B$.

We will now show that $i_S(\Sigma_S)$ is immersed in T^*B. For this purpose we introduce local coordinates (q,p,ξ,ζ) in T^*E. Consider the graph L_S of dS

$$L_S = \{(q,p,\xi,\zeta) \in T^*E : p = \frac{\partial S}{\partial q}.\zeta = \frac{\partial S}{\partial \xi}\}$$

and the subbundle Q of T^*E consisting of forms annihilating all tangent vectors to the fibers.

$$Q = \{(q,p,\xi,\zeta) \in T^*E : \zeta = 0\}.$$

There is a fibering of Q over T^*B with fibers

$$Q^{\perp}_{qp} = \{(q', p'. \xi. 0) \in Q : q' = q. p' = p\}$$

and a projection along the fibers $pr : Q \to Q/Q^{\perp}$. The space $Q/Q^{\perp}$ can be identified with T^*B. The tangent space TL_S of L_S is spanned by the $n + k$ column vectors of the matrix

$$\begin{pmatrix} I_n & 0 \\[6pt] \dfrac{\partial^2 S}{\partial q^2} & \dfrac{\partial^2 S}{\partial \xi \partial q} \\[6pt] 0 & I_k \\[6pt] \dfrac{\partial^2 S}{\partial q \partial \xi} & \dfrac{\partial^2 S}{\partial \xi^2} \end{pmatrix}$$

If S is nondegenerate, then $T(Q \cap L_S) = TQ \cap TL_S$ and $L_S \cap Q$ is a manifold, as we have already seen. The kernel of the differential of the projection pr restricted to $L_S \cap Q$ is given by

$$\ker d(pr|_{L_S \cap Q}) = TL_S \cap TQ^{\perp},$$

where $TQ^{\perp}$ is the tangent space to the fibers ($TQ^{\perp}$ is orthogonal to TQ with respect to the symplectic form on T^*E).

Since S is nondegenerate, $TL_S \cap TQ^{\perp} = \{0\}$ on $L_S \cap Q$. It follows that $pr|_{L_S \cap Q}$ is an immersion into T^*B. The procedure described is a special case of the reduction construction for Lagrange manifolds (see [107, lecture 3]).

Example 1. The one-dimensional bundle $E = \mathbf{R}^n \times \mathbf{R}$ with the function

$$\begin{aligned} S(q, \xi) &= \frac{\xi^3}{3} - \xi q^2 \\ \Sigma_S &= \{(q, \xi) : \xi^2 - q^2 = 0\}. \end{aligned}$$

Here, S is degenerate at $(q, \xi) = 0$ and Σ_S is not a manifold.

Example 2. The trivial bundle $E = \mathbf{R} \times \mathbf{R}$ with

$$\begin{aligned} S(q, \xi) &= \frac{\xi^3}{3} + \xi(q^2 - 1) \\ \Sigma_S &= \{(q, \xi) : \xi^2 + q^2 - 1 = 0\} \\ \left(\frac{\partial^2 S}{\partial q\, \partial \xi}, \frac{\partial^2 S}{\partial \xi^2}\right) &= (2q, 2\xi) \neq 0 \end{aligned}$$

on Σ_S. In this example, S is nondegenerate, Σ_S is the unit circle, and

$$i_S(\Sigma_S) = \{(q, p) :\ p = 2\xi q = \pm 2q\sqrt{1 - q^2}\}$$

is the figure "8" in the (q, p)-plane (see Figure 1).

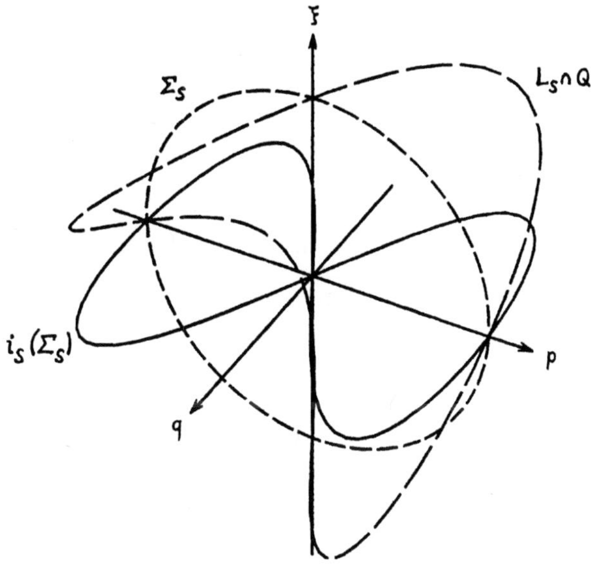

Figure 1

Definition 3.10 *The real valued nondegenerate function S defined on the vector bundle E over B is a* generating function *for the Lagrange manifold $L \subset T^{\cdot}B$ if*

$$L = i_S(\Sigma_S).$$

Observe that i_S is an exact Lagrange immersion: If $p\,dq$ is the Liouville form on $T^{*}B$, then

$$
\begin{aligned}
p\,dq &= \frac{\partial}{\partial q} S(q,\xi)\,dq \\
&= dS \quad \text{at points in } \Sigma_S, \text{ where } \tfrac{\partial S}{\partial \xi} = 0 \\
i^{*}(p\,dq) &= i^{*}\,dS = d\,i^{*}S.
\end{aligned}
$$

Therefore, $p\,dq$ restricted to L is exact and a fortiori $\omega = d(p\,dq)$ restricted to L is zero:

$$i^{*}\omega = d\,i^{*}(p\,dq) = d^2\,i^{*}\,S = 0.$$

3.4 The action functional considered as a generating function

(This section is based on [101] and [104], see also [94].) Consider the Hamiltonian flow φ_t on $\mathbf{R}^{2n} = T^*\mathbf{R}^n$ given by a C^2-Hamiltonian $H = H(q, p, t)$ with bounded second derivatives. Set $L_0 = \mathbf{R}^n$ and $L_1 = \varphi_1(L_0) \subset \mathbf{R}^{2n}$. The action functional is defined by

$$A(\gamma) \;=\; \int_0^1 (p\dot{q} - H(q, p, t))dt,$$
$$\text{where} \quad \gamma \;=\; (q. p) : [0, 1] \to \mathbf{R}^{2n}$$

are curves in the Sobolev space $H^{1,2}([0, 1], R^{2n})$. We restrict A to

$$\mathcal{P} = \{\gamma = (q, p) \in H^{1,2} : p(0) = 0\}$$

(note that $\gamma \in H^{1,2}$ is continuous). This gives rise to the fibration $\pi : \mathcal{P} \to \mathbf{R}^n$ determined by $\pi(\gamma) = q(1)$. A vector space structure on the fibers $\mathcal{P}_{q(1)}$ is provided by the space of derivatives: any $\gamma \in \mathcal{P}_{q(1)}$ is determined by its derivative $\dot{\gamma} \in L^2([0, 1], \mathbf{R}^{2n})$

$$\gamma = (q(t), p(t)) = \left(q(1) - \int_t^1 \dot{q}(s)\, ds, \int_0^t \dot{p}(s)\, ds\right).$$

We use the notation $\gamma = \gamma(q(1), \dot{\gamma})$. A formal argument indicates that A is an "infinite dimensional" generating function:

$$dA(\gamma)(\delta\gamma) \;=\; \int_0^1 [p\,\delta\dot{q} + \dot{q}\,\delta p - (\nabla H(\gamma, t), \delta\gamma)]\, dt$$
$$=\; p(1)\,\delta q(1) - \int_0^1 (J\dot{\gamma} + \nabla H, \delta\gamma)\, dt.$$

The "critical manifold" is given by

$$\Sigma_A = \{\gamma \in \mathcal{P} : dA(\gamma)(\delta\gamma) = 0 \;\; \forall\, \delta\gamma \in T_\gamma \mathcal{P} \cap \ker d\pi\}.$$

The elements $\gamma \in \Sigma_A$ thus satisfy

$$\int_0^1 (J\dot{\gamma} + \nabla H, \delta\gamma)dt = 0$$

for all $\delta\gamma$ with $\delta q(1) = 0$ ($\delta\gamma \in \ker d\pi$) and $\delta p(0) = 0$ ($\delta\gamma \in T_\gamma \mathcal{P}$). These curves $\gamma \in \Sigma_A$ are therefore solutions of the Hamilton equation

$$-J\dot{\gamma} = \nabla H(\gamma, t).$$

On the other hand,

$$\frac{\partial A}{\partial q(1)} = p(1),$$

which shows that A is a generating function for the immersion $\Sigma_A \to \mathbf{R}^{2n}$ given by $\gamma \mapsto (q(1).p(1)) = \varphi_1(q(0), p(0))$. The Lagrangian manifold $L_1 = \varphi_1(L_0)$. $L_0 = \mathbf{R}^n$, has the action functional A as a generating function.

The reduction method of Amann and Zehnder [4] will reduce the infinite dimensional variational problem involving the action functional A to a finite dimensional one. In the present context, A can be replaced by the generating function $a(q(1), u)$ (see below) which is defined on a finite dimensional vector bundle over $\mathbf{R}^n$.

The variational equation for A shows that

$$\nabla A(\gamma) = -J\dot{\gamma} - \nabla H(\gamma. t)$$

is the L^2-gradient of the action functional in direction of the fiber $\mathcal{P}_{q(1)}$. The elements in $L^2([0, 1], \mathbf{R}^{2n})$ can be represented by their Fourier series

$$z(t) = \sum_{k \in \mathbf{Z}} e^{2\pi kt J} z_k, \qquad z_k \in \mathbf{R}^{2n}.$$

Define the projection operators $P = P_N$ and $Q = Q_N$ by

$$Pz = \sum_{|k| \leq N} e^{2\pi kt J} z_k$$

$$Qz = \sum_{|k| > N} e^{2\pi kt J} z_k$$

and decompose ∇A into its P- and Q-components

$$\nabla_P A(\gamma) = -PJ\dot{\gamma} - P\nabla H(\gamma),$$
$$\nabla_Q A(\gamma) = -QJ\dot{\gamma} - Q\nabla H(\gamma).$$

Assume now that there is a C^1-mapping w from $\mathbf{R}^n \times E = \mathbf{R}^n \times PL^2([0, 1], \mathbf{R}^{2n})$ into $F = QL^2([0, 1], \mathbf{R}^{2n})$ assigning to each $(q(1). u) \in \mathbf{R}^n \times E$ the unique value $w = w(q(1). u) \in F$ such that $\nabla_Q A(\gamma(q(1), u + w)) = 0$. Then the variational function

$$a(q(1), u) = A(\gamma(q(1), u + w))$$

will have $(q(1), u)$ as a critical point if and only if $\gamma = \gamma(q(1), u + w(q(1), u))$ is a critical point for A. The formal calculation justifying this statement is given by

$$\nabla a(q(1), u) = \nabla_P A(\gamma(q(1), u + w)) + \nabla_Q A(\gamma(q(1), u + w))w'(q(1), u).$$

The defining property for the mapping w asserts that the last term in this sum vanishes. The variations for a and A then vanish simultaneously.

In the remainder of this section, rigorous arguments will be given which will prove that the mapping w exists (for N sufficiently big) and that $a(q(1), u)$ is a generating function for $L_1 = \varphi_1(L_0)$. Fix $(q(1), u) \in \mathbf{R}^n \times E$ and determine γ in dependence of $v \in F:\ \gamma = \gamma(q(1), u + v)$.

Lemma 3.11 *If $H \in C^2(\mathbf{R}^{2n} \times [0,1])$ with $|H''| \leq C$, then for sufficiently big N, the mapping*

$$v \mapsto QJ\nabla H(\gamma)$$

is contracting in $F = Q_N L^2([0,1], \mathbf{R}^{2n})$. The unique fixed point $w = w(q(1), u)$ depends differentiably on $(q(1), u) \in \mathbf{R}^n \times E$.

Proof. If $v_1, v_2 \in F$, then $\gamma(q(1), u + v_2) - \gamma(q(1), u + v_1)$ is given by

$$-\int_t^1 (v_2 - v_1)\, ds.$$

With the Fourier series representation

$$\sum_{|k|>N} e^{2\pi k t J} z_k$$

for the difference $v = v_2 - v_1$, the integral can be expressed as

$$-\int_t^1 v\, ds = z_0 - \sum_{|k|>N} \frac{1}{2\pi k} e^{2\pi k t J} J z_k$$

with the constant

$$\begin{aligned}
z_0 &= \int_0^1 \left(-\int_t^1 v\, ds\right) dt = -\int_0^1 v\, t\, dt \\
&= \int_0^1 v\, Q(-t)\, dt.
\end{aligned}$$

This gives the estimate $|z_0| \leq c_0(N)\, \|v\|_2$ with $\lim_{N \to \infty} c_0(N) = 0$. It follows that

$$\left\| \int_t^1 v\, ds \right\|_2 \ \leq\ c_0(N)\, \|v\|_2 + \left(\sum_{|k|>N} k^{-2} |z_k|^2 \right)^{1/2}$$

$$\leq\ \left(c_0(N) + \frac{1}{N}\right) \|v\|_2.$$

The curves $\gamma_i = \gamma(q(1), u + v_i)\ (i = 1, 2)$ thus satisfy the uniform estimate

$$\|\gamma_2 - \gamma_1\|_2 \leq c(N)\|v_2 - v_1\|_2$$

with $\lim\limits_{N \to \infty} c(N) = 0$. Since H'' is uniformly bounded it follows that

$$\|QJ\nabla H(\gamma_2) - QJ\nabla H(\gamma_1)\|_2 \leq C \cdot c(N) \|v_2 - v_1\|_2.$$

The integer N must now be chosen such that $C \cdot c(N) < 1$. The implicit function theorem (see e.g. [65, p. 17]) then tells us that $v = w(q(1), u)$ defined by the equation

$$G(q(1), u, v) - v = 0$$
$$\text{with} \quad G(q(1), u, v) = QJ\nabla H(\gamma(q(1). u + v))$$

is continuously differentiable at $(q(1), u)$ if G is continuously differentiable in a neighbourhood of $(q(1), u, w(q(1), u))$ with $D_v G - I$ invertible.

Now G is in fact continuously differentiable and at the fixed point $w = w(q(1). u)$ one has

$$\|D_v G(q(1), u, w)\| \leq C \cdot c(N) < 1.$$

Therefore $D_v G - I$ is invertible in a neighbourhood of $(q(1), u, w(q(1), u))$ and the assumptions for the implicit function theorem are satisfied. This proves that w is continuously differentiable. $\square$

Lemma 3.12 *If γ is critical for A, then $(q(1). u)$ with $u = P\dot\gamma$ is critical for*

$$a(q(1), u) = A\big(\gamma(q(1), u + w(q(1). u))\big)$$

and conversely, if $(q(1), u))$ is critical for a, then $\gamma = \gamma(q(1), u + w(q(1), u))$ is critical for A.

By saying that γ is critical for A we mean that

$$dA(\gamma)\,\delta\gamma = 0$$

for all $\delta\gamma \in T_\gamma P \cap \ker d\pi$ (in particular $\delta q(1) = 0$) and similarly for a.

Let us first calculate the variation of a:

$$da(q(1), u)\,(\delta q(1), \delta u) = dA\Big(\gamma\big(q(1). u + w(q(1), u)\big)\Big)\,\delta\gamma$$

with $\delta\gamma$ determined by $\delta q(1)$ and $\delta\dot\gamma = \delta u + w'(q(1), u)\,(\delta q(1), \delta u)$. Hence

$$\begin{aligned}
da(q(1), u)\,(\delta q(1), \delta u) &= p(1)\,\delta q(1) - \int_0^1 (J\dot\gamma + \nabla H(\gamma). \delta\gamma)\,dt \\
&= p(1)\,\delta q(1) - \int_0^1 (PJ\dot\gamma + P\nabla H(\gamma), \delta\gamma)\,dt
\end{aligned}$$

if γ is determined as $\gamma(q(1), u + w(q(1), u))$. Note that in this final expression, $\delta\gamma$ can be taken to be determined by $\delta q(1)$ and $\delta\dot\gamma = \delta u$. Assume now that

$$dA(\gamma)\,\delta\gamma = 0$$

for all $\delta\gamma \in T_\gamma \mathcal{P} \cap \ker d\pi$. Then γ satisfies

$$J\dot\gamma + \nabla H(\gamma) = 0$$

and a fortiori

$$QJ\dot\gamma + Q\nabla H(\gamma) = 0.$$

The function $v = Q\dot\gamma$ is thus given by $w(q(1), u), u = P\dot\gamma$, and it follows that

$$da(q(1), u)(0, \delta u) = 0$$

for all $\delta u \in E$.

Conversely, if $(q(1), u)$ is critical for a, then $\gamma = \gamma(q(1), u + w(q(1), u))$ satisfies

$$\int_0^1 (PJ\dot\gamma + P\nabla H(\gamma), \delta\gamma)\, dt = 0$$

for all $\delta\gamma$ with $\delta q(1) = 0$ and $\delta\dot\gamma = \delta u$ (see the remark above). It follows that γ is critical for A. The expression for the variation of $a(q(1), u)$ shows that

$$\frac{\partial a}{\partial q(1)} = p(1).$$

This completes the proof that a is a generating function for $L_1 = \varphi_1(L_0)$.

3.5　Critical points for generating functions

Given a real valued C^2-function f on a compact manifold X, set

$$X^b = \{x \in X : f(x) \le b\}, \quad b \in \mathbf{R}.$$

Morse theory asserts that the topological type of X^b in dependence of b is unchanged outside of critical values of f. If $\alpha \in H^*(X), \alpha \neq 0$ is a cohomology class of X and

$$i_\lambda : X^\lambda \hookrightarrow X$$

the imbedding, then $i_\lambda^* \alpha = 0$ for sufficiently small λ. The value

$$c(\alpha, f) = \inf\{\lambda : i_\lambda^* \alpha \neq 0\}$$

will then be critical for f (see section 3.1).

The same situation will now be considered for functions S defined on a vector bundle $\pi : E \to B$ over the compact manifold B. In the applications these are the generating functions for some Lagrangian manifolds in T^*B.

Typically these functions S will be "quadratic at infinity". In the particular case where S is a finite dimensional reduction of the action functional $A(\gamma)$, the quadratic term is due to the part

$$\int_0^1 p\dot{q}\,dt$$

in A.

If q is a nondegenerate quadratic function on E then the *negative bundle* $E^-(q)$ of q is defined as follows: In local coordinates, defined in a neighbourhood U, q can be represented in the form

$$q(x,\xi) = |\xi^+|^2 - |\xi^-|^2 \qquad (x,\xi) \in U \times \mathbf{R}^n$$

with the orthogonal decomposition $\xi = \xi^+ \oplus \xi^-, \xi^+ \in \mathbf{R}^m, \xi^- \in \mathbf{R}^k$ (use a linear change of variables in each fiber). The dimension k of the "negative subspace" is constant, since q is nondegenerate. The system of coordinates $\{U \times \mathbf{R}^k\}$ then defines the negative bundle $E^-(q)$ of q.

Definition 3.13 *A function $S \in C^2(E, \mathbf{R})$ is called* quadratic at infinity *if it coincides outside a compact set in E with a nondegenerate quadratic function q whose negative bundle $E^-(q)$ is trivial over B.*

We will consider relative homology and cohomology H_* and H^* with coefficients in some fixed field. For functions S quadratic at ∞ set

$$E^\lambda = \{x \in E : S(x) \le \lambda\} \qquad \lambda \in \mathbf{R}$$

and denote by $D(E^-)$ and $S(E^-)$ the disk and sphere bundles associated to $E^- = E^-(q)$.

For sufficiently large λ the homotopy types of (E^λ, E^μ) and $(E^\mu, E^{-\lambda})$ are independent of λ. This justifies the notations E^∞ for E^λ and $E^{-\infty}$ for $E^{-\lambda}$ for large λ. Observe then that by homotopy

$$H^*(E^\infty, E^{-\infty}) \cong H^*(D(E^-), S(E^-)).$$

The Thom isomorphism $H^*(B) \to H_c^*(E^-)$ is an isomorphism between the cohomology of B and compact cohomology (in the direction of the fibers) of the vector bundle E^- over B, shifting the grading by $k = \operatorname{rank} E^-$. There exists a k-form t on E^- (with compact support in direction of the fibers) such that the isomorphism can be described by

$$\alpha \mapsto t \wedge \pi^* \alpha \qquad \alpha \in H^*(B).$$

Compact cohomology on E^- can be identified with relative cohomology on $(D(E^-), S(E^-))$. The cohomology $H^*(E^\infty, E^{-\infty})$ is therefore isomorphic (with a shift by k in the grading) to $H^*(B)$. The resulting isomorphism

$$T : H^*(B) \longrightarrow H^*(E^\infty, E^{-\infty})$$

will still be referred to as the Thom isomorphism.

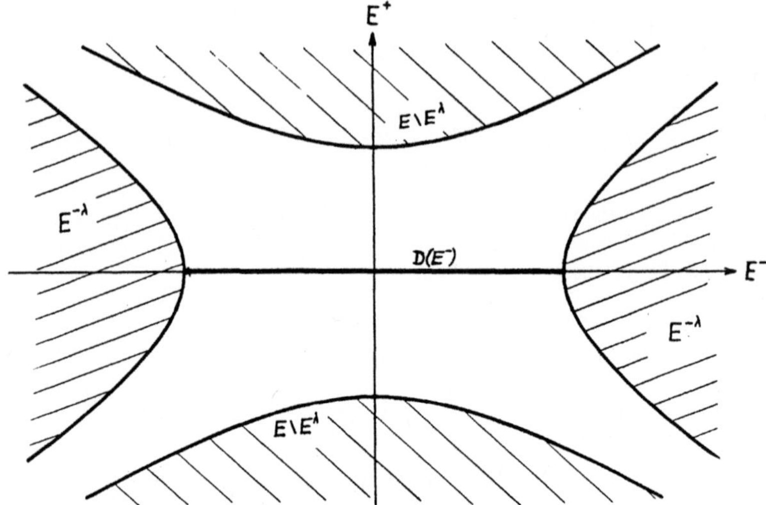

Figure 2: The drawing illustrates that $(E^\lambda, E^{-\lambda})$ retracts onto $(D(E^-),$ $S(E^-))$.

Definition 3.14 *For $S \in C^2(E, \mathbf{R})$ quadratic at ∞ and for $\alpha \in H^*(B)$, set*

$$c(\alpha, S) = \inf\{\lambda : i_\lambda^* T\alpha \neq 0\},$$

where $i_\lambda : E^\lambda \hookrightarrow E$ is the inclusion. By Lusternik-Schnirelman theory (see section 3.1), $c(\alpha, S)$ is a critical value of S. We will thus refer to $c(\alpha, S)$ as the critical value of S determined by α.

The product $\alpha \wedge \beta$ of a closed form β with an exact form $\alpha = d\gamma$ will give an exact form: $d(\gamma \wedge \beta) = d\gamma \wedge \beta + 0 = \alpha \wedge \beta$. This shows that the inequality in the following proposition is correct. Lusternik-Schnirelman theory provides a finer statement in case of equality.

Proposition 3.15 ([103, Prop. 2.2]) *If S is quadratic at ∞ and $\alpha, \beta \in H^*(B)$, then*

$$c(\alpha \wedge \beta, S) \geq c(\alpha, S)$$

If equality holds with $\beta \notin H^0(B)$, then $T\beta$ induces a nonzero cohomology class in any neighbourhood of the set K_c of critical points on the level $c = c(\alpha, S)$

$$K_c = \{e \in E : dS(e) = 0 \text{ and } S(e) = c\}.$$

Theorem 3.16 (Hofer) *If $L_1 = \varphi_1(L_0)$, $L_0 = 0_B \cong B$, then the number of intersection points of L_1 with L_0 is at least $cl(B) (\geq 2)$.*

Proof. The action functional provides a generating function S quadratic at ∞. Theorem 3.5 and Proposition 3.7 show that

$$\#\{e \in E : dS(e) = 0\} \geq cl(B).$$

But the critical points of S are in bijective correspondence to points in $L_1 \cap 0_B$. (L_1 is an embedded Lagrangian manifold in T^*B). $\qquad\square$

3.6 Stabilization

There are two obvious ways of changing a generating function S (quadratic at infinity) such that the modified function describes the same Lagrangian manifold $L \subset T^*B$.

1. **Stabilization.** Let $q : E' \to \mathbf{R}$ be a nondegenerate quadratic form on the fibers of $E' \to B$. Then

$$S_1(b, \xi, \eta) = S(b, \xi) + q(\eta)$$

is called a *stabilization* of S. The functions $S : E \to \mathbf{R}$ and $S_1 : E \oplus E' \to \mathbf{R}$ generate the same Lagrangian manifold.

2. **Equivalence.** Two generating functions quadratic at infinity $S_1 : E_1 \to \mathbf{R}$ and $S_2 : E_2 \to \mathbf{R}$ are *equivalent,* if there exists a fibre preserving diffeomorphism $\varphi : E_1 \to E_2$ and a constant C such that

$$S_1 = S_2 \circ \varphi + C.$$

Equivalent generating functions will generate the same Lagrangian manifold. The following result is proved in [103, Prop. 1.5].

Theorem 3.17 *Assume that φ_t is a Hamiltonian isotopy of T^*B (B a compact manifold). If S_1 and S_2 are generating functions quadratic at infinity which are both generating for the embedded Lagrangian manifold $\varphi_1(0_B)$, then after stabilization S_1 is equivalent to S_2.*

The quantities $c(\alpha, S)$ will change if the generating function S is replaced by $S + \mathrm{const}$. Therefore either a normalization is required or the differences

$$c(\alpha, S) - c(\beta, S)$$

have to be considered. But if S_1 and S_2 are equivalent generating functions quadratic at infinity and if there is no additive constant:

$$S_1 = S_2 \circ \varphi,$$

then $c(\alpha. S_1) = c(\alpha, S_2)$, since E_1^λ and E_2^λ are diffeomorphic. In the case that S_2 is a stabilization of S_1, it is still true that $c(\alpha, S_1) = c(\alpha, S_2)$ (see [103]).

If the considerations are restricted to Lagrangian manifolds $L \subset T^*B$ arising as images of $B \cong 0_B$ under a Hamiltonian isotopy, then up to normalization the critical levels $c(a, S)$ are invariantly assigned to the Lagrangian manifold L; they are independent of the particular generating function. In the presence of a normalization, we write $c(\alpha, L)$ instead of $c(\alpha, S)$.

3.7 Symplectic diffeomorphisms of $\mathbf{R}^{2n}$

Consider the group $\mathcal{H} = \mathcal{H}(\mathbf{R}^{2n})$ of time-one mappings of Hamiltonian flows associated to time dependent compactly supported Hamiltonians $H(x,t)$: $\mathbf{R}^{2n} \times [0,1] \to \mathbf{R}$. For $\psi \in \mathcal{H}$ its graph Γ_ψ is a Lagrangian submanifold of $(\mathbf{R}^{2n} \times \mathbf{R}^{2n}, -\omega_0 \oplus \omega_0)$ which coincides with the diagonal Δ outside a compact set. Under the symplectic mapping (see the introduction to this chapter)

$$h : (x, y, X, Y) \mapsto (\frac{x + X}{2}, \frac{y + Y}{2}, y - Y, X - x)$$

Δ is identified with $0_{\mathbf{R}^{2n}}$. Under the one point compactification of $\mathbf{R}^{2n}$, $S^{2n} = \mathbf{R}^{2n} \cup \{\infty\}$, $T^*\Delta$ embeds into T^*S^{2n} and $h(\Gamma_\psi)$ is compactified by the addition of the point $(\infty, 0) \in T^*S^{2n}$. In this way one obtains a Lagrange submanifold $\tilde{\Gamma}_\psi \subset T^*S^{2n}$ with the additional property that

$$\tilde{\Gamma}_\psi \cap T^*U = U$$

in a neighbourhood U of ∞ in S^{2n}. The generating functions $S : S^{2n} \times \mathbf{R}^k \to \mathbf{R}$ for $\tilde{\Gamma}_\psi$ (quadratic at infinity) will then be normalized by the condition that the critical value associated with the point $(\infty, 0) \in 0_{S^{2n}} \cap \tilde{\Gamma}_\psi$ is zero. The cohomology of S^{2n} is generated by $1 \in H^0(S^{2n})$ and the orientation class $\mu \in H^{2n}(S^{2n})$.

Definition 3.18

$$c_-(\psi) = -c(\mu, \tilde{\Gamma}_\psi), \quad c_+(\psi) = -c(1, \tilde{\Gamma}_\psi), \quad c(\psi) = c_+(\psi) - c_-(\psi)$$

Remarks 1. Since $\mu = 1 \wedge \mu$, it follows that $c_- \leq c_+$.

2. It is a consequence of the normalization (see [103, Prop. 4.2]) that

$$c_- \leq 0 \leq c_+.$$

3. The graph $\Gamma_{\psi^{-1}}$ of the inverse function is obtained by reflecting Γ_ψ on the diagonal. This symmetry shows up again in the generating functions: If S is generating for $\tilde{\Gamma}_\psi$, then $-S$ will generate $\tilde{\Gamma}_{\psi^{-1}}$. Poincaré duality in combination with this symmetry will then lead to the equality

$$c_+(\psi^{-1}) = c_-(\psi)$$

(see [103, Prop. 4.2]).

4. $c(\psi) = 0$ if and only if $\psi = id$.

The fixed points of the mapping ψ are the points in $\Gamma_\psi \cap \Delta$. They correspond to the points in $\tilde{\Gamma}_\psi \cap 0_{S^{2n}}$ and are therefore in one to one correspondence with the critical points of the generating functions S for $\tilde{\Gamma}_\psi \subset T^* S^{2n}$. If (x_0, ξ_0) and (x_1, ξ_1) are critical points of the generating function S and if $\tilde{\gamma} : [0, 1] \to \tilde{\Gamma}_\psi$ is any path in $\tilde{\Gamma}_\psi$ connecting x_0 to x_1, then

$$S(x_1, \xi_1) - S(x_0, \xi_0) = \int_{\tilde{\gamma}} \tilde{p}\, d\tilde{q}.$$

since on $L = \tilde{\Gamma}_\psi$ one has $dS = \tilde{p}\, d\tilde{q}$. In this notation. $\tilde{q} = (\frac{q+Q}{2}, \frac{p+P}{2})$ and $\tilde{p} = (p-P, Q-q)$ are the coordinates on $T^* S^{2n}$. Transforming the integral by the linear symplectomorphism $h(q, p, Q, P) = (\tilde{q}, \tilde{p})$ and using partial integration, this becomes

$$S(x_1, \xi_1) - S(x_0, \xi_0) = \int_\gamma p\, dq + \int_{\psi \circ \gamma^{-1}} p\, dq.$$

The graph of the curve γ is mapped onto $\tilde{\gamma}$ by the mapping h.

In the general setting, critical points (x_α, ξ_α) for the generating function S can be chosen such that the critical values are $c(\alpha, S)$, $\alpha \in H^*(B)$. In the present context there exist critical points $(x_\pm, \xi_\pm)$ of S such that $S(x_\pm, \xi_\pm) = c_\pm$. Choose any curve $\gamma_\pm$ connecting $x_\pm$ with ∞. Then

$$\begin{aligned} c_\pm &= S(x_\pm, \xi_\pm) = S(x_\pm, \xi_\pm) - S(\infty, 0) \\ &= -\left(\int_{\gamma_\pm} p\, dq + \int_{\psi \circ \gamma_\pm^{-1}} p\, dq \right). \end{aligned}$$

Since by assumption $\psi = id$ in a neighbourhood U of ∞. the point at infinity can be replaced by any point $z \in U$. Similarly, if γ is a curve connecting x_- to x_+, then

$$c = c_+ - c_- = \int_\gamma p\, dq + \int_{\psi \circ \gamma^{-1}} p\, dq.$$

If the flow ψ_t is given by the Hamiltonian H, then the tangent vectors to the flow lines in extended phase space are in the kernel of the form $d(p\, dq - H\, dt)$. Therefore. if $\Psi : [0, 1]^2 \to \mathbf{R}^{2n}$ is the mapping $\Psi(t, s) = \psi_t(\gamma^{(s)})$. then

$$\int_{\Psi[0,1]^2} d(p\, dq - H\, dt) = 0.$$

As above take γ to be a curve connecting x_+ or x_- to z. By Stokes' theorem

$$0 = \int_\gamma p\, dq + \int_{\psi_t(z)} (p\, dq - H\, dt) - \int_{\psi \circ \gamma} p\, dq - \int_{\psi_t(x_\pm)} (p\, dq - H\, dt).$$

Therefore

$$c_\pm = -\left(\int_\gamma p\,dq + \int_{\psi\circ\gamma^{-1}} p\,dq\right)$$

$$= \int_{\psi_t(z)} (p\,dq - H\,dt) - \int_{\psi_t(x_\mp)} (p\,dq - H\,dt) = -\int_{\psi_t(x_\pm)} (p\,dq - H\,dt).$$

The last equality holds since $\psi_t(z) \equiv z$ and $H(t,z) \equiv 0$.

Proposition 3.19 *Let φ_t, $t \in [0,1]$, be a flow of conformally symplectic mappings, $\varphi_t^* \omega = \lambda(t)\omega$, $\varphi_o = id$. Then for $\psi \in \mathcal{H}$*

$$c_\pm(\varphi_t \circ \psi \circ \varphi_t^{-1}) = \lambda(t)\, c_\pm(\psi).$$

Proof. If x is a fixed point of ψ, then $\varphi_t(x)$ is a fixed point of $\varphi_t \circ \psi \circ \varphi_t^{-1}$. Connect two fixed points x_0 and x_1 by a path γ and consider the closed path

$$g_t = \varphi_t \circ (\gamma \cup \psi \circ \gamma^{-1}) = \varphi_t \circ \gamma \cup \varphi_t \circ \psi \circ \varphi_t^{-1} \circ (\varphi_t \circ \gamma^{-1})$$

Then

$$\int_{g_t} p\,dq = \lambda(t) \int_{\gamma \cup \psi \circ \gamma^{-1}} p\,dq,$$

since φ_t is conformally symplectic. The integral

$$\int_{\gamma \cup \psi \circ \gamma^{-1}} p\,dq$$

is the difference of two critical values of the generating function for Γ_ψ. These differences make up a totally disconnected set D of real numbers, in fact D has zero Lebesgue measure. On the other hand, $c_\pm(\varphi_t \circ \psi \circ \varphi_t^{-1})$ is expressed as

$$\int_{\gamma_t \cup (\varphi_t \circ \psi \circ \varphi_t^{-1}) \circ \gamma_t^{-1}} p\,dq$$

for a curve γ_t connecting two fixed points of the mapping $\varphi_t \circ \psi \circ \varphi_t^{-1}$. Such curves are of the form $\varphi_t \circ \gamma$ with γ a curve connecting two fixed points of ψ. The above equality therefore implies that the function

$$\frac{1}{\lambda(t)} c_\pm(\varphi_t \circ \psi \circ \varphi_t^{-1})$$

has its values in the discrete set D. But this function is continuous and therefore it must be constant. $\qquad\square$

Theorem 3.20 *For all $\varphi, \psi \in \mathcal{H}$*

$$c_+(\varphi \circ \psi) \le c_+(\varphi) + c_+(\psi), \quad c_-(\varphi \circ \psi) \ge c_-(\varphi) + c_-(\psi)$$

and consequently

$$c(\varphi \circ \psi) \le c(\varphi) + c(\psi).$$

For the proof we refer to [103, Prop. 4.4].

Corollary 3.21 *Let $\varphi, \psi \in \mathcal{H}$ and assume that the isotopy φ_t for φ satisfies $\operatorname{supp} \varphi_t \subset U$ for all $t \in [0,1]$. If $\psi(U) \cap U = \emptyset$, then*

$$c_\pm(\psi \circ \varphi) = c_\pm(\psi)$$

and consequently

$$c_+(\varphi) \leq c(\psi)$$
$$c_-(\varphi) \geq -c(\psi).$$

For the proof consider a fixed point x of $\psi \circ \varphi_t$. It cannot lie in U since $\psi(U) \cap U = \emptyset$. Therefore it must be a fixed point of ψ. Take then two fixed points x and y of $\psi \circ \varphi_t$ and choose a path γ joining them. Then

$$\gamma \cup (\psi \circ \varphi_t \circ \gamma^{-1}) = \gamma \cup (\psi \circ \gamma^{-1}) \cup (\psi \circ \gamma) \cup (\varphi \circ \varphi_t \circ \gamma^{-1})$$
$$= (\gamma \cup \psi \circ \gamma^{-1}) \cup \psi \circ (\gamma \cup \varphi_t \circ \gamma^{-1})$$

Since ψ is symplectic,

$$\int_{\gamma \circ \varphi_t \circ \gamma^{-1}} p\, dq = \int_{\psi \circ (\gamma \circ \varphi_t \circ \gamma^{-1})} p\, dq$$

and since φ_t is an isotopy, this integral equals

$$\int_{\varphi_t(x)} (p\, dq - H\, dt) - \int_{\varphi_t(y)} (p\, dq - H\, dt).$$

But $H(t,x) = H(t,y) = 0$ and $\varphi_t(x) = x$, $\varphi_t(y) = y$ for all $t \in [0,1]$, therefore this integral vanishes. This then proves that

$$\int_{\gamma \circ \psi \circ \varphi_t \circ \gamma^{-1}} p\, dq$$

is independent of t. By the same argument as before this implies that $c_\pm(\psi \circ \varphi_t)$ is independent of t. By the theorem,

$$c_+(\varphi) \leq c_+(\psi^{-1}) + c_+(\psi \circ \varphi)$$
$$= -c_-(\psi) + c_+(\psi) = c(\psi).$$

3.8 The Viterbo capacities

Definition 3.22 *The* Viterbo capacities c and γ *are defined by*

$$c(U) = \sup_{\varphi \in \mathcal{H}} \{c_+(\varphi) : \operatorname{supp} \varphi_t \subset U\}, \quad \gamma(U) = \inf_{\psi \in \mathcal{H}} \{c(\psi) : \psi(U) \cap U = \emptyset\}$$

for compact sets $U \subset \mathbf{R}^n$. For arbitrary sets $V \subset \mathbf{R}^n$ set $\gamma(V) = \sup\{\gamma(U) : U \subset V,\ U \text{ compact}\}$.

Both invariants γ and c are symplectic capacities for subsets of $\mathbf{R}^{2n}$ (in the sense of definition 4.12). Monotonicity is obvious and conformality is a direct consequence of Proposition 3.19. The above corollary shows that $c(U) \leq \gamma(U)$. Clearly $\gamma(u) < \infty$ and hence $c(U) < \infty$. The capacities are thus nontrivial. Local nontriviality follows from the fact that $\gamma(U) \geq c(U) > 0$ if $U^\circ \neq \emptyset$. The capacities of the unit ball $B \subset \mathbf{R}^{2n}$ and of the cylinder $Z = \{(p,q) \in \mathbf{R}^{2n} : p_1^2 + q_1^2 < 1\}$ can in fact be calculated explicitly:

$$c(B) = c(Z) = \pi, \qquad \gamma(B) = \gamma(Z) = \pi$$

Proof. We will show that $c(B) \geq \pi$ and $\gamma(z) \leq \pi$. Take a radial Hamiltonian function $H(x) = h(|x|)$ with support in B such that $0 \leq h'(r) \leq (z\pi + \varepsilon)r$ on $[0,1]$ and $h(0) = \pi - \varepsilon$. Then the integral curves of the flow φ_t are Hopf fibers on the spheres Σ_r and 0 is the only fixed point for $\varphi = \varphi_1$ in B°. The invariant $c_+(\varphi)$ is therefore given by the action

$$c_+ = -\int_{\varphi_t(0)} (p\,dq - H\,dt) = h(0) \int_0^1 dt = \pi - 2\varepsilon.$$

whereas $c_- = 0$. This shows that $c(B) \geq \pi$.

In order to show that $\gamma(z) \leq \pi$ choose an area preserving mapping $\alpha_1 : \mathbf{R}^2 \to \mathbf{R}^2$ which maps the unit disc onto the half disc

$$D^+ = \{q_1^2 + p_1^2 \leq \sqrt{2}, p \geq 0\}.$$

The product mapping $\alpha : \mathbf{R}^{2n} \to \mathbf{R}^{2n}$, which in the (q_1, p_1)-coordinates is given by α_1 and in the other coordinates by the identity, is symplectic. The cylindrical Hamiltonian $H(z) = h(|z_1|)$ with support in $Z(\sqrt{2}) = B^2(\sqrt{2}) \times \mathbf{R}^{2n-2}$ such that $0 \geq h'(r) \geq -\pi r$ and

$$h'(r) = -\pi r \quad \text{on} \quad (\varepsilon, \sqrt{2} - \varepsilon)$$

generates a flow φ_t such that $\varphi = \varphi_1$ turns the slightly smaller half cylinder $Z_\varepsilon^+ = \{z \in Z^+ = D^+ \times \mathbf{R}^{2n-2} : d(z, \partial Z^+) \geq \varepsilon\}$ by an angle π. In this way

$$\varphi(Z_\varepsilon^+) \cap Z_\varepsilon^+ = \emptyset.$$

As before, $c_-(\varphi) = 0$ and $c_+(\varphi) = h(0) \leq \pi$. The construction has to be modified, since $\mathrm{supp}\, H$ is not compact. This can easily be done by multiplying H with a C^∞-function $\chi \geq 0$ with compact support, which equals 1 in a big ball and whose derivative is arbitrarily small. Letting ε tend to zero, it then follows that

$$\gamma(Z) = \gamma(Z^+) \leq \pi.$$

$$\square$$

4 Symplectic Capacities

The first symplectic invariant appeared in Gromov's famous paper [50]. He called it the symplectic radius and used it in the proof of the famous squeezing theorem. When similar invariants were discovered later by Ekeland, Hofer, Viterbo, Zehnder and others, it became clear that their existence has many useful consequences, e.g. on the rigidity of symplectic mappings. Therefore, the properties of these invariants were axiomatized and the notion of symplectic capacity was introduced.

These invariants are closely related to the existence of closed Hamiltonian trajectories on energy surfaces. That is why, in section 4.1, we shall first derive the existence results we shall need for the actual construction of a capacity in section 4.2. These existence results are of interest in their own right. The proof uses a variation principle to identify the closed trajectories with the critical points of some functional. The Palais-Smale condition of this functional and Leray-Schauder degree theory are then used to find a suitable critical point.

After the definition of a symplectic capacity in section 4.2, we shall first derive the squeezing theorem and some rigidity results as consequences of the existence of a capacity. Then we shall construct the Hofer-Zehnder capacity [58], which is the one most directly related to Hamiltonian trajectories. For comparision, the Ekeland-Hofer capacity [24] will also be described briefly.

4.1 Existence of Periodic Solutions

Let $(\mathbf{R}^{2n}, \omega)$ be euclidean space with the standard symplectic structure. If $S = H^{-1}(0)$ is a regular hypersurface in $\mathbf{R}^{2n}$, i.e. $\nabla H \neq 0$ on S, then the unparametrized orbits on S of the Hamiltonian vector field $X = J\nabla H$ are independent of the choice of H (only the length of the vector X depends on H, but not its direction). One can therefore ask which hypersurfaces carry a periodic orbit. A general result was given in 1978 by P. Rabinowitz [87] who proved that every star-like hypersurface carries a periodic orbit. Later, C. Viterbo [100] showed the same holds more generally for hypersurfaces of contact type (cf. below). In this section, following H. Hofer and E. Zehnder [57], we shall prove an even more general theorem which will imply Viterbo's result.

Definition 4.1 *A parametrized family of compact hypersurfaces in* $\mathbf{R}^{2n}$*, modelled on a compact hypersurface S is a diffeomorphism $\psi : (-1, 1) \times S \to \mathbf{R}^{2n}$ onto a bounded open neighbourhood of S, so that $\psi(0, x) = x$ for $x \in S$. We abbreviate $\psi(\{\varepsilon\} \times S)$ by S_ε.*

Example. Let $S \subset \mathbf{R}^{2n}$ be a smooth compact hypersurface. Choose a smooth function H on $\mathbf{R}^{2n}$ such that $S = H^{-1}(0)$ and $\nabla H(x) \neq 0$ on S. Denote the (local) flow associated to $c\frac{\nabla H}{|\nabla H|^2}$ by φ^t. If the constant c is small enough, one can define a parametrized family of hypersurfaces $\psi : (-1, 1) \times S \to \mathbf{R}^{2n}$ by $\psi(t, x) = \varphi^t(x)$. By definition it satisfies $H(\psi(t, x)) = ct$.

For a loop x with period T the *action* is defined by

$$A(x) = \frac{1}{2} \int_0^T \langle -J\dot{x}, x \rangle dt = \int_{[0,T]} x^* \lambda,$$

where $\lambda = \sum p_i dq_i$ is the canonical 1-form satisfying $d\lambda = \omega$.

Theorem 4.2 (Hofer-Zehnder) *Let $S \subset \mathbf{R}^{2n}$ be a smooth compact hypersurface and ψ an associated parametrized family of hypersurfaces. Then there is a constant $d = d(\psi)$ so that there exist arbitrarily small ε for which S_ε carries a periodic orbit x and $0 < A(x) < d$.*

Before giving the proof, let us derive Viterbo's result as a corollary. For further applications see [57]. Recall that a hypersurface is *of contact type* if there exists a vector field X in a neighbourhood of S which is transverse to S and is homothetic, i.e. satisfies $\mathcal{L}_X \omega = \omega$. For an equivalent definition see section 8.2. Note that every starlike hypersurface in $\mathbf{R}^{2n}$ is of contact type. The homothetic vector field is given by $X(x) = \frac{1}{2}x$.

Corollary 4.3 *Every smooth compact hypersurface in $(\mathbf{R}^{2n}, \omega)$ which is of contact type has a periodic orbit.*

Proof. Let X be the vector field transverse to S with $\mathcal{L}_X \omega = \omega$. The flow φ^t generated by X allows to define a special parametrized family of hypersurfaces modelled on S: $\psi(\varepsilon, x) = \varphi^\varepsilon(x)$ for $x \in S$. Now, since $\frac{d}{dt}(\varphi^t)^* \omega = (\varphi^t)^* \mathcal{L}_X \omega = (\varphi^t)^* \omega$, we get

$$(\varphi^t)^* \omega = e^t \omega.$$

It follows that y is a periodic orbit on S_ε if and only if $x = \varphi^{-\varepsilon} \circ y$ is a periodic orbit of $S = S_0$. (Note that y is a Hamiltonian orbit on S_ε if and only if $\omega(\dot{y}(t), v) = 0 \ \forall v \in T_{y(t)} S_\varepsilon = \varphi^\varepsilon_* T_{x(t)} S$.) The claim now follows from the theorem. $\square$

Proof of Theorem 4.2. Let $\psi : (-1, 1) \times S \to \mathbf{R}^{2n}$ be the parametrized family of hypersurfaces and let $\delta \in (0, 1)$ be arbitrary. We must show there exists $\varepsilon \in (-\delta, \delta)$ so that S_ε carries a periodic orbit. Since S is cooriented, the complement of $\psi([-\delta, \delta] \times S)$ has two components, provided δ is small enough. Let A be the unbounded and B the bounded component. Without loss of generality we may assume $0 \in B$ and $S_\varepsilon \subset B$ for $-1 < \varepsilon < -\delta$. We set $U = \psi((-1, 1) \times S)$ and $\gamma = \operatorname{diam}(U)$.

Now Hofer and Zehnder construct a special Hamiltonian function $H \in C^\infty(\mathbf{R}^{2n}, \mathbf{R})$ for their variational argument. Choose r and b so that

$$\gamma < r < 2\gamma, \quad \frac{3}{2}\pi r^2 < b < 2\pi r^2. \tag{4.1}$$

Choose $f \in C^\infty((-1,1), \mathbf{R})$ so that

$$f|_{(-1,-\delta]} = 0, \quad f|_{[\delta,1)} = b, \quad f'(s) > 0 \text{ for } -\delta < s < \delta$$

and $g \in C^\infty((0,\infty), \mathbf{R})$ such that

$$\begin{aligned}
g(s) &= b, & s &\leq r \\
g(s) &= \tfrac{3}{2}\pi s^2, & s &\text{ large} \\
g(s) &\geq \tfrac{3}{2}\pi s^2, & s &> r \\
0 &< g'(s) \leq 3\pi s, & s &> r.
\end{aligned}$$

Now define

$$H(x) = \begin{cases}
0, & x \in B \\
f(\varepsilon), & x \in S_\varepsilon \text{ for some } \varepsilon \in [-\delta, \delta] \\
b, & x \in A \text{ and } |x| \leq r \\
g(|x|), & |x| > r
\end{cases}$$

It follows that

$$-b + \frac{3}{2}\pi|x|^2 \leq H(x) \leq b + \frac{3}{2}\pi|x|^2 \quad \forall\, x \in \mathbf{R}^{2n}.$$

The 1-periodic solutions of $\dot{x} = J\nabla H(x)$ are the critical points of the functional

$$\Phi(x) = \frac{1}{2}\int_0^1 \langle -J\dot{x}, x\rangle dt - \varphi(x), \quad \varphi(x) = \int_0^1 H(x(t))dt, \tag{4.2}$$

defined on the space of loops $x : [0,1] \to \mathbf{R}^{2n}$ with $x(0) = x(1)$.

Lemma 4.4 *If x is a 1-periodic solution of $\dot{x} = J\nabla H(x)$ with $\Phi(x) > 0$, then the image of x is contained in S_ε for some $\varepsilon \in (-\delta, \delta)$.*

Proof. If x is constant, then $\Phi(x) \leq 0$, since H is nonnegative. Assume now x to be nonconstant and $|x(0)| > r$. Then $|x(t)|$ is constant and x satisfies $-J\dot{x} = \frac{g'(|x|)}{|x|}\, x$. Hence by the definition of H,

$$\begin{aligned}
\Phi(x) &= \frac{1}{2}g'(|x(0)|)\,|x(0)| - H(x(0)) \\
&\leq \frac{3}{2}\pi|x(0)|^2 - \frac{3}{2}\pi|x(0)|^2 = 0
\end{aligned}$$

and the statement follows. $\qquad\square$

For the proof of the theorem, it thus suffices to find a 1-periodic solution x with $\Phi(x) > 0$. For this, Leray-Schauder degree theory or a 'linking argument' will be used.

Let $E = H^2_{\frac{1}{2}}(S^1, \mathbf{R}^{2n})$ be the Hilbert space of all $x \in L^2(S^1, \mathbf{R}^{2n})$ whose Fourier coefficients $x_k \in \mathbf{R}^{2n}$,

$$x(t) = \sum_{k=-\infty}^{\infty} e^{2\pi k t J} x_k,$$

satisfy

$$\sum_{k=-\infty}^{\infty} |k||x_k|^2 < \infty.$$

The scalar product is

$$(x, y) = \langle x_0, y_0 \rangle + 2\pi \sum_{k=-\infty}^{\infty} |k|\langle x_k, y_k \rangle.$$

Note that the elements of E need not be continuous. In this section, unless otherwise specified. $(\,.\,)$ and $\|\,\|$ will always refer to the product and norm of E. We use the obvious decomposition $E = E^- \oplus E^0 \oplus E^+$ and let P^-, P^0, P^+ denote the corresponding orthogonal projection operators. An easy calculation yields

$$\Phi(x) = \frac{1}{2}((-P^- + P^+)x, x) - \varphi(x). \tag{4.3}$$

Lemma 4.5 *The embedding $E \hookrightarrow L^p(S^1, \mathbf{R}^{2n})$ is compact for $1 \leq p < \infty$.*

Proof. This is easy to show for $p = 2$; hence it immediately follows for $1 \leq p \leq 2$. To get it for $p > 2$. note that by [114, Thm. XII.5.24], one has a continuous imbedding

$$H^2_\alpha(S^1, \mathbf{R}^{2n}) \hookrightarrow L^p(S^1, \mathbf{R}^{2n})$$

for $\alpha = \frac{1}{2} - \frac{1}{p}$. So it suffices to show one has a compact embedding

$$H^2_{\frac{1}{2}}(S^1, \mathbf{R}^{2n}) \hookrightarrow H^2_\alpha(S^1, \mathbf{R}^{2n})$$

for $-\frac{1}{2} \leq \alpha < \frac{1}{2}$.

Assume (x^ν) is a sequence in $H^2_{\frac{1}{2}}$ such that $\|x^\nu\|_{\frac{1}{2},2} \leq 1$ for all ν. Of course, the norm $\|.x\|$ of E is equivalent to

$$\|x\|_{\frac{1}{2},2} = \left(\sum_{k=-\infty}^{\infty} (1 + |k|^2)^{\frac{1}{2}} |x_k|^2 \right)^{\frac{1}{2}}.$$

Hence $(1 + |k|^2)^{\frac{1}{4}}|x^\nu_k| \leq 1$ for all k and all ν. Using a diagonal process we can choose a subsequence. which we again denote by (x^ν), such that all the Fourier coefficients converge:

$$\lim_{\mu,\nu\to\infty} |x^\nu_k - x^\mu_k| = 0 \quad \forall k \in \mathbf{Z}.$$

Given $\varepsilon > 0$ we now choose K so that

$$\frac{(1 + |k|^2)^{\alpha/2}}{(1 + |k|^2)^{1/4}} < \varepsilon \quad \text{for } |k| \geq K.$$

Then

$$\|x^\nu - x^\mu\|_{\alpha,2} = \left(\sum_{k=-\infty}^{\infty} (1 + |k|^2)^\alpha |x_k^\nu - x_k^\mu|^2 \right)^{\frac{1}{2}}$$

$$\leq \varepsilon \|x^\nu - x^\mu\|_{\frac{1}{2},2} + \left(\sum_{|k|<K} (1 + |k|^2)^\alpha |x_k^\nu - x_k^\mu|^2 \right)^{\frac{1}{2}}.$$

Here the first summand is less than 2ε and the second converges to zero when $\mu, \nu \to \infty$. showing that (x^ν) is Cauchy in $H_\alpha^2(S^1 . \mathbf{R}^{2n})$. $\qquad\square$

The gradient $\varphi'(x) \in E$ is defined by

$$(\varphi'(x). y) = d\varphi_x(y) = \int_0^1 \langle \nabla H(x), y \rangle dt = (\nabla H(x). y)_{L^2} \quad \forall y \in E.$$

Lemma 4.6 *The action* $\Phi : E \to \mathbf{R}$ *is smooth and the map* $E \to E, x \mapsto \varphi'(x)$
is compact.

Proof. It suffices to show $\varphi : E \to \mathbf{R}$ is smooth. Using the Taylor series for H
with remainder. we get

$$\varphi(x + u) = \varphi(x) + \varphi'_x(u) + \cdots + \frac{1}{n!} \varphi_x^{(n)}(u^{[n]}) + R_n(x, u),$$

where

$$\varphi_x^{(n)}(u^{[n]}) = \int_0^1 H_{x(t)}^{(n)}(u(t)^{[n]}) \, dt,$$

$$R_n(x, u) = \int_0^1 \int_0^1 \frac{(1 - \tau)^n}{n!} H_{x(t)+\tau u(t)}^{(n+1)}(u(t)^{[n+1]}) \, d\tau dt.$$

Hence,

$$|R_n(x, u)| \leq \int_0^1 \int_0^1 \frac{(1 - \tau)^n}{n!} |H_{x(t)+\tau u(t)}^{(n+1)}(u(t)^{[n+1]})| \, d\tau dt.$$

Now, $H(x) = \frac{3}{2}\pi|x|^2 + h(x), h \in C_c^\infty$, implies

$$|H_{x(t)+\tau u(t)}^{(n+1)}(u(t)^{[n+1]})| \leq c|u(t)|^{n-1},$$

whence

$$|R_n(x, u)| \leq c' \int_0^1 |u(t)|^{n+1} dt = c'\|u\|_{L^{n+1}}^{n+1} \leq c''\|u\|^{n+1}.$$

To show that $x \mapsto \varphi'(x)$ is compact, we first note that $L^2 \to L^2$, $x \mapsto \nabla H(x)$ is continuous. This follows from $\nabla H(x) = 3\pi x + k(x)$, where k has compact support, hence is Lipschitz.

Now let (x_i) be a bounded sequence in E. By the compact embedding $E \hookrightarrow L^2$ there is a subsequence such that $x_i \to x$ in $L^2(S^1, \mathbf{R}^{2n})$ and by continuity, $\nabla H(x_i) \to \nabla H(x)$ in L^2. It follows that $\varphi'(x_i)$ is a Cauchy sequence:

$$\|\varphi'(x_i) - \varphi'(x_j)\| = \sup_{\|u\|=1} (\nabla H(x_i) - \nabla H(x_j), u)_{L^2} \leq \|\nabla H(x_i) - \nabla H(x_j)\|_{L^2},$$

using the Cauchy-Schwartz inequality and $\|u\|_{L^2} \leq \|u\|$. $\qquad\square$

Lemma 4.7 *There exist $\alpha \in (0,1)$ and $\beta > 0$ such that*

$$\Phi|_\Gamma \geq \beta \qquad for \qquad \Gamma = \{x \in E^+ : \|x\| = \alpha\}.$$

Proof. This follows immediately from $\Phi(0) = 0$, $\Phi'(0) = 0$, $\Phi''(0) = -P^- + P^+$. (Note that $H''(0) = 0$ implies $\varphi''(0) = 0$.) $\qquad\square$

Let $e \in E^+$ be the function $e(t) = (2\pi)^{-1/2} e^{2\pi t J} a$ for some fixed unit vector $a \in \mathbf{R}^{2n}$. Note that $\|e\| = 1$. We set $\hat{E} = E^- \oplus E^0 \oplus \mathbf{R}e$ and

$$\Sigma = \{x^- + x_0 + se \in \hat{E} : \|x^- + x_0\| \leq \rho, 0 \leq s \leq \rho\},$$

where $\rho > 1$ will be chosen later. Let $\partial\Sigma$ denote the boundary of Σ in $\hat{E}$.

Lemma 4.8 *For ρ sufficiently large, $\Phi|_{\partial\Sigma} \leq 0$.*

Proof. From $H(x) \geq -b + \frac{3}{2}\pi|x|^2$ we get $\varphi(x) \geq -b + \frac{3}{2}\pi\|x\|_{L^2}^2$. Hence for $x = x^- + x^0 + se$

$$\Phi(x) \leq \frac{1}{2}((-P^- + P^+)x, x) + b - \frac{3\pi}{2}\|x\|_{L^2}^2 \leq -\frac{1}{2}\|x^-\|^2 - \frac{s^2}{4} - \frac{3\pi}{2}\|x^0\|^2 + b,$$

where $\|x\|_{L^2}^2 \geq \|x^0\|^2 + \frac{s^2}{2\pi}$ was used. On the other hand, by (4.3), $\Phi|_{E^- + E^0} \leq 0$ and the result follows. $\qquad\square$

The gradient flow of Φ,

$$\frac{d}{dt}x = -\Phi'(x), \qquad x : \mathbf{R} \to E,$$

is defined globally, because Φ is smooth and $\Phi'(x) = (-P^- + P^+)x - \varphi'(x)$ grows linearly for large x:

$$\|\varphi'(x)\| = \sup_{\|u\|=1} (\nabla H(x), u)_{L^2}$$
$$\leq \|\nabla H(x)\|_{L^2} = \|3\pi x + k(x)\|_{L^2} \leq c\|x\|_{L^2} \leq c\|x\|,$$

where k is Lipschitz and $k(0) = 0$. For simplicity we shall denote the flow by $(x, t) \mapsto x \cdot t$.

Since Φ can only decrease along trajectories of the flow, the two preceding lemmas imply

$$(\partial \Sigma) \cdot t \cap \Gamma = \emptyset \qquad \forall\, t \geq 0. \tag{4.4}$$

The decisive lemma now is the following; here Leray-Schauder degree theory or a 'linking argument' will be used.

Lemma 4.9 $\Sigma \cdot t \cap \Gamma \neq \emptyset \quad \forall\, t \geq 0.$

Proof. The statement is equivalent to

$$\forall\, t \geq 0 \; \exists\, x \in \Sigma : \quad (e^{-t} P^- + P^0)(x \cdot t) = 0, \; \|x \cdot t\| = \alpha. \tag{4.5}$$

Variation of the constant yields for the flow

$$x \cdot t = e^t x^- + x_0 + e^{-t} x^+ + B(t, x).$$
$$B(t, x) = \int_0^t (e^{t-s} P^- + P^0 + e^{-(t-s)} P^+)\, \varphi'(x \cdot s)\, ds.$$

The map $B : \mathbf{R} \times E \to E$ is smooth and compact. With this notation and $x = x^- + x^0 + se$, (4.5) can be written

$$x + T_t(x) = 0 \quad \text{with} \quad T_t(x) = -(s + \alpha - \|x \cdot t\|)\, e + (e^{-t} P^- + P^0)\, B(t, x).$$

We must show this equation has a solution for all $t \geq 0$. This will be done by Leray-Schauder theory (see e.g. [19, chapter VII.B.3] or [92, chapter III]).

Note that the map $(t, x) \mapsto T_t(x)$ is compact because the first summand is one-dimensional, B is compact and $e^{-t} P^- + P^0$ is continuous. Let Ω be the interior of Σ in $\hat{E}$. By (4.4) there is no solution on $\partial \Omega = \partial \Sigma$, hence by invariance of degree under compact homotopies,

$$\deg\,(id + T_t, \Omega, 0) = \deg\,(id + T_0, \Omega, 0).$$

Since $B(0, x) = 0$, $T_0(x) = -(s + \alpha - \|x\|)\, e$. Hence $T_0 : \bar{\Omega} \to \hat{E}$ is compactly homotopic to the constant map $-\alpha e$. The homotopy is given by $R_\sigma(x^- + x^0 + se) = -(s + \alpha - \|\sigma(x^- + x^0) + se\|)\, e$. Since $\tau > \alpha$, we have $x + R_\sigma(x) \neq 0$ for $x \in \partial \Omega, \sigma \in [0, 1]$. Therefore,

$$\deg\,(id + T_t, \Omega, 0) = \deg\,(id - \alpha e, \Omega, 0) = 1 \quad (\alpha e \in \Omega),$$

which means there is a solution for all $t \geq 0$ as claimed. $\qquad\square$

Lemma 4.10 Φ *satisfies the* **Palais-Smale** *condition, i.e. every sequence* (x_i) *in E such that $\Phi(x_i)$ is bounded and $\Phi'(x_i) \to 0$ has a convergent subsequence. (The limit is automatically a critical point of Φ).*

Proof. Assume first (x_i) is bounded. Then by Lemma 4.6, $\varphi'(x_i)$ has a convergent subsequence. From $\Phi'(x_i) = -P^- x_i + P^+ x_i - \varphi'(x_i) \to 0$ it now follows that $(P^- x_i)$ and $(P^+ x_i)$ have convergent subsequences. The same holds for $P^0 x_i$ since E^0 is one-dimensional.

Assume now $\|x_i\| \to \infty$. We shall show this case cannot occur. First we claim that $z_i = \frac{1}{\|x_i\|} \varphi'(x_i)$ is a Cauchy sequence in E. Set $v_i = x_i/\|x_i\|$. Since $E \hookrightarrow L^2$ is compact, we may assume v_i converges in L^2. Using $\nabla H(x) = 3\pi x + k(x)$, $k \in C_c^\infty(\mathbf{R}^{2n}, \mathbf{R}^{2n})$ once again, we get

$$
\begin{aligned}
\|z_i - z_j\| &= \sup_{\|u\|=1} \left(\frac{1}{\|x_i\|} \nabla H(x_i) - \nabla H(x_j), u \right)_{L^2} \\
&= \sup_{\|u\|=1} \left(3\pi(v_i - v_j) + \frac{1}{\|x_i\|} k(x_i) - \frac{1}{\|x_j\|} k(x_j), u \right)_{L^2} \\
&\leq 3\pi \|v_i - v_j\|_{L^2} + \frac{1}{\|x_i\|} \|k(x_i)\|_{L^2} + \frac{1}{\|x_j\|} \|k(x_j)\|_{L^2},
\end{aligned}
$$

proving the claim.

Together with $-P^- v_i + P^+ v_i - \varphi'(x_i)/\|x_i\| \to 0$ it follows that $P^\pm v_i$ and hence v_i have a subsequence converging in E, say $v_i \to v \in E$. As in the above estimate one sees that

$$
\frac{1}{\|x_i\|} \varphi'(x_i) \to \varphi'_\infty(v) = 3\pi v
$$

in E, where $\varphi_\infty(x) = \int_0^1 \frac{3\pi}{2} \|x(t)\|^2 dt$. This means that v is a solution of the problem

$$
-J\dot{v} = 3\pi v, \quad v(0) = v(1).
$$

But the self-adjoint operator on the left hand side has spectrum $2\pi \mathbf{Z}$, contradicting $\|v\| = 1$. $\qquad\square$

The next (and last) lemma provides a critical point x with $\Phi(x) > 0$, hence a periodic solution in S_ε by Lemma 4.4.

Lemma 4.11 *The limit* $c = \lim_{t \to +\infty} \sup \Phi(\Sigma \cdot t)$ *exists,* $c \geq 3$ *and* c *is a critical value for* Φ.

Proof. Since $0 \leq \varphi(x) \leq b + \frac{3\pi}{2} \|x\|^2$, $\Phi(\Sigma)$ is bounded in $\mathbf{R}$. Hence $\sup \Phi(\Sigma \cdot t)$ is non-increasing and $\geq \beta$ by Lemmas 4.9 and 4.7.

Assume c is not a critical level. We claim there exists a positive ε such that for large τ

$$
\Phi^{c+\varepsilon} \cdot \tau \subseteq \Phi^{c-\varepsilon},
$$

where $\Phi^a = \{x \in E : \Phi(x) \leq a\}$. By the Palais-Smale condition the set C of critical values of Φ is closed. Hence we can choose ε such that $C \cap [c-\varepsilon, c+\varepsilon] = \emptyset$.

By the Palais-Smale condition there exists $\delta > 0$ such that $\|\Phi'(x)\| \geq \delta > 0$ for $x \in \Phi^{-1}[c - \varepsilon, c + \varepsilon]$ and the claim follows.

By the definition of c we can pick $t > 0$ such that $\Sigma \cdot t \subseteq \Phi^{c+\varepsilon}$. Then, $\Sigma \cdot (t + \tau) = (\Sigma \cdot t) \cdot \tau \subseteq \Phi^{c+\varepsilon} \cdot \tau \subseteq \Phi^{c-\varepsilon}$, contradicting the definition of c. $\qquad\square$

To finish the proof of Theorem 4.2 it only remains to estimate the action $A(x) = \Phi(x) + \varphi(x)$. By the definition of H, $0 < H(x(t)) \leq b$, hence $0 < \varphi(x) \leq b$. By the definition of c, $\beta \leq \Phi(x) \leq \sup \Phi(\hat{E}) \leq b$ (see the estimate in the proof of Lemma 4.8). Thus, $0 < \beta \leq A(x) \leq 2b < 16\gamma^2$, where $\gamma = \mathrm{diam}(\psi(-1, 1) \times S)$ is independent of δ. $\qquad\square$

Remarks. 1. The same proof (replacing the interval $(-1, 1)$ by $(\varepsilon_0 - \delta, \varepsilon_0 + \delta)$ for arbitrary $\varepsilon_0 \in (-1, 1)$) shows that the set of ε values for which S_ε contains a periodic orbit is dense in $(-1, 1)$.

2. A clever refinement of the proof due to M. Struwe [99] shows that the set actually has full Lebesgue measure.

3. More generally, this holds for compact hypersurfaces S in symplectic manifolds (M, ω), provided S bounds a compact symplectic manifold and $c_{HZ}(M, \omega) < \infty$, where c_{HZ} is the Hofer-Zehnder capacity to be introduced in the next section (see [59]).

4. It is not known whether every compact surface in (R^{2n}, ω_0) carries a closed characteristic.

4.2 Symplectic Capacities

A symplectic invariant for symplectic manifolds was first constructed by Gromov [50] who called it the *symplectic radius*

$$\mathrm{rad}(M) = \sup \{r : \exists \text{ a symplectic embedding } B(r) \to M\},$$

where $B(r)$ is the euclidean ball with radius r and standard symplectic structure ω_0. Similar invariants were constructed by Ekeland and Hofer [24, 25], Hofer and Zehnder [58], Viterbo and others [46, 103]. Such invariants are now called symplectic capacities. As we shall see, they are closely related to the existence of closed Hamiltonian trajectories.

In this section, after the general definition of capacities and some consequences of their existence, a complete construction of the Hofer-Zehnder capacity will be given. Finally, the Ekeland-Hofer capacity will be described briefly.

Definition 4.12 *A symplectic capacity of dimension $2n$ is a map c assigning to every symplectic $2n$-manifold (M, ω) (possibly with boundary) an element of $[0, +\infty]$ and satisfying the following four axioms.*

C1. Monotonicity. If $\varphi : (M, \omega) \to (M', \omega')$ is a symplectic embedding, then $c(M, \omega) \leq c(M', \omega')$.

C2. Conformality. For every $a \in \mathbf{R} \setminus \{0\}$, $c(M, a\omega) = |a| \, c(M, \omega)$.

C3. Local nontriviality. For the open unit ball $B \subset \mathbf{R}^{2n}$,

$$c(B, \omega_0) > 0.$$

C4. Nontriviality. For the open cylinder $Z = \{z \in \mathbf{C}^n : |z_1| < 1\}$,

$$c(Z, \omega_0) < \infty.$$

Often symplectic capacities are defined only for subsets of symplectic vector spaces, as e.g. the Ekeland-Hofer capacity [24]. Of course, axiom C1 implies c is a *symplectic invariant*: if $\varphi : (M, \omega) \to (M', \omega')$ is a symplectic diffeomorphism, then $c(M, \omega) = c(M', \omega')$.

Note that a capacity cannot assume the value zero. Indeed, by the Darboux Theorem 2.3 one can symplectically embed an ε-ball $(B(\varepsilon), \omega_0)$ into (M, ω) for some $\varepsilon > 0$. Axioms C1 to C3 then imply

$$c(M, \omega) \geq c(B(\varepsilon), \omega_0) = \varepsilon^2 c(B, \omega_0) > 0.$$

It can be shown that the square of Gromov's radius is a capacity. Only the verification of C4 is nontrivial. It follows from Gromov's *squeezing theorem*: if $\varphi : B(r) \to Z(r')$ is a symplectic embedding, then $r' \geq r$. In fact, the so-called *Darboux width*

$$\underline{c} = \pi \, \mathrm{rad}^2$$

is the smallest capacity such that B and Z have capacity π. Analogously, one can define the largest capacity such that B and Z have capacity π:

$$\bar{c}(M) = \inf \left\{ \pi r^2 : \exists \text{ a symplectic embedding } M \to Z(r) \right\}.$$

The squeezing theorem is actually equivalent to the existence of a capacity satisfying $c(B) = c(Z)$. Indeed,

$$r^2 c(B) = c(B(r)) \leq c(Z(r')) = r'^2 c(Z) = r'^2 c(B).$$

(For another proof of the squeezing theorem using holomorphic curves, see section 6.3.6.)

Before constructing a capacity explicitly, we give some more consequences of the existence of a capacity. An *ellipsoid* in $\mathbf{R}^{2n}$ is a set

$$E = \{x \in \mathbf{R}^{2n} : q(x) < 1\}$$

for some positive definite quadratic form q. By C3 and C4, we have $0 < c(E, \omega_0) < \infty$.

The following lemma is an easy consequence of the axioms (cf. [24, Theorem 3]).

Lemma 4.13 *If a linear map* $\varphi : \mathbf{R}^{2n} \to \mathbf{R}^{2n}$ *preserves the capacities of all ellipsoids, then*

$$\varphi^* \omega_0 = \omega_0 \quad or \quad \varphi^* \omega_0 = -\omega_0.$$

Arguing as in [24, Theorems 5 and 6] one deduces the following C^0-rigidity result in $(\mathbf{R}^{2n}, \omega_0)$.

Theorem 4.14 *Let* $\varepsilon > 0$ *and let* $\varphi_i : B(\varepsilon) \to \mathbf{R}^{2n}$ *be a sequence of symplectic embeddings converging uniformly to a continuous map* φ. *If* φ *is differentiable at* 0, *then* $\varphi'(0)$ *is symplectic.*

Using locally Darboux charts one concludes the following.

Corollary 4.15 *On a compact symplectic manifold* (M, ω) *the set of symplectic* C^∞-*diffeomorphisms is* C^0-*closed in the set of* C^∞-*maps on* M.

A different proof of this result is outlined in [50].

We shall now construct the Hofer-Zehnder capacity. Let (M, ω) be a symplectic manifold. Roughly, the capacity will measure how large a positive Hamiltonian $H : M \to \mathbf{R}$ can get when it is zero on an open set and does not allow nonconstant periodic solutions with period less than one. To make this precise, let $\mathcal{H}(M, \omega)$ denote the set of smooth functions $H : M \to [0, \infty)$ satisfying the following two properties.

P1. For every $H \in \mathcal{H}(M, \omega)$ there is a compact subset $K \subset M \setminus \partial M$ and a constant m_H such that

$$H(M \setminus K) = m_H \qquad \text{and} \qquad 0 \le H(x) \le m_H \quad \forall\, x \in M.$$

P2. For every H there is a subset $U \subset M \setminus \partial M$ with nonempty interior such that $H|_U = 0$.

A function $H \in \mathcal{H}(M, \omega)$ is called *admissible* if all T-periodic solutions of the Hamiltonian equation $-J\dot{x} = \nabla H(x)$ on M with period $T \le 1$ are constant.

Definition 4.16 *The* Hofer-Zehnder capacity *is defined by*

$$c_{HZ}(M, \omega) = \sup\{m_H : H \in \mathcal{H}(M, \omega),\ H \text{ admissible}\}.$$

Theorem 4.17 c_{HZ} *is a symplectic capacity and* $c_{HZ}(Z) = c_{HZ}(B) = \pi$.

Proof. We have to verify C1–C4.

1. If $\varphi : M \to M'$ is a symplectic embedding and $H \in \mathcal{H}(M)$ is admissible, then the function

$$H'(x) = \begin{cases} H(\varphi^{-1}(x)) & \text{if } x \in \varphi(M) \\ m_H & \text{otherwise} \end{cases}$$

belongs to $\mathcal{H}(M')$ and is admissible.

2. We claim that $H \in \mathcal{H}(M,\omega)$ is admissible if and only if $H_a = |a|H$ is admissible in $\mathcal{H}(M,a\omega)$. Let X and Y denote the Hamiltonian vector fields given by H on $(M.\omega)$ and by H_a on $(M,a\omega)$, respectively. Then.

$$i(Y)\,a\omega = dH_a = |a|\,dH = |a|\,i(X)\,\omega,$$

hence $Y = \text{sign}(a)X$ and the claim follows. Since $m_{H_a} = |a|m_H$. we conclude $c_{HZ}(M,a\omega) = |a|c_{HZ}(M,\omega)$.

3. We show

$$c_{HZ}(B,\omega_0) \geq \pi. \tag{4.6}$$

For $\varepsilon > 0$ small choose a C^∞-function $f : [0,1] \to [0,\pi)$ such that $0 \leq f'(x) < \pi$, f equals 0 near 0 and $\pi - \varepsilon$ near 1. Set $H(x) = f(|x|^2)$. Obviously, $H \in \mathcal{H}(B,\omega_0)$. The estimate (4.6) will follow if we show H is admissible, since $m_H = \pi - \varepsilon$ and $\varepsilon > 0$ was arbitrary.

On a solution of the Hamiltonian equation

$$-J\dot{x} = \nabla H(x) = 2f'(|x|^2)\,x$$

$|x|$ and hence $\delta := 2f'(|x|^2)$ are constant. By the choice of f, $0 \leq \delta < 2\pi$. If $\delta = 0$ the solutions are constant, while if $\delta \neq 0$ they have period $2\pi/\delta > 1$. Thus H is admissible.

4. It remains to prove $c_{HZ}(Z,\omega_0) \leq \pi$. We shall show every $H \in \mathcal{H}(Z,\omega_0)$ with $m_H > \pi$ has a nonconstant 1-periodic solution.

Let $K \subset Z$ be a compact set such that $H(x) = m_H$ for $x \in Z \setminus K$. We may assume H vanishes in a neighborhood of the origin. We want to extend $H|_K$ to a function $\bar{H}$ defined an $\mathbf{R}^{2n}$. Choose $\varepsilon \in (0,\frac{1}{2})$ so that $m_H > \pi + \varepsilon$ and pick a smooth function $f : [0,\infty) \to \mathbf{R}$ satisfying

$$\begin{aligned}
f(s) &= m_H & 0 &\leq s \leq 1, \\
f(s) &\geq (\pi + \varepsilon)\,s^2 & s &\geq 0, \\
f'(s) &\leq 2(\pi + \varepsilon)\,s & s &\geq 0, \\
f(s) &= (\pi + \varepsilon)\,s^2 & &\text{for } s \text{ large.}
\end{aligned}$$

Choose $R > 1$ so that $K \subset B(R)$ and pick a smooth function $g : [0,\infty) \to \mathbf{R}$ such that

$$\begin{aligned}
g(s) &= 0 & 0 &\leq s \leq R, \\
0 &\leq g'(s) < 2\pi s & s &\geq 0, \\
g(s) &= \tfrac{\pi}{2}s^2 & &\text{for } s \text{ large.}
\end{aligned}$$

We decompose $x = (x_1.x_r)$, where $x_1 \in \mathbf{R}^2$ and $x_r \in \mathbf{R}^{2n-2}$ and set

$$\bar{H}(x) = \begin{cases} H(x) & x \in Z \cap B(R) \\ f(|x_1|) + g(|x_r|) & \text{otherwise.} \end{cases}$$

Clearly, $\bar{H} \in C^\infty(\mathbf{R}^{2n})$ and $\bar{H} = H$ on K. The 1-periodic solutions of $-J\dot{x} = \nabla\bar{H}(x)$ are the critical points of the functional (4.2) with H replaced by $\bar{H}$.

Lemma 4.18 *Every 1-periodic solution of* $-J\dot{x} = \nabla \bar{H}(x)$ *with* $\Phi(x) > 0$ *satisfies*

$$\{x(t) : t \in \mathbf{R}\} \subset K \quad and \quad -J\dot{x} = \nabla H(x).$$

Proof. We have to show $\Phi(x) \leq 0$ for every 1-periodic solution not contained in K. This is clear if x is constant, since $\bar{H} \geq 0$. If the solution x is not constant, it lies completely in the complement of K. It therefore satisfies $-J\dot{x} = \nabla \bar{H}(x)$, which splits into the independent equations

$$-J\dot{x}_1 = \frac{f'(|x_1|)}{|x_1|}\, x_1, \quad -J\dot{x}_r = \frac{g'(|x_r|)}{|x_r|}\, x_r.$$

Clearly, $|x_1|$ and $|x_r|$ are independent of t. From $0 \leq \frac{g'(|x_r|)}{|x_r|} < 2\pi$ we infer $\dot{x}_r = 0$. Thus,

$$\Phi(x) = \frac{1}{2}\int_0^1 \langle -J\dot{x}_1, x_1\rangle dt - f(|x_1|) - g(|x_r|).$$

By the choice of f and g we get

$$\begin{aligned}
\Phi(x) &= \frac{1}{2}f'(|x_1|)\,|x_1| - f(|x_1|) - g(|x_r|) \leq \\
&\leq (\pi + \varepsilon)\,|x_1|^2 - (\pi + \varepsilon)\,|x_1|^2 = 0,
\end{aligned}$$

which proves the lemma. $\qquad\qquad\qquad\qquad\qquad\qquad\qquad\qquad\square$

To finish the proof of Theorem 4.17, it only remains to show there is a 1-periodic solution of $-J\dot{x} = \nabla \bar{H}(x)$ with $\Phi(x) > 0$. This is done exactly as in the proof of Theorem 4.2. $\qquad\qquad\qquad\qquad\qquad\qquad\qquad\qquad\square$

With the same methods one can show (see [58, Proposition 4]).

Proposition 4.19 *Assume* $S = \partial C$ *is the smooth compact boundary of a convex region in* $(\mathbf{R}^{2n}, \omega_0)$. *Then there exists a periodic Hamiltonian trajectory* x^* *on* S *such that*

$$c_{HZ}(C) = A(x^*) = \min\{|A(x)| : x \text{ a periodic trajectory on } S\}.$$

Remember the action A is defined by

$$A(x) = \frac{1}{2}\int_0^1 \langle -J\dot{x}, x\rangle dt.$$

The same representation formula holds for the Ekeland-Hofer capacity, so that the two capacities agree on convex regions with smooth compact boundary.

For the convenience of the reader we now shortly reproduce the definition of the Ekeland-Hofer capacity [24]. For any $2n$-dimensional symplectic vector space (V, ω) there is a linear symplectic diffeomorphism $\psi : (V, \omega) \to (\mathbf{R}^{2n}, \omega_0)$. Setting

$$c_{EH}(S) = c_{EH}(\psi(S))$$

for any subset S of V, we can restrict the discussion to the standard space $(\mathbf{R}^{2n}, \omega_0)$. For a bounded subset $S \subset \mathbf{R}^{2n}$ define $\mathcal{F}(S)$ as the set of smooth Hamiltonians $H : \mathbf{R}^{2n} \to \mathbf{R}$ such that

$$
\begin{aligned}
H &\geq 0 && \text{on } \mathbf{R}^{2n}, \\
H &= 0 && \text{in an open neighborhood of the closure of } S \\
H(x) &= a|x^2| && \text{for large } |x| \text{ and some } a > \pi,\ a \notin \mathbf{Z}\pi.
\end{aligned}
$$

Further, we need the Hilbert space $E = E^- \oplus E^0 \oplus E^+$ introduced in section 4.1. Let Γ denote the group of homeomorphisms $h : E \to E$ such that

$$h(x) = e^{\gamma^+(x)} P^+ x + P^0 x + e^{\gamma^-(x)} P^- x + k(x).$$

where $\gamma^\pm : E \to \mathbf{R}$ are continuous and map bounded sets to bounded sets, $k : E \to E$ is continuous and maps bounded sets into compact sets. and moreover there is a $\rho > 0$ such that

$$A(x) \leq 0 \quad \text{or} \quad \|x\| \geq \rho \quad \text{implies} \quad h(x) = x \ \ (\text{i.e. } k(x) = 0.\ \gamma^\pm(x) = 0).$$

Set $S^+ = \{x \in E^+ : \|x\| = 1\}$. The *Ekeland-Hofer capacity* of the bounded set S is defined by

$$c_{EH}(S) = \inf_{H \in \mathcal{F}(S)} \ \sup_{h \in \Gamma} \ \inf_{x \in S^+} \ \Phi_H(h(x)).$$

where

$$\Phi_H(x) = A(x) - \int_0^1 H(x(t))dt.$$

For an unbounded set $S \subset \mathbf{R}^{2n}$ one defines

$$c_{EH}(S) = \sup\{c_{EH}(T) : T \subset S \text{ bounded}\}.$$

The Ekeland-Hofer capacity also satisfies $c_{EH}(B) = c_{EH}(Z) = \pi$. In [25] a sequence of capacities $c_1 \leq c_2 \leq \ldots$ is defined which do not satisfy $c_j(B) = c_j(Z)$.

5 Floer Homology

In this chapter we sketch Floer's proof of the Arnol'd conjecture [38, 39, 40, 41, 42]. The conjecture is concerned with the minimal number of fixed points of symplectomorphisms on compact manifolds. Floer's construction is now known as Floer homology. There will not be enough space for all the details, nevertheless we hope to give a clear overview. In the proof that Floer homology reproduces singular homology we mainly follow Salamon and Zehnder [91]. Another good reference is the review article by McDuff [71].

As a preparation, in the first section we shall present the classical Morse homology using (nonclassical) methods that generalize to the Floer homology case [90]. In particular, it will be shown that Morse homology agrees with the singular homology of the underlying manifold and that this fact implies the classical Morse inequalities. In the second section, the Arnol'd conjecture is discussed and the exact statement of Floer's results is given. The following four sections will be used to develop Floer's homology theory. In the last section we shall briefly sketch the construction of a symplectic homology theory due to Floer and Hofer [45].

5.1 Morse Homology and the Conley Index

In the infinite-dimensional case the classical methods of Morse theory (see e.g. [77, 14, 54]) can only be used if the Morse function has a well-behaved gradient flow with respect to a suitable metric and all the critical points have finite index. Unfortunately, this is not the case for the action functional (on the infinite dimensional loop space) we will have to consider in section 5.3. However, there is another approach to Morse theory (on finite-dimensional manifolds) which does generalize to the infinite dimensional situation. This is what we are going to explain in this section. First we will review some results on gradient dynamical systems, in particular the Conley index. Then we will define the Morse complex (or Morse-Smale-Witten complex) and show that its homology, called Morse homology, agrees with the usual singular homology of the underlying manifold. Finally we will see how this implies the (classical) Morse inequalities. Another, very detailed reference for Morse homology is [93].

Let M be a compact n-dimensional smooth manifold (without boundary) and $f : M \to \mathbf{R}$ a smooth function. A critical point p of f is called *nondegenerate* if for some (hence for any) local coordinate system at p the matrix $((\partial^2 f / \partial x_i \partial x_j)(p))$ is nonsingular. This matrix is called the *Hessian* of f at p. The number of its negative eigenvalues is independent of the choice of local coordinates and is called the *(Morse) index* $\mathrm{ind}(p)$ of f at p. A *Morse function* is a smooth function such that all its critical points are nondegenerate. Assume now we have a Morse function f on our compact manifold M. Choose a Riemannian metric g on M and consider the negative gradient flow φ^t of

f (associated to the vector field $-\nabla_g f$). The critical points of f are then hyperbolic fixed points of φ^t. The following lemma yields local coordinates near critical points which determine the local picture of the flow.

Lemma 5.1 (Morse) *If p is a nondegenerate critical point of f of index k, then there is a coordinate map $h : U \to \mathbf{R}^n$ on a neighborhood U of p such that*

$$f \circ h^{-1}(x) = f(p) - \sum_{i=1}^{k} x_i^2 + \sum_{i=k+1}^{n} x_i^2$$

For a proof see e.g. [77, p. 6] or [54, 6.1.1]. Using these local coordinates and the flow, it is clear that the sets

$$W^s(p) = \{x \in M : \lim_{t \to \infty} \varphi^t(x) = p\}, \quad W^u(p) = \{x \in M : \lim_{t \to -\infty} \varphi^t(x) = p\}$$

are submanifolds of M of dimension $n - \mathrm{ind}(p)$ and $\mathrm{ind}(p)$, respectively, diffeomorphic to open euclidean balls of the respective dimension.

Now we define the Conley index (or homotopy index) of an isolated invariant set. A subset S of M is *invariant* if $\varphi^t(S) = S$ for every $t \in \mathbf{R}$. It is *isolated* if there exists an 'isolating neighborhood' N of S such that

$$S = \bigcap_{t \in \mathbf{R}} \varphi^t(N)$$

i.e. S agrees with the maximal invariant set in N. An *index pair* for an isolated invariant set S is a pair (N, L) of compact sets $L \subset N \subseteq M$ such that S lies in the interior of $N \setminus L$, $\mathrm{cl}(N \setminus L)$ is an isolating neighborhood for S, and

(1) $x \in L, \varphi^{[0,t]}(x) \subset N \Rightarrow \varphi^t(x) \in L$,

(2) $x \in N \setminus L \Rightarrow \exists t > 0 : \varphi^{[0,t]}(x) \subset N$.

Condition (1) says that L is forward invariant in N and (2) means that the orbits can leave N only through L, which is therefore called an *exit set* for N. In [20, 88] it is shown that every isolated invariant set has an index pair such that the topological quotient N/L has the homotopy type of a finite polyhedron. Moreover, the homotopy type is independent of the choice of the index pair.

Lemma 5.2 (Conley) *If (N, L) and (N', L') are two index pairs for S, then the index spaces N/L and N'/L' are homotopy equivalent.*

For a proof, see [20] or [90]. The *Conley index* of S is by definition the homotopy type of the pointed space N/L. If L is a neighborhood deformation retract in N, then the homology of N/L agrees with $H_*(N, L)$. Hence the latter is independent of the choice of index pair.

We give two important examples of index pairs

Example 1. If p is a nondegenerate critical point of index k, then an index pair for the isolated invariant set $\{p\}$ can be given by

$$N = h^{-1}(D^k \times D^{n-k}), \quad L = h^{-1}(\partial D^k \times D^{n-k})$$

where h is a Morse chart near p and D^k is the closed unit ball in $\mathbf{R}^k$. The Conley index of $\{p\}$ is the homotopy type of a pointed k-sphere and in this sense, the Conley index generalizes the Morse index $\mathrm{ind}(p) = k$. Note also that $H_*(N, L) = H_*(D^k, \partial D^k)$.

Example 2. If S is the set of critical points of f on the (critical) level c, then an index pair is given by $N = M^b$, $L = M^a$, where $M^a = \{x \in M : f(x) \leq a\}$ and a, b are regular values of f such that c is the only critical value in the interval (a, b).

If the metric g is *generic*, i.e. belongs to a certain dense set of metrics, then the flow is of *Morse-Smale type*: for every pair (x, y) of critical points the unstable submanifold $W^u(x)$ intersects the stable submanifold $W^s(y)$ transversely, see [96]. We can identify the space of trajectories of the gradient flow connecting x to y

$$M(x, y) = \{\gamma : \mathbf{R} \to M : \dot{\gamma} = -\nabla_g f \circ \gamma, \ \lim_{t \to -\infty} \gamma(t) = x, \ \lim_{t \to \infty} \gamma(t) = y\}$$

with the intersection

$$M(x, y) = W^u(x) \cap W^s(y).$$

Observe that by the Morse lemma, $\dim W^u(x) = \mathrm{ind}(x)$ and $\mathrm{codim}\, W^s(y) = \mathrm{ind}(y)$. Since for transversal intersections codimensions add, we get

$$\dim M(x, y) = \mathrm{ind}(x) - \mathrm{ind}(y)$$

provided $M(x, y)$ is not empty. Because every trajectory $\gamma : \mathbf{R} \to M$ has a one-dimensional family of reparametrizations $\gamma_\tau(t) = \gamma(t + \tau)$, $\tau \in \mathbf{R}$, the *manifold of unparametrized trajectories* from x to y, $\hat{M}(x, y) = M(x, y)/\mathbf{R}$, has dimension

$$\dim \hat{M}(x, y) = \mathrm{ind}(x) - \mathrm{ind}(y) - 1.$$

In particular, connecting orbits from x to y exist only if $\mathrm{ind}(x) > \mathrm{ind}(y)$. We can identify $\hat{M}(x, y)$ with $M(x, y) \cap \{z \in M : f(z) = a\}$ for some a between $f(y)$ and $f(x)$. It follows that $\hat{M}(x, y)$ is compact. Indeed, if $\gamma_j(t)$ is a sequence of orbits connecting x to y, then a subsequence converges (in an appropriate sense) to a finite family of orbits γ^i connecting x^{i-1} to x^i for $i = 1, \ldots, m$, where $x^0 = x$, $x^m = y$ and $\mathrm{ind}(x^i) < \mathrm{ind}(x^{i-1})$.

Let us now define the *Morse complex* for coefficients in $\mathbf{Z}$. (Note that for coefficients in $\mathbf{Z}_2 = \mathbf{Z}/2\mathbf{Z}$ the following discussion simplifies significantly because signs do not matter, so that one does not have to consider orientations.)

For every critical point x of f we choose an orientation of the vector space $E^u(x) = T_x W^u(x)$ and denote by $\langle x \rangle$ the pair consisting of x and this orientation. For $k = 0, 1, \ldots, n = \dim M$ let C_k be the free group generated by the set of critical points of index k:

$$C_k = \bigoplus_{x:\, \mathrm{ind}(x)=k} \mathbf{Z}\langle x \rangle$$

Since f is of Morse-Smale type, if $\mathrm{ind}(x) = \mathrm{ind}(y) + 1$, then $\hat{M}(x.y)$ consists of finitely many connecting orbits, since it is zero-dimensional and compact. To every connecting orbit γ from x to y we can assign a sign $n_\gamma = \pm 1$ as follows. The orthogonal complement of $v = \lim_{t \to -\infty} \dot{\gamma}(t)/|\dot{\gamma}(t)|$ in $E^u(x)$ inherits an orientation from $\langle x \rangle$. The tangent flow induces an isomorphism from this orthogonal complement onto $E^u(y)$ and we define n_γ to be $+1$ or -1 according to whether this map is orientation preserving or not. Set

$$n(x, y) = \sum_\gamma n_\gamma$$

where the sum runs over all orbits connecting x to y. The boundary operator $\bar{\partial}_k : C_{k+1} \to C_k$ of the Morse complex is defined by

$$\bar{\partial}\langle x \rangle = \sum_y n(x, y)\langle y \rangle \tag{5.1}$$

where the sum runs over all critical points of index k. The significance of this construction rests on the following theorem, which is due to R. Thom, S. Smale, C. Conley, E. Witten.

Theorem 5.3 $(C_k, \bar{\partial})$ *is a chain complex:* $\bar{\partial}_{k-1} \circ \bar{\partial}_k = 0$ *and its homology reproduces the singular homology of* M: $\ker \bar{\partial}_{k-1}/\mathrm{im}\, \bar{\partial}_k \simeq H_k(M; \mathbf{Z})$.

Remark. Using the universal coefficient theorem, the Morse complex can be defined over any abelian group, and Theorem 5.3 remains valid if $\mathbf{Z}$ is replaced by any principal ideal domain (see e.g. [41] or [90]).

For the proof of Theorem 5.3 we will follow Salamon in [90]. First we have to describe Conley's connection matrix for the special case of a Morse-Smale gradient flow. For every critical point x of f let (N_x, L_x) be an index pair as in Example 1 above. Since an orientation of $E^u(x)$ determines a generator of $H_k(N_x, L_x) \simeq H_k(D^k, \partial D^k) \simeq \mathbf{Z}$ if $k = \mathrm{ind}(x)$, the group C_k can be identified with

$$C_k = \bigoplus_{x:\, \mathrm{ind}(x)=k} H_k(N_x, L_x)$$

If $\mathrm{ind}(x) = \mathrm{ind}(y) + 1$, then

$$S(x, y) = M(x, y) \cup \{x, y\}$$

is an isolated invariant set. Let (N_2, N_0) be an index pair for $S(x, y)$ and set $N_1 = N_0 \cup (N_2 \cap \{f \leq a\})$, where $f(y) < a < f(x)$. Then (N_2, N_1) is an index pair for $\{x\}$ and (N_1, N_0) is an index pair for $\{y\}$. Define the homomorphism

$$\Delta_k(x, y) : H_{k+1}(N_x, L_x) \to H_k(N_y, L_y)$$

to be the composition

$$H_{k+1}(N_x, L_x) \xrightarrow{\simeq} H_{k+1}(N_2, N_1) \xrightarrow{\partial} H_k(N_1, N_0) \xrightarrow{\simeq} H_k(N_y, L_y)$$

where the isomorphisms are induced by the homotopy equivalence of Lemma 5.2. This determines a homomorphism $\Delta_k : C_{k+1} \to C_k$ which is a special case of Conley's connection matrix and agrees with the boundary operator (5.1).

Lemma 5.4 $\bar{\partial} = \Delta$.

Proof. We restrict to the case where x and y are the only critical points of f in $f^{-1}([a, b])$, where $a = f(y)$ and $b = f(x)$. The general case can be reduced to this one by changing f outside an isolating neighborhood of $S(x, y)$, see [90, p. 119] and [88]. For $c \in (a, b)$, $\varepsilon > 0$ small and $T > 0$ large, define the index pairs

$$
\begin{aligned}
N_x &= \{z \in M : f(\varphi^{-T}(z)) \leq b + \varepsilon, \ f(z) \geq c\}, \\
L_x &= \{z \in N_x : f(z) = c\}, \\
N_y &= \{z \in M : f(\varphi^T z) \geq a - \varepsilon, \ f(z) \leq c\}, \\
L_y &= \{z \in N_y : f(\varphi^T z) = a - \varepsilon\}
\end{aligned}
$$

(The reader should make a picture of these sets, e.g. for $n = 3$, $\mathrm{ind}(y) = 1$.) An index triple for the repeller-attractor pair (x, y) in the isolated invariant set $S(x, y)$ is then given by

$$N_2 = N_x \cup N_y, \quad N_1 = N_y \cup L_x, \quad N_0 = L_y \cup \mathrm{cl}\,(L_x \setminus N_y)$$

Note that letting $T \to \infty$, N_x contracts onto $W^u(x) \cap \{f \geq c\}$ and N_y contracts onto $W^s(y) \cap \{f \leq c\}$. Since $W^u(x)$ and $W^s(y)$ intersect transversely, it follows that $N_y \cap W^u(x) \cap \{f = c\}$ consists of finitely many components $V_1, \ldots, V_m$, one for every orbit connecting x to y, and each V_j contains a unique point $z_j \in M(x, y) \cap V_j$. More precisely, there exists a diffeomorphism

$$\psi : N_y \to D^k \times D^{n-k}, \qquad k = \mathrm{ind}\,(y) = \dim W^u(y)$$

such that $\psi(L_y) = \partial D^k \times D^{n-k}$, $\psi(W^s(y) \cap N_y) = \{0\} \times D^{n-k}$ and $\psi(V_j) = D^k \times \{\theta_j\}$ for some $\theta_j \in \partial D^{n-k}$. Hence the map $\psi_1 = \pi_1 \circ \psi : N_y \to D^k$ and the diffeomorphism $\psi_1|V_j : V_j \to D^k$ induce isomorphisms in homology

$$H_k(N_y, L_y) \simeq H_k(D^k, \partial D^k) \simeq H_k(V_j, W_j)$$

where $W_j = V_j \cap L_y$ is the boundary of V_j. The orientation of $E^u(y)$ given by $\langle y \rangle$ determines a generator α of $H_k(N_y, L_y) \simeq \mathbf{Z}$. Under the above isomorphism it is mapped to a generator α_j of $H_k(V_j, W_j)$. The homology class α_j is determined by the orientation $T_{z_j} V_j$ inherits from $E^u(y)$ via the isomorphism $T_{z_j} V_j \to E^u(y)$ induced by the flow. This orientation agrees with the one inherited from $W^u(x)$ via the injection

$$T_{z_j} V_j = T_{z_j} W^u(x) \cap \nabla f(z_j)^\perp \hookrightarrow T_{z_j} W^u(x)$$

(choosing $-\nabla f(z_j)$ as the first basis vector) if and only if $n_j = 1$, where $n_j \in \{1, -1\}$ is the sign associated to the connecting orbit $\gamma_j(t) = \varphi^t(z_j)$.

Now choose a triangulation of the k-manifolds V_j and extend it to one of the $k+1$-manifold $W^u(x) \cap \{f \geq c\}$ with boundary $W^u(x) \cap \{f = c\}$. Together with the orientation of $W^u(x)$ given by $\langle x \rangle$ this determines a generator

$$\beta \in H_{k+1}(W^u(x) \cap N_x, W^u(x) \cap L_x) \simeq H_{k+1}(N_x, L_x)$$

The homology class $\partial_j \beta \in H_k(W^u(x) \cap L_x, \mathrm{cl}(W^u(x) \cap L_x \setminus V_j)) \simeq H_k(V_j, W_j)$ is represented by the original trangulation of V_j together with the orientation inherited from $W^u(x)$ and therefore agrees with $n_j \alpha_j$. Using the above isomorphisms $H_k(N_y, L_y) \simeq H_k(V_j, W_j)$, we obtain

$$\Delta_k(x, y)\,\beta = \sum_{j=1}^{m} n_j\, \alpha = n(x, y)\, \alpha \in H_k(N_y, L_y)$$

Identifying β with $\langle x \rangle$ and α with $\langle y \rangle$ (as in the definition of Δ_k) and summing over all $y \in C_k$ connected to x, yields $\Delta_k \langle x \rangle = \sum n(x, y) \langle y \rangle = \bar{\partial} \langle x \rangle$. $\square$

Proof of Theorem 5.3. For $j \leq k$ let S_{kj} be the union of the sets $M(x, y)$ over all pairs (x, y) of critical points of f with $j \leq \mathrm{ind}(y) \leq \mathrm{ind}(x) \leq k$. These sets are compact if the gradient flow of f is Morse-Smale (by the same argument as used before to show that $\hat{M}(x, y)$ is compact). It follows that S_{kj} is an isolated invariant set for $j \leq k$. In particular, $S_{n0} = M$ and S_{kk} consists of the critical points of index k. By [20] there exists an *index filtration* $N_0 \subset N_1 \subset \ldots \subset N_n = M$ such that (N_k, N_{j-1}) is an index pair for S_{kj} where $j \leq k$ and $N_{-1} = \emptyset$.

By Lemma 5.4 there exists a commutative diagram

$$
\begin{array}{ccccc}
C_{k+1} & \xrightarrow{\bar{\partial}_k} & C_k & \xrightarrow{\bar{\partial}_{k-1}} & C_{k-1} \\
\simeq \downarrow & & \simeq \downarrow & & \simeq \downarrow \\
H_{k+1}(N_{k+1}, N_k) & \xrightarrow{\partial_k} & H_k(N_k, N_{k-1}) & \xrightarrow{\partial_{k-1}} & H_{k-1}(N_{k-1}, N_{k-2})
\end{array}
\qquad (5.2)
$$

Hence $\bar{\partial}_{k-1} \circ \bar{\partial}_k = 0$.

By Example 1, $H_j(N_k, N_{k-1}) \simeq H_j(D^k, \partial D^k) = 0$ if $j \neq k$. Thus the middle map in the homology exact sequence

$$H_{j+1}(N_{k+1}, N_k) \xrightarrow{\partial} H_j(N_k) \to H_j(N_{k+1}) \to H_j(N_{k+1}, N_k)$$

is an isomorphism for $j \neq k, k+1$. It follows that the composed map $H_j(N_k) \to H_j(M)$ is an isomorphism for $j < k$ and $H_j(N_k) = 0$ for $j > k$. In the commutative diagram

$$
\begin{array}{ccccccc}
 & & & & & & H_{k-1}(N_{k-2}) = 0 \\
 & & & & & & \downarrow \\
0 = H_k(N_{k-1}) & \longrightarrow & H_k(N_k) & \longrightarrow & H_k(N_k, N_{k-1}) & \xrightarrow{\ \partial\ } & H_{k-1}(N_{k-1}) \\
 & & & & & \mathrel{\underset{\partial}{\searrow}} & \downarrow \\
 & & & & & & H_{k-1}(N_{k-1}, N_{k-2})
\end{array}
$$

the horizontal and vertical sequences are exact. Hence the kernels of the two boundary homomorphisms agree and are isomorphic to $H_k(N_k)$. By (5.2) they are also isomorphic to $\ker \bar{\partial}_{k-1} \subseteq C_k$. We conclude that the homology exact sequence

$$H_{k+1}(N_{k+1}, N_k) \xrightarrow{\partial_k} H_k(N_k) \longrightarrow H_k(N_{k-1}) \longrightarrow 0$$

is isomorphic to the sequence

$$C_{k+1} \xrightarrow{\bar{\partial}_k} \ker \bar{\partial}_{k-1} \longrightarrow H_k(M) \longrightarrow 0$$

The latter is therefore exact, too, and $H_k(M) \simeq \ker \bar{\partial}_{k-1} / \operatorname{im} \bar{\partial}_k$. $\square$

We can now get the *Morse inequalities* as a corollary of Theorem 5.3.

Corollary 5.5 *Let f be a Morse function on the compact n-manifold M. If c_k denotes the number of critical points of index k and β_k denotes the Betti number $\operatorname{rank} H_k(M; R)$ for coefficients in a principal ideal domain R, then*

$$c_k - c_{k-1} + \ldots \pm c_0 \geq \beta_k - \beta_{k-1} + \ldots \pm \beta_0$$

for $k \geq 0$. and equality holds for $k \geq n$.

In particular, $c_k \geq \beta_k$ ($\forall\, k$), hence the number of critical points of f is at least the sum of the Betti numbers of M.

Proof. Let $(C_k(R), \partial_k(R))$ be the Morse complex with coefficients in R. Note that $c_k = \dim C_k(R)$ and set $d_k = \operatorname{rank} \partial_k(R) = \dim \operatorname{im} \partial_k(R)$. Since

$$\operatorname{im} \partial_{k-1}(R) \simeq C_k(R) / \ker \partial_{k-1}(R)$$

we have $d_{k-1} = c_k - \dim \ker \partial_{k-1}(R)$.

The isomorphism $H_k(M; R) \simeq \ker \partial_{k-1}(R)/\operatorname{im} \partial_k(R)$ now implies

$$\beta_k = \dim \ker \partial_{k-1}(R) - d_k = c_k - d_{k-1} - d_k$$

Hence

$$\beta_k - \beta_{k-1} + \ldots \pm \beta_0 = c_k - c_{k-1} + \ldots \pm c_0 - d_k$$

The claim on equality follows from the fact that $C_k(R) = \{0\}$ for $k > n$. $\quad\square$

5.2   The Arnol'd Conjecture

Without further notice we always assume that (M, ω) is a compact smooth symplectic manifold without boundary. We are interested in the dynamics associated to a time-dependent Hamiltonian vector field X_t:

$$\omega(X_t, .) = dH_t, \quad H \in C^\infty(\mathbf{R} \times M, \mathbf{R}).$$

The time-1 map $\varphi = \varphi_1$ of such a vector field is called an *exact (symplectic) diffeomorphism* or a *Hamiltonian map*. In [8] it is shown that the set of Hamiltonian maps is the commutator subgroup of the group of symplectomorphisms, and a proper subgroup if $H^1(M, \mathbf{R}) \neq 0$. Considered as dynamical systems, exact diffeomorphisms should have interesting special properties. Concerning the simplest object of dynamical theory, the fixed point set, V. I. Arnol'd conjectured that the *Lefschetz fixed point theory* (which applies to general diffeomorphisms) should be replaced here by a *Morse-type theory*. This is obviously the case if H is time independent, since then every critical point of H is a fixed point of the flow (i.e. of φ_t for all t). Moreover, it was proven in [6] and [108] that every exact diffeomorphism which is C^1-close to the identity has at least as many fixed points as a smooth function on M has critical points. The question was raised in [7] whether this holds for general exact diffeomorphisms and this has become known as the *Arnol'd conjecture*. In 1982, Conley and Zehnder [21] proved the conjecture for the torus T^{2n} with the standard symplectic structure. They used a variational principle similar to the one described in the next section and then could reduce the problem to finite dimension.

Floer's famous contribution to the problem gave rise to what is now called *Floer homology*. Actually, Floer's method did not yield a proof of the Arnol'd conjecture in exactly the (general) form stated above. To describe what has been proved we need to make some definitions.

Definition. A diffeomorphism $\varphi : M \to M$ *satisfies the cup-length estimate* with respect to a ring R if $\#\mathrm{Fix}(\varphi)$ is at least the *cup-length* $\mathrm{cl}(M, R)$, i.e. the largest integer k such that there exist cohomology classes $\alpha_1, \ldots, \alpha_{k-1} \in H^*(M; R)$ of positive dimension with $\alpha_1 \cup \ldots \cup \alpha_{k-1} \neq 0$.

By Lusternik-Schnirelman theory (see Corollary 3.8), a smooth function on M has at least $\mathrm{cl}(M; \mathbf{Z})$ critical points. An exact diffeomorphism which is

C^1-close to the identity satisfies the cup-length estimate with respect to any ring, since it can be identified via its graph with an exact 1-form df in such a way that its fixed points correspond to the critical points of f (cf. chapter 3).

Definition. If the fixed points of φ are nondegenerate, then $\mathrm{Fix}(\varphi)$ is finite and we say that φ *satisfies the ungraded Morse inequalities* with respect to the ring R if there exists a homomorphism $\partial : F_* \to F_*$ of the free R-module over $\mathrm{Fix}(\varphi)$ with

$$\partial \circ \partial = 0 \quad \text{and} \quad \ker \partial / \mathrm{im}\, \partial \cong H_*(M; R).$$

In particular, if this holds for some principal ideal domain R, then as in Corollary 5.5 we get

$$\#\mathrm{Fix}(\varphi) \geq \sum_k \beta_k.$$

where $\beta_k = \dim H_k(M; R)$ is the Betti number.

Influenced by ideas of Witten, Gromov, and Conley, Floer defined a 'Conley index' by elliptic methods. Instead, following Salamon and Zehnder we shall use the Maslov type index introduced in section 1.4. These indexes are easiest to use when the symplectic form ω vanishes on all 2-sheres in M: $\omega|_{\pi_2(M)} = 0$. More generally, since TM has a complex structure which is well defined up to homotopy, it has *Chern classes* $c_i(M) \in H^{2i}(M; \mathbf{R})$ (we will only need the first Chern class c_1; see § 20 of [15]) and we can define:

Definition. The symplectic manifold (M, ω) is *monotone* if there exists $k \geq 0$ such that the (de Rham) cohomology class $[\omega] - k\, c_1(M)$ vanishes on every 2-sphere in M.

We are now ready to cite Floer's results ([40. Theorem 1], [42], cf. [71, Theorem 3.3.1]):

Theorem 5.6 *Let* (M, ω) *be a compact symplectic manifold.*

(i) If both $[\omega]$ *and* $c_1(M)$ *vanish on* $\pi_2(M)$. *then every exact diffeomorphism satisfies the cup-length estimates for every coefficient ring* R.

(ii) If (M, ω) *is monotone, then every exact diffeomorphism with nondegenerate fixed points satisfies the ungraded Morse inequalities with respect to any ring* R.

In the remaining sections of this chapter we will sketch the proof of (ii) in the case $R = \mathbf{Z}_2$, i.e.

$$\#\mathrm{Fix}(\varphi) \geq \sum_k \dim H_k(M; \mathbf{Z}_2).$$

For symplicity we will also assume that $[\omega]$ and $c_1(M)$ vanish on $\pi_2(M)$. The first condition means $\int_{S^2} u^*\omega = 0$ for all smooth maps $u : S^2 \to M$. This implies that there are no nonconstant holomorphic spheres in M, because they satisfy

$$\int_{S^2} u^*\omega = \frac{1}{2} \int_{S^2} |\nabla u|^2.$$

The second condition will be used in the definition of a Morse-type index.

5.3   The Variational Setup

Let (M,ω) be a compact symplectic manifold and φ an exact diffeomorphism (see the previous section). By reparametrizing the Hamiltonian H_t with respect to time we may assume that H_t depends (smoothly) on $t \in S^1 \equiv \mathbf{R}/\mathbf{Z}$. The fixed points $x \in M$ of φ correspond bijectively to the closed paths $\varphi_t(x)$, $0 \leq t \leq 1$, that is to periodic solutions of the Hamiltonian equation

$$\dot{u}(t) = X_t(u(t)), \tag{5.3}$$

where X_t is the Hamiltonian vector field. As in [21], we introduce the space $\Omega M = C^\infty_{\mathrm{contr}}(S^1, M)$ of smooth *contractible* loops in M and the space

$$\mathcal{P} = \mathcal{P}(H) \subset \Omega M$$

of contractible periodic solutions of (5.3). As we shall see below, the elements of $\mathcal{P}$ are precisely the critical points of an action functional $A : \Omega M \to \mathbf{R}$. Hence we can count them using 'Morse theory' on the infinite dimensional space ΩM. Note that we are ignoring all fixed points corresponding to noncontractible loops.

The action functional is defined by

$$A(u) = - \int_{D^2} \tilde{u}^*\omega - \int_{S^1} H_t(u(t))\, dt, \qquad u \in \Omega M. \tag{5.4}$$

where $\tilde{u} : D^2 \to M$ is a smooth extension of $u : S^1 \to M$. Because $\omega|_{\pi_2(M)} = 0$, $A(u)$ is independent of the choice of $\tilde{u}$. (In general, it depends only on the homotopy class of $\tilde{u}$ with fixed boundary.) The tangent space to ΩM at u consists of C^∞-sections $\zeta(t)$ of the pull-back bundle $u^*(TM)$. The derivative of A in the direction of ζ is defined by

$$dA(u)\zeta = \frac{\partial}{\partial s}\big|_{s=0} A(u_s), \qquad \text{where} \quad u_0 = u, \quad \frac{\partial}{\partial s}\big|_{s=0} u_s = \zeta.$$

The second term in A clearly contributes $-\int_{S^1} H_t(u(t))\zeta(t)\, dt$ to $dA(u)\zeta$. For the first term we use Stokes' theorem to get

$$\int_{D^2} \frac{\partial}{\partial s}\big|_{s=0} \tilde{u}_s^*\omega \;=\; \int_{D^2} \tilde{u}_0^* \mathcal{L}_{\tilde\zeta}\omega \;=\; \int_{D^2} \tilde{u}^* d\,i(\tilde\zeta)\omega$$

$$\hspace{2cm} =\; \int_{S^1} u^* i(\zeta)\omega \;=\; \int_{S^1} \omega(\zeta(t), \dot{u}(t))\, dt.$$

Collecting terms, the derivative of A is given by

$$dA(u)\zeta = \int_{S^1} \{\omega(\dot{u}(t),\zeta(t)) - dH_t(u(t))\,\zeta(t)\}\,dt. \tag{5.5}$$

Because

$$dH_t(u(t))\,\zeta(t) = \omega(X_t(u(t)).\zeta(t)),$$

the right hand side of (5.5) vanishes for all ζ if and only if u satisfies the Hamiltonian equation, i.e. $u \in \mathcal{P}$, as claimed.

Now, choose an ω-compatible almost complex structure J to get a Riemannian metric $g = g_J$ defined by

$$g(\xi,\zeta) = \omega(\xi, J(x)\zeta)$$

for $\xi, \zeta \in T_x M$. (In other words, g and J determine an almost Kähler structure on M.) Consider the gradient of A with respect to the inner product

$$\langle \xi, \zeta \rangle = \int_{S^1} g_J(\xi(t), \zeta(t))\,dt$$

on ΩM. By definition, $\langle \mathrm{grad}\, A(u), \zeta \rangle = dA(u)\zeta$, hence if ∇H_t denotes the g_J-gradient of H, then by (5 5) we get

$$\mathrm{grad}\, A(u)(t) = J(u(t))\,\dot{u}(t) - \nabla H_t(u(t)). \tag{5.6}$$

Note that our signs, which are those of [71], differ from Floer's in [40] since his metric g satisfies $g(.,.) = \omega(J.,.)$, while they differ from those of [91], because there X_t is defined by $\omega(X_t,.) = -dH_t$. By (5.6) the trajectory $u_s \in \Omega M$ of the 'gradient flow'

$$\frac{\partial}{\partial s} u_s = -\mathrm{grad}\, A(u_s)$$

is a map $u : \mathbf{R} \times S^1 \to M$, $u(s,t) = u_s(t)$, satisfying

$$\frac{\partial u}{\partial s} + J(u)\frac{\partial u}{\partial t} - \nabla H_t(u) = 0. \tag{5.7}$$

Note that for $H = 0$ the solutions of (5.7) are holomorphic curves from the complex manifold $\mathbf{R} \times S^1 = \mathbf{C}/i\mathbf{Z}$ into the almost complex manifold (M, J). Since the elliptic equation (5.7) does not have a well-posed Cauchy problem (for given $u(0,.)$), it does *not* define a flow. But as Floer had noticed, to build a complex analogous to the Morse complex it is enough to have well-behaved spaces of trajectories which go from one critical point to another.

The fundamental regularity and compactness result for holomorphic curves remains valid in the presence of a Hamiltonian term [90]. Denote by

$$\mathcal{M} = \mathcal{M}(H, J)$$

the space of *bounded solutions* of (5.7), i.e. smooth functions $u : \mathbf{C}/i\mathbf{Z} \to M$ which are contractible, satisfy (5.7) and have finite *flow energy*

$$\Phi(u) := \frac{1}{2} \int_{-\infty}^{\infty} \int_0^1 (|\frac{\partial u}{\partial s}|^2 + |\frac{\partial u}{\partial t} - X_t(u)|^2)\, dt\, ds < \infty.$$

Since M is compact and $\omega|_{\pi_2(M)} = 0$, the space $\mathcal{M}$ is *compact* in the topology of uniform convergence with all derivatives on compact sets ([40, 90]). Moreover for every $u \in \mathcal{M}$ there exists a pair $x, y \in \mathcal{P}$ such that

$$\lim_{s \to -\infty} u(s, t) = x(t), \qquad \lim_{s \to +\infty} u(s, t) = y(t). \tag{5.8}$$

The convergence is uniform in t and $\frac{\partial u}{\partial s} \to 0$ uniformly in t as $|s| \to \infty$. In fact, if x and y are non-degenerate, then u converges exponentially with all derivatives as $|s| \to \infty$. A loop $x \in \mathcal{P}$ is called *non-degenerate* if $\det[I - d\varphi_1(x(0))] \neq 0$.

Given $x, y \in \mathcal{P}(H)$ denote by

$$\mathcal{M}(x, y) = \mathcal{M}(x, y; H, J)$$

the space of all $u \in \mathcal{M}$ which satisfy (5.8). This is the set of absolute minima of the energy functional Φ subject to the asymptotic boundary condition (5.8) as the following calculation shows

$$\frac{1}{2} \int_{-\infty}^{\infty} \int_0^1 |\frac{\partial u}{\partial s} + J(u)\frac{\partial u}{\partial t} - \nabla H(t, u)|^2 dt\, ds$$

$$= \quad \Phi(u) + \int_{-\infty}^{\infty} \int_0^1 \langle \frac{\partial u}{\partial s}, J(u)\frac{\partial u}{\partial t} - \nabla H(t, u)\rangle dt\, ds$$

$$= \quad \Phi(u) + \int_{-\infty}^{\infty} \frac{d}{ds} A(u(s, .)) ds \quad = \quad \Phi(u) + A(y) - A(x).$$

Hence if u belongs to $\mathcal{M}(x, y)$, then

$$\Phi(u) = A(x) - A(y).$$

For a generic choice of the Hamiltonian H, the space $\mathcal{M}(x, y)$ is a finite-dimensional manifold. As in chapter 6 this is seen by linearizing (5.7) in the direction of a vector field $\xi \in C^\infty(u^*TM)$ along u. This yields the first order linear differential operator

$$F(u)\xi = \nabla_s \xi + J(u)\nabla_t \xi + \nabla_\xi J(u)\frac{\partial u}{\partial t} - \nabla_\xi \nabla H(t, u) \tag{5.9}$$

where $\nabla_s\ (= \nabla_{u_*\frac{\partial}{\partial s}})$, ∇_t and ∇_ξ denote the covariant derivative with respect to the metric $g = g_J$. If u satisfies (5.8) and $x, y \in \mathcal{P}$ are non-degenerate, then $F(u)$ is a *Fredholm operator* between appropriate Sobolev spaces (cf. section 1.4).

The pair $(H.J)$ (J compatible with ω) is called *regular* if every contractible 1-periodic solution of (5.3) is non-degenerate and $F(u)$ is onto for $u \in \mathcal{M}$. As in chapter 6 one can show that the set $(\mathcal{H} \times \mathcal{J})_{\mathrm{reg}}$ of regular pairs is *dense* with respect to the C^∞-topology (see [91, §8]). For regular pairs it follows from an implicit function theorem (e.g. [65, p. 17]) that $\mathcal{M}(x.y)$ is a finite-dimensional manifold with local dimension near u equal to the Fredholm index of $F(u)$.

5.4 Definition of Floer Homology

In the regular case, the dimension of $\mathcal{M}(x,y)$ was used by Floer to define a *relative Morse index* for pairs of critical points $x, y \in \mathcal{P}$. Namely, if $c_1|_{\pi_2(M)} = 0$, then the Fredholm index of $F(u)$ depends only on the boundary condition (5.8) and we denote it by $m(x,y)$. For every triple $x.y.z \in \mathcal{P}$ one has

$$m(x,y) + m(y,z) = m(x.z).$$

Hence there exists a function $m : \mathcal{P} \to \mathbf{Z}$ (defined up to an additive integer) such that the Fredholm index

$$\operatorname{ind} F(u) = m(x) - m(y)$$

for every smooth map $u : \mathbf{C}/i\mathbf{Z} \to M$ which satisfies (5.8).

In particular, if $m(x) - m(y) = 1$, then $\mathcal{M}(x.y)$ is a 1-dimensional manifold. As in the case of Morse homology, there is an $\mathbf{R}$-action on $\mathcal{M}(x,y)$ by shift in the s-variable. Thus, it follows from Gromov's compactness that $\hat{\mathcal{M}}(x,y) = \mathcal{M}(x,y)/\mathbf{R}$ is a finite set of isolated orbits if $m(x) - m(y) = 1$.

To define Floer homology for $\mathbf{Z}_2$-coefficients. let $C_k = C_k(M; H)$ be the vector space over $\mathbf{Z}_2$ generated by $\{x \in \mathcal{P} : m(x) = k\}$, $k \in \mathbf{Z}$. For any pair $x, y \in \mathcal{P}$ set

$$\langle \partial x, y \rangle = \#\hat{\mathcal{M}}(x,y) \qquad \mathrm{mod}\ 2.$$

The boundary operator $\partial_k = \partial_k(M; H, J) : C_{k+1} \to C_k$ is defined by

$$\partial_k x = \partial x = \sum_{y \in \mathcal{P}, m(y)=k} \langle \partial x. y \rangle y$$

for $x \in \mathcal{P}.\ m(x) = k + 1$.

As Floer has proven in [40], this operator satisfies $\partial \circ \partial = 0$ so that one gets a chain complex. Its homology

$$H_*^F(M; H, J) = \ker \partial / \operatorname{im} \partial$$

is the *Floer homology* of the pair $(H, J) \in (\mathcal{H} \times \mathcal{J})_{reg}$.

We sketch the proof of $\partial \circ \partial = 0$. One has to show that for $m(x) - 1 = m(y) = m(z) + 1$, the matrix elements vanish:

$$\langle \partial \partial x, z \rangle = \sum_y \langle \partial x, y \rangle \langle \partial y, z \rangle \equiv 0 \quad \mathrm{mod}\ 2.$$

This is equivalent to the number of elements in $\bigcup_y \hat{\mathcal{M}}(x, y) \times \hat{\mathcal{M}}(y, z)$ ("broken trajectories") being even if $m(x) - m(z) = 2$. This follows from a *gluing argument*. Any pair $(u, v) \in \mathcal{M}(x, y) \times \mathcal{M}(y, z)$ can be glued together at y to yield a one-parameter family $\{\tilde{w}_R\}$ of approximate solutions of (5.7) such that

$$\tilde{w}_R(s, t) = \begin{cases} u(s + R, t), & s \le -1 \\ v(s - R, t), & s \ge 1 \end{cases}$$

For R large enough there exists a genuine solution $w_R \in \mathcal{M}(x, z)$ close to $\tilde{w}_R$. This follows from an iteration procedure which was first applied to partial differential equations by C. Taubes. It uses the linearized operator $F(\tilde{w}_R)$ from (5.9) and quadratic estimates of higher order terms [40]. (Floer's original paper is not easy to read, but there seems to be no better reference.)

For $R \to \infty$, $w_R(s - R, t)$ and $w_R(s + R, t)$ converge to $u(s, t)$ and to $v(s, t)$, respectively, in the C^∞_{loc}-topology. (This obviously holds for $\tilde{w}_R$ and $w_R, \tilde{w}_R$ approach each other exponentially when $R \to \infty$.) Moreover, any connecting orbit in $\mathcal{M}(x, z)$ which is sufficiently close to the pair (u, v) must be a member of the family $\{w_R\}$, up to reparametrization. In other words, a pair in $\hat{\mathcal{M}}(x, y) \times \hat{\mathcal{M}}(y, z)$ uniquely determines a one-parameter family in $\hat{\mathcal{M}}(x, z)$. Since $\hat{\mathcal{M}}(x, z)$ is a one-dimensional manifold, this determines an interval component of $\hat{\mathcal{M}}(x, z)$, whose other end converges to another broken trajectory from x to z. Thus the broken trajectories are paired and their number is even.

The following theorem shows that the Floer homology is independent of (H, J) and nontrivial. In particular, the space $\mathcal{M}$ of bounded solutions is not empty.

Theorem 5.7 (Floer) *(a) If (H, J) and (H', J') are regular pairs then there exists a natural chain homomorphism which induces an isomorphism of Floer homology*

$$H^F_*(M; H, J) \cong H^F_*(M; H', J').$$

(b) There is a natural isomorphism between the Floer homology and the singular homology of M

$$H^F_*(M; H, J) \cong H_*(M; \mathbf{Z}_2).$$

Note that this implies the diffeomorphism $\varphi = \varphi_1$ described by $H = H_t$ satisfies the ungraded More inequalities with respect to $\mathbf{Z}_2$. The grading of the Floer homology groups is only defined up to an additive integer. As Salamon and Zehnder have shown [91], this ambiguity can be removed by using a Maslow type index $\mu : \mathcal{P} \to \mathbf{Z}$ which is a symplectic invariant obtained from the linearized Hamiltonian flow along the loop $x \in \mathcal{P}$ (cf. section 1.4).

In terms of this Maslov index one gets (Theorem 1.15)

$$\operatorname{ind} F(u) = \mu(x) - \mu(y)$$

and μ defines a natural grading for the Floer homology groups. With this grading one has an isomorphism

$$H_k^F(M; H, J) \cong H_{k+n}(M; \mathbf{Z}_2), \qquad -n \leq k \leq n.$$

Remarks. 1. If $[\omega]$ does not vanish on $\pi_2(M)$, then A takes values in S^1. If c_1 does not vanish on $\pi_2(M)$, then the Maslow index μ is only defined modulo an integer N.

2. To define the Floer homology with integer coefficients (and hence with coefficients in any abelian group) one has to assign an integer $+1$ or -1 to every $u \in \mathcal{M}(x, y)$ with $\mu(x) - \mu(y) = 1$ [40, 90]. To determine these signs one has to consistently orient the moduli spaces $\hat{\mathcal{M}}(x, y)$. For the details see [43].

5.5 Continuation of Floer Homology

In this section, following [91], we sketch the proof of part (a) of Theorem 5.7 using the Maslov index of section 1.4. More precisely we shall show the following.

Proposition 5.8 *Let* (H^α, J^α), (H^β, J^β) *and* (H^γ, J^γ) *denote regular pairs (cf. end of section 5.3). There exist natural isomorphisms*

$$h_k^{\beta\alpha} : H_k^F(M; H^\alpha, J^\alpha) \to H_k^F(M; H^\beta, J^\beta)$$

satisfying

$$h_*^{\alpha\alpha} = id, \quad h_*^{\gamma\beta} h_*^{\beta\alpha} = h_*^{\gamma\alpha}. \tag{5.10}$$

Proof. To construct the homomorphisms $h_*^{\beta\alpha}$ consider a smooth *homotopy from* (H^α, J^α) to (H^β, J^β), i.e. smooth homotopies $H^{\alpha\beta} : \mathbf{R} \times S^1 \times M \to \mathbf{R}$ and $J^{\alpha\beta} : \mathbf{R} \times M \to \mathrm{End}\,(TM)$ such that $H^{\alpha\beta}(s, t, x)$ and $J^{\alpha\beta}(s, x)$ are independent of s for large $|s|$ and satisfy

$$\lim_{s \to -\infty} H^{\alpha\beta}(s, t, x) = H^\alpha(t, x), \quad \lim_{s \to +\infty} H^{\alpha\beta}(s, t, x) = H^\beta(t, x),$$

$$\lim_{s \to -\infty} J^{\alpha\beta}(s, x) = J^\alpha(x), \qquad \lim_{s \to +\infty} J^{\alpha\beta}(s, x) = J^\beta(x)$$

For $x^\alpha \in \mathcal{P}(H^\alpha)$, $x^\beta \in \mathcal{P}(H^\beta)$ denote by

$$\mathcal{M}(x^\alpha, x^\beta; H^{\alpha\beta}, J^{\alpha\beta})$$

the space of solutions $u : \mathbf{C}/i\mathbf{Z} \to M$ of

$$\frac{\partial u}{\partial s} + J(s, u)\frac{\partial u}{\partial t} - \nabla H(s, t, u) = 0 \tag{5.11}$$

satisfying the boundary conditions

$$\lim_{s \to -\infty} u(s, t) = x^\alpha(t), \qquad \lim_{s \to +\infty} u(s, t) = x^\beta(t). \tag{5.12}$$

As with the s-independent case (5.7), a solution of (5.11) has bounded flow energy if and only if the limits (5.12) exist.

The linearization of (5.11) in the direction of a vector field $\xi \in C^{\infty}(u^*TM)$ along u leads to the differential operator $F(u) : W^{1,2}(u) \to L^2(u)$.

$$F(u)\xi = \nabla_s\xi + J(s,u)\nabla_t\xi + \nabla_\xi J(s,u)\frac{\partial u}{\partial t} + \nabla_\xi \nabla H(s,t,u), \qquad (5.13)$$

where $W^{1,2}(u) = \{\xi \in L^2(u) : \nabla_s\xi, \nabla_t\xi \in L^2(u)\}$ and $L^2(u)$ is the Hilbert space of vector fields ξ along u such that $\xi(s,t+1) = \xi(s,t)$ with norm

$$\|\xi\|^2 = \int_{-\infty}^{\infty}\int_0^1 |\xi(s,t)|^2\, dt\, ds < \infty.$$

As in the s-independent case, $F(u)$ is a Fredholm operator whenever (5.12) is fulfilled and $x^\alpha \in \mathcal{P}(H^\alpha)$, $x^\beta \in \mathcal{P}(H^\beta)$ are non-degenerate. As before, the Fredholm index is

$$\operatorname{ind} F(u) = \mu(x^\alpha; H^\alpha) - \mu(x^\beta; H^\beta).$$

The pair $(H^{\alpha\beta}, J^{\alpha\beta})$ is called a *regular homotopy* if both (H^α, J^α) and (H^β, J^β) are regular and if the operator (5.13) is onto for every solution u of (5.11) and (5.12) with non-degenerate x^α, x^β.

If (H^α, J^α) and (H^β, J^β) are regular, then the space of regular homotopies is dense in the space of all smooth homotopies from (H^α, J^α) to (H^β, J^β) with respect to the topology of uniform convergence with all derivatives on compact sets (see [44] and [91, §8]).

It follows from the index formula above and the surjectivity of $F(u)$ that the spaces $\mathcal{M}(x^\alpha, x^\beta; H^{\alpha\beta}, J^{\alpha\beta})$ are manifolds of dimension

$$\dim \mathcal{M}(x^\alpha, x^\beta; H^{\alpha\beta}, J^{\alpha\beta}) = \mu(x^\alpha; H^\alpha) - \mu(x^\beta; H^\beta).$$

provided the homotopy $(H^{\alpha\beta}, J^{\alpha\beta})$ is regular. By compactness it follows that $\mathcal{M}(x^\alpha, x^\beta; H^{\alpha\beta}, J^{\alpha\beta})$ is a finite set if $\mu(x^\alpha; H^\alpha) = \mu(x^\beta; H^\beta)$. Thus one can define a homomorphism $\varphi^{\beta\alpha} : C_k(M; H^\alpha) \to C_k(M; H^\beta)$.

$$\varphi^{\beta\alpha}(x^\alpha) = \sum_{\mu(x^\beta; H^\beta)=k} \langle \varphi^{\beta\alpha}x^\alpha, x^\beta \rangle x^\beta.$$

setting the matrix element $\langle \varphi^{\beta\alpha}x^\alpha, x^\beta \rangle$ equal to the number of connecting orbits in $\mathcal{M}(x^\alpha, x^\beta; H^{\alpha\beta}, J^{\alpha\beta})$ modulo 2.

The maps $\varphi^{\beta\alpha}$ just defined (for regular $(H^{\alpha\beta}, J^{\alpha\beta})$) are chain homomorphisms, i.e.

$$\partial^\beta \varphi^{\beta\alpha} = \varphi^{\beta\alpha}\partial^\alpha.$$

The proof is omitted (see [91, Lemma 6.2]), since it is very similar to that of $\partial \circ \partial = 0$. The induced homomorphisms of Floer homology turn out to be independent of the choice of the regular homotopy $(H^{\alpha\beta}, J^{\alpha\beta})$. They are the $h^{\beta\alpha}$ appearing in Proposition 5.8.

To prove the claimed independence one chooses a smooth homotopy of homotopies $(H^{\alpha\beta}_\lambda, J^{\alpha\beta}_\lambda)$ from $(H^{\alpha\beta}_0, J^{\alpha\beta}_0)$ to $(H^{\alpha\beta}_1, J^{\alpha\beta}_1)$ and introduces the spaces

$$\{(u, \lambda) : 0 \leq \lambda \leq 1, u \in \mathcal{M}(x^\alpha, y^\beta : H^{\alpha\beta}_\lambda, J^{\alpha\beta}_\lambda)\}.$$

For a generic homotopy of homotopies this is a manifold (with boundary) of dimension $\mu(x^\alpha; H^\alpha) - \mu(x^\beta; H^\beta) + 1$. This can be used to define homomorphisms $\Phi^{\beta\alpha} : C_k(M; H^\alpha) \to C_{k+1}(M; H^\beta)$ with the chain homotopy property:

$$\partial^\beta \Phi^{\beta\alpha} + \Phi^{\beta\alpha} \partial^\alpha = \varphi^{\beta\alpha}_1 - \varphi^{\beta\alpha}_0.$$

Hence $\varphi^{\beta\alpha}_1$ and $\varphi^{\beta\alpha}_0$ induce the same homomorphism $h^{\beta\alpha}$ in homology and the claim is proved. (For more details see [91, Lemma 6.3].)

It remains to show (5.10). The first property, $h^{\alpha\alpha}_* = id$, follows immediately by choosing the constant homotopy. To show the second property, assume $(H^{\alpha\beta}, J^{\alpha\beta})$ and $(H^{\beta\gamma}, J^{\beta\gamma})$ are regular homotopies from (H^α, J^α) to (H^β, J^β) and from (H^β, J^β) to (H^γ, J^γ), respectively. Then for $R > 0$ large enough one can construct a regular homotopy from (H^α, J^α) to (H^γ, J^γ) by

$$H^{\alpha\gamma}_R(s, t, x) = \begin{cases} H^{\alpha\beta}(s + R, t, x), & s \leq 0 \\ H^{\beta\gamma}(s - R, t, x), & s \geq 0 \end{cases} \qquad J^{\alpha\gamma}_R(s, x) = \begin{cases} J^{\alpha\beta}(s + R, x), & s \leq 0 \\ J^{\beta\gamma}(s - R, x), & s \geq 0 \end{cases}$$

Let $\varphi^{\gamma\alpha}_R$ be the associated chain homomorphism.

Similarly as in the proof of $\partial \circ \partial = 0$, every pair

$$u^{\alpha\beta} \in \mathcal{M}(x^\alpha, x^\beta; H^{\alpha\beta}, J^{\alpha\beta}), \quad u^{\beta\gamma} \in \mathcal{M}(x^\beta, x^\gamma; H^{\beta\gamma}, J^{\beta\gamma})$$

with $\mu(x^\alpha; H^\alpha) = \mu(x^\beta; H^\beta) = \mu(x^\gamma; H^\gamma)$ can be glued together to give a connecting orbit $u^{\alpha\gamma}_R \in \mathcal{M}(x^\alpha, x^\gamma; H^{\alpha\gamma}_R, J^{\alpha\gamma}_R)$. Conversely, it follows from Gromov's compactness that for R large enough, these glued orbits comprise all of $\mathcal{M}(x^\alpha, x^\gamma; H^{\alpha\gamma}_R, J^{\alpha\gamma}_R)$. Thus, for $\mu(x^\alpha; H^\alpha) = \mu(x^\gamma; H^\gamma) = k$, we get

$$\langle \varphi^{\gamma\alpha}_R x^\alpha, x^\gamma \rangle = \sum_{\mu(x^\beta; H^\beta) = k} \langle \varphi^{\beta\alpha} x^\alpha, x^\beta \rangle \langle \varphi^{\gamma\beta} x^\beta, x^\gamma \rangle.$$

This means that $\varphi^{\gamma\alpha}_R = \varphi^{\gamma\beta} \circ \varphi^{\beta\alpha}$, hence $h^{\gamma\alpha} = h^{\gamma\beta} \circ h^{\beta\alpha}$.

Setting $\gamma = \alpha$ one concludes that the $h^{\alpha\beta}$ are isomorphisms and the proof is finished. $\qquad\square$

5.6 Floer Homology Equals Singular Homology

In this section we sketch the proof of part (b) of Theorem 5.7 as given in [91, §7].

Consider first the case of a Hamiltonian H which is independent of t and is a Morse function on M. Since M is compact, there exists $\varepsilon > 0$ such that the period of every nonconstant periodic solution of the Hamiltonian equation is larger than ε. Replacing H by εH, we may assume that $\mathcal{P}(H)$ equals the set of critical points of H, considered as constant loops

$$\mathcal{P}(H) = \{x(t) \equiv x \in M : dH(x) = 0\}.$$

The Maslow index of a constant orbit $x(t) \equiv x$ of the Hamiltonian equation is related to its Morse index $\operatorname{ind}_H(x)$. (To define the Maslow index one chooses a fixed almost complex structure on M which is compatible with ω, cf. section 1.4).

Lemma 5.9 *There exists $\varepsilon > 0$ such that the Morse function $H : M \to \mathbf{R}$ satisfies $\|H\|_{C^2} < \varepsilon$, then*

$$\mu(x; H) = \operatorname{ind}_H(x) - n$$

for every critical point x of H.

For the proof, see [91, Lemma 7.2]. The lemma has also been proved independently by Viterbo [102].

For H independent of t, every solution u of (5.7) which is also independent of t satisfies

$$\frac{du}{ds} = \nabla H(u(s)) \tag{5.14}$$

and is therefore a gradient flow line of H. For an open dense set of almost complex structures J (compatible with ω and determining the metric $g_J(.,.) = \omega(., J.)$) the flow (5.14) is of Morse-Smale type (see e.g. [91, Theorem 8.1]). It can be shown that if H is again multiplied by a sufficiently small constant, then one can choose J so that at the same time $(H, J) \in (\mathcal{H} \times \mathcal{J})_{reg}$, (5.14) is a Morse-Smale flow and every solution u of (5.7) with bounded flow energy is independent of t, see [91, Theorem 7.3] and [90].

In this case, the Floer complex agrees with the Morse complex of section 5.1 with the grading shifted by n. Since the homology of the latter agrees with singular homology by Theorem 5.3, we get the isomorphism

$$H_k^F(M; H, J) \cong H_{n+k}(M; \mathbf{Z}_2)$$

claimed in Theorem 5.7.

This also finishes the proof of the special case of part (ii) of Theorem 5.6 announced at the end of section 5.2.

5.7 Symplectic Homology

In this section we shall sketch the symplectic homology theory of Floer and Hofer. For more details we refer to [45, 44, 47, 59]. Essentially, for fixed generic H and J one builds Floer homology groups using only periodic solutions x whose action functional $A(x)$ lies in a given interval. Taking a direct limit over suitable pairs (H, J) one gets a homology group which only depends on the chosen interval.

Let (M, ω) be a compact symplectic manifold with boundary ∂M of contact type (or empty). For simplicity assume $\pi_2(M) = 0$ and restrict to homology with coefficients in $\mathbf{Z}_2$. For a generic pair (H, J) and $a \in \mathbf{R} \cup \{\infty\}$ define the free groups

$$C_k^a(H, J) = \operatorname{span}_{\mathbf{Z}_2}\{x \in \mathcal{P}(H) : \mu(x) = k, A(x) < a\}, \quad C^a(H, J) = \oplus C_k^a(H, J).$$

If u is a solution of (5.7) with asymptotic boundary data $x^+, x^- \in \mathcal{P}(H)$, then $A(x^+) \leq A(x^-)$, hence $\partial = \partial_{(H,J)}$ maps C_k^a into C_{k-1}^a. Defining the quotient groups

$$C_k^{[a,b)}(H, J) = C_k^b(H, J)/C_k^a(H, J) \quad (-\infty \leq a \leq b \leq \infty),$$

∂ induces an operator $\partial_k : C_k^{[a,b)}(H, J) \to C_{k-1}^{[a,b)}(H, J)$ and we can define the Floer homology group

$$F_k^{[a,b)}(H, J) = \ker \partial_k / \operatorname{im} \partial_{k+1}$$

which depends on (H, J).

We now define admissible pairs (H, J). Let $H : S^1 \times M \to [0, +\infty)$ be smooth with compact support in $S^1 \times (M \setminus \partial M)$ and such that all contractible solutions x of X_H satisfying $A(x) < 0$ are nondegenerate. (This allows to define $F^{[a,b)}(H, J)$ for $-\infty \leq a \leq b \leq 0$.) The almost complex structure J can be chosen so that every solution $u : \mathbf{R} \times S^1 \to M$ of (5.7) and with boundary data x^+, x^- satisfies $u(\mathbf{R} \times S^1) \subset M \setminus \partial M$, provided x^+ and x^- are contained in $M \setminus \partial M$. The boundary ∂M is then called J-pseudoconvex. A pair (H, J) satisfying all these conditions is called *admissible*. The set of admissible pairs becomes a directed set with the partial ordering

$$(H, J) \leq (H', J') \; :\Longleftrightarrow \; H(t, x) \leq H'(t, x) \quad \forall (t, x) \in S^1 \times M.$$

For such pairs one defines a *monotone homotopy* $(\tilde{H}, \tilde{J})$, where $\tilde{H} : \mathbf{R} \times S^1 \times M \to \mathbf{R}$ is smooth, $\tilde{J}(s, t, x)$ is compatible with ω, $(\tilde{H}, \tilde{J}) = (H, J)$ for $s \leq -s_0$, $(\tilde{H}, \tilde{J}) = (H', J')$ for $s \geq s_0$, and $\partial \tilde{H}/\partial s \geq 0$. Using the solutions of

$$\frac{\partial u}{\partial s} + \tilde{J}(s, t, u)\frac{\partial u}{\partial t} - (\nabla_j \tilde{H})(s, t, u) = 0,$$

$$u(s, .) \to x \in \mathcal{P}(H) \quad (s \to -\infty), \quad u(s, .) \to y \in \mathcal{P}(H') \quad (s \to +\infty)$$

as in section 5.5 one constructs a natural homomorphism

$$F^{[a,b)}(H,J) \to F^{[a,b)}(H',J')$$

of $\mathbf{Z}_2$-vector spaces independent of the chosen generic monotone homotopy. Passing to the direct limit yields the *symplectic homology group*

$$S^{[a,b)} = \varprojlim_{(H,J)} F^{[a,b)}(H,J).$$

For an open set $U \subset M$ we can restrict the direct limit to those pairs (H,J) such that the support of H is contained in $S^1 \times (U \setminus \partial M)$ and get a symplectic homology group of U:

$$S^{[a,b)}(U) = \varprojlim_{\mathrm{supp}H \subset S^1 \times (U \setminus \partial M)} F^{[a,b)}(H,J).$$

We briefly mention the main properties of these homology groups. There is a natural homomorphism

$$S^{[a,b)}(U) \to S^{[a,b)}$$

and an exact sequence for $-\infty \le a \le b \le c \le 0$:

$$\ldots \to S_k^{[a,b)}(U) \to S_k^{[a,c)}(U) \to S_k^{[b,c)}(U) \xrightarrow{\partial} S_{k-1}^{[a,b)}(U) \to \ldots$$

If $U \subseteq V$, the inclusion induces a homomorphism

$$\sigma_{V,U} : S^{[a,b)}(U) \to S^{[a,b)}(V).$$

A symplectomorphism $\psi : M \to M$ induces a natural isomorphism

$$\psi_\# S^{[a,b)}(U) \xrightarrow{\approx} S^{[a,b)}(\psi(U)).$$

For $\psi(U) \subseteq V$ define

$$\psi_* = \sigma_{V,\psi(U)} \circ \psi_\# : S^{[a,b)}(U) \to S^{[a,b)}(V).$$

One can show that $(\varphi \circ \psi)_* = \varphi_* \circ \psi_*$ and the identity of M induces the identity on $S^{[a,b)}(U)$.

An important property is isotopy invariance: if $\psi_s : M \to M, 0 \le s \le 1$, is a smooth family of symplectomorphisms satisfying $\psi_s(U) \subseteq V \ \forall s$, then $(\psi_s)_* : S^{[a,b)}(U) \to S^{[a,b)}(V)$ is independent of s.

In general, $S^{[a,b)}$ is difficult to calculate, see [45, 47] for computations and applications. As an example, the computation of the symplectic homology groups of polydisks leads to a proof of a conjecture of Gromov on the symplectic classification of open polydisks: For $r = (r_1,\ldots,r_n)$ with the convention $0 < r_1 \le r_2 \le \ldots \le r_n$, let $D(r)$ be the polydisk $B^2(r_1) \times \ldots \times B^2(r_n)$. Then $D(r)$ and $D(r')$ are symplectomorphic if and only if $r = r'$. For a proof see [45].

6 Pseudoholomorphic Curves

6.1 Introduction

With this note we want to give a short introduction to the theory of pseudo-holomorphic curves considered as a tool in symplectic geometry. The invention and the whole foundation of this theory are due to M. Gromov. In the following only some of the simplest arguments of Gromov's seminal paper [50] will be presented but we will try to give all the details of the proofs. We develop the theory as far as to be able to prove Gromov's famous "squeezing theorem" [50, corollary 0.3A]. Further information on and applications of pseudoholomorphic curves may be found in [50, 70, 73, 75, 86, 112].

6.1.1 Linear algebra of symplectic and almost complex structures

A symplectic structure on a real vector space V is an alternating bilinear form ω, which is nondegenerate, i.e. $\ker \omega = \{ v \in V \mid \omega(v.V) = 0 \} = \{0\}$. An almost complex structure on V is an endomorphism $J \in \mathrm{End}\,(V)$ with $J^2 = -1_V$. A *hermitian metric* g for J is a euclidean scalar product, such that J is antisymmetric with respect to g. Given J and a hermitian g we get by

$$\omega(x, y) = g(Jx, y)$$

a symplectic structure.

Definition 6.1 *An almost complex structure J is called* compatible *with a symplectic structure ω if:*

(i) $\omega(x. Jx) > 0$ for $x \neq 0$

(ii) $\omega(Jx. y) + \omega(x, Jy) = 0$,

or equivalently: $g(x, y) = \omega(x, Jy)$ is a hermitian metric.
The set of all ω-compatible almost complex structures is denoted by $\mathcal{J}(\omega)$.

The set $\mathcal{J}(\omega)$ is a real analytic subvariety of the vector space $\mathrm{End}(V)$. The following two propositions show that this space is very simple.

Proposition 6.2 *The space of ω-compatible structures is not empty and contractible.*

Proof. Given any metric g on V, we get an antisymmetric endomorphism $A \in \mathrm{End}(V)$ with $g(Ax, y) = \omega(x, y)$. Then

$$g(-A^2 x, x) = -\omega(Ax, x) = \omega(x, Ax) = g(Ax. Ax) > 0,$$

for $x \neq 0$. Consequently, $-A^2$ has its spectrum $\sigma(-A^2)$ in $\mathbf{C} \setminus (-\infty, 0]$. On $\mathbf{C} \setminus (-\infty, 0]$ the square root has a unique branch with $\sqrt{1} = 1$. On the open set

$$\mathcal{P} = \{B \in \operatorname{End}(V) \mid \sigma(B) \subset \mathbf{C} \setminus (-\infty, 0]\}$$

there exists a real analytic function $r : \mathcal{P} \to \operatorname{End}(V)$ with the following properties: (i) $r(B)^2 = B$, (ii) $r(CBC^{-1}) = Cr(B)C^{-1}$ for $C \in \operatorname{Aut}(V)$ arbitrary, (iii) $BC = CB \Longrightarrow r(B)C = Cr(B)$ for C arbitrary, (iv) $\sigma(r(B)) = \sqrt{\sigma(B)}$. When B is symmetric with respect to a scalar product, then $r(B)$ will be symmetric with respect to the same scalar product (r is most conveniently constructed with the Cauchy integral).
If we set $J = r(-A^2)^{-1}A$, the endomorphism J will be antisymmetric and $J^2 = -1_V$. Further

$$\omega(x, Jx) = g(Ax. \, r(-A^2)^{-1}Ax) \geq 0.$$

We got a real analytic mapping $g \mapsto J(g)$ producing ω-compatible structures. This proves existence which we can get also by considering the normal form of ω.

But when J_1, J_2 are two ω-compatible structures, we get two scalar products $g_i(x, y) = \omega(x. J_i y)$, $i = 1, 2$. Then $J(g_i) = J_i$. The linear homotopy $g_t = tg_1 + (1 - t)g_2$ of euclidean scalar products gives us a homotopy $J(g_t)$ from J_1 to J_2. This proves that $\mathcal{J}(\omega)$ is contractible. $\square$

Let us denote by $Sp(V, \omega)$ the group of automorphisms of the couple (V, ω). This is a Lie group of dimension $3n^2 + n$, if $2n = \dim V$. With a fixed metric g the group of endomorphisms leaving invariant ω and g, $U(V, g, \omega) = Sp(V, \omega) \cap SO(V, g)$, is isomorphic to the unitary group in dimension $\frac{\dim V}{2} = n$.

Proposition 6.3 *The group $Sp(V, \omega)$ acts on $\mathcal{J}(\omega)$ by conjugation. Fixing a scalar product gives the following presentation*

$$\mathcal{J}(\omega) = \frac{Sp(V. \omega)}{U(V, g, \omega)}.$$

Proof. The first assertion is easily verified.

Given two compatible almost complex structures, J_1 and J_2, we find a symplectic automorphism conjugating the first to the second:
All complex vector spaces of a fixed dimension are isomorphic, such that we can find a $\varphi \in \operatorname{End}(V)$ with $J_2 = \varphi J_1 \varphi^{-1}$. Let $g_i(x, y) = \omega(x. J_i y)$, $i = 1, 2$. Then $\tilde{g}(x, y) = g_2(\varphi x, \varphi y)$ will be a J_1-hermitian metric. All hermitian metrics on a fixed complex vector space are equivalent, so there is a $\psi \in \operatorname{End}(V)$ commuting with J_1 and with $\tilde{g}(x, y) = g_1(\psi x, \psi y)$. It is immediatly verified that $\omega(\psi\varphi^{-1}x, \psi\varphi^{-1}y) = \omega(x, y)$, i.e. $\psi\varphi^{-1} \in Sp(V. \omega)$ and we have $(\psi\varphi^{-1})J_2(\psi\varphi^{-1})^{-1} = \psi J_1 \psi^{-1} = J_1$. So the group $Sp(V, \omega)$ acts transitively

on $\mathcal{J}(\omega)$. The isotropy group in $Sp(V,\omega)$ of a ω-compatible almost complex structure J is equal to the isotropy group of the metric $g(x,y) = \omega(x,Jy)$. $\quad\square$

The Lie algebra of the matrix group $Sp(V.\omega)$ is:

$$sp(V,\omega) = \{A \in \mathrm{End}(V) \mid \omega(Ax,y) + \omega(x,Ay) = 0, \forall x, y\}.$$

If we fix an ω-compatible almost complex structure J. then the corresponding unitary group has the following Lie algebra:

$$
\begin{aligned}
u(V,J) &= \{A \in \mathrm{End}(V) \mid A^t = -A, \; JA = AJ\} \\
&= \{A \in \mathrm{End}(V) \mid A^t = -A, \; \omega(Ax,y) + \omega(x,Ay) = 0, \; \forall x, y\}
\end{aligned}
$$

and

$$sp(V,\omega) = \{A \in \mathrm{End}(V) \mid A^t J = -JA\}.$$

Here the tranposition is defined with respect to the hermitian metric which belongs to ω, J. The involution $I : A \mapsto -A^t$ leaves $sp(V,\omega)$ invariant and decomposes this Lie algebra into a symmetric and an antisymmetric part:

$$sp(V,\omega) = \ker(I - 1) \oplus \ker(I + 1) = u \oplus h.$$

Here we have set

$$h(V,J) = \{A \in \mathrm{End}(V) \mid AJ = -JA. \; A^t = A\}$$

and this is a J-stable subspace of (real) dimension $n(n+1)$.

Corollary 6.4 *If we choose an ω-compatible structure J, the space $\mathcal{J}(\omega)$ gets diffeomorphic to $h \subset sp(V,\omega)$ via the map $A \mapsto e^A J e^{-A}$.*

Proof. Because $\mathcal{J}(\omega)$ is a homogeneous space as described in the foregoing proposition, we know by the elementary theory of symmetric spaces (which is explained for example in [22]), that $\mathcal{J}(\omega)$ is a nonpositively curved space such that the curves $t \mapsto e^{tA} J e^{-tA}$ are the geodesics starting at J. We know also that $\mathcal{J}(\omega)$ is simply connected, such that there is a unique geodesic joining J and an arbitrary element of $\mathcal{J}(\omega)$. $\quad\square$

The variation of the function $J(A) = e^A J e^{-A}$ at a point $A_0 \in h(V,J)$ can be very easily described as follows. For A_0 we have the diffeomorphism

$$\Phi : h(V,J) \xrightarrow{\sim} \mathcal{J}(\omega) \xleftarrow{\sim} h(V, J(A_0)),$$

$$e^A J e^{-A} = e^{\Phi(A)} J(A_0) e^{-\Phi(A)}.$$

This map has a differential at A_0

$$\varphi_{A_0} = d\Phi(A_0) : h(V,J) \xrightarrow{\sim} h(V. J(A_0)).$$

With these definitions we get for $\dot{A} \in h(V, J)$:

$$
\begin{aligned}
\frac{d}{dt}\big|_{t=0} J(A_0 + t\dot{A}) &= \frac{d}{dt}\big|_{t=0} e^{A_0 + t\dot{A}} J e^{-(A_0 + t\dot{A})} \\
&= \frac{d}{dt}\big|_{t=0} e^{\Phi(A_0 + t\dot{A})} J(A_0) e^{-\Phi(A_0 + t\dot{A})} \\
&= [\frac{d}{dt}\big|_{t=0} \Phi(A_0 + t\dot{A}) . J(A_0)] \\
&= [\varphi_{A_0}(\dot{A}), J(A_0)].
\end{aligned}
$$

The definitions of symplectic, almost complex and hermitian structures on vector spaces give immediately corresponding definitions on vector bundles. We never used a basis in the proofs of the propositions about these structures, so all of them go over to the relative situation:

Corollary 6.5 *Let $E \to X$ be a smooth symplectic vector bundle with symplectic structure ω. The space $\mathcal{J}(\omega)$ of ω-compatible almost complex structures on E is parametrized by the space of sections of the vector bundle $h(E, J)$ of symmetric anti-J-linear endomorphisms of E, once we have chosen an almost complex structure in $\mathcal{J}(\omega)$.*

Remark 6.6: The variation of the function

$$
\alpha \in C^\infty(X, h(E, J)) \mapsto J(\alpha) = e^\alpha J e^{-\alpha} \in \mathcal{J}(\omega) \subset C^\times(X, \mathrm{End}(E))
$$

at the point α_0 can be described as follows: There is a smooth vector bundle isomorphism $\varphi_{\alpha_0} : h(E, J) \xrightarrow{\sim} h(E, J(\alpha_0))$, such that for $\dot{\alpha} \in C^\infty(X, h(E, J))$ we get

$$
\frac{d}{dt}\big|_{t=0} J(\alpha_0 + t\dot{\alpha}) = [\varphi_{\alpha_0}(\dot{\alpha}), J(\alpha_0)].
$$

6.1.2　Definition of pseudoholomorphic curves

A symplectic manifold is a manifold with a symplectic structure ω on its tangent bundle, with the further condition that ω must be a closed 2-form. All symplectic manifolds of a fixed dimension are locally isomorphic. To distinguish one from the other we must look at some global object attached to them. The idea we want to pursue is to look at some imitation of holomorphic object allied to a compatible almost complex structure.

Let Σ be a compact Riemann surface and M an almost complex manifold. This means that on M there is given an automorphism J of TM with $J^2 = -1$. The complex structure of Σ gives a canonical almost complex structure on it.

Definition 6.7 *Let $f : \Sigma \to M$ be a smooth map. It is called a pseudoholomorphic curve (or holomorphic curve for short) if its differential*

$$
df : T\Sigma \to TM
$$

is (i, J)-linear, i.e. $df \circ i = J \circ df$.

The strategy for the following development is to show the existence of a holomorphic curve under some appropriate conditions. This will be done by a subtle version of the continuity method. If we have a compatible almost complex structure for which we can find all holomorphic curves, then we can show that the set of holomorphic curves cannot disappear under a deformation of the almost complex structure because a certain invariant of it, its cobordism class. is constant along a deformation. For this we need to provide the set of holomorphic curves with the structure of a smooth manifold and to show that this manifold is compact.

There are other global invariants that can be attached to a symplectic manifold (see section 4.2) and the result which we will give as an illustration of the method of holomorphic curves could also be obtained by them.

6.1.3 Regularity of holomorphic curves

As for functions, the condition of being holomorphic is an elliptic equation, so that the smoothness required in the definition is a consequence of much weaker requirements. Let us write $H_{loc}^{k,p}$ for the space of functions which are locally L^p-integrable and whose distributional derivatives up to order k are still L_{loc}^p functions.

Proposition 6.8 *Set $D = \{z \in \mathbf{C} \mid |z| < 1\}$ and let V be a real vector space, $U \subset V$ open and $J : U \to \mathrm{End}(V)$ a map of class C^{k+1} with $J^2 = -1$. A map $f : D \to U \subset V$ of class $H_{loc}^{1,p}$ with $p > 2$ and $df \circ i = J \circ df$ is already in $C^k(D, V)$.*

Proof. For the moment we consider a smooth f, but without the condition on the differential. We write $h = df + J \circ df \circ i$

$$\frac{\partial f}{\partial x} + J(f(z))\frac{\partial f}{\partial y} = h_1(f)$$

$$\frac{\partial f}{\partial y} - J(f(z))\frac{\partial f}{\partial x} = h_2(f)$$

where (x, y) are standard coordinates on D. Then

$$\frac{\partial^2 f}{\partial x^2} + \frac{\partial^2 f}{\partial y^2} + J(f(z))_x\frac{\partial f}{\partial y} - J(f(z))_y\frac{\partial f}{\partial x} = h_{1x} + h_{2y},$$

where we have abbreviated $\frac{\partial}{\partial x}J(f(z)) = J(f(z))_x$, etc..

With a scalar product on V, the corresponding L^2-scalar product on functions with values in V and the Laplace operator $\Delta = -\frac{\partial^2}{\partial x^2} - \frac{\partial^2}{\partial y^2}$, we get for a function $g \in C_0^\infty(D, V)$ with compact support:

$$\langle \Delta f, g \rangle - \langle J_x\frac{\partial f}{\partial y} - J_y\frac{\partial f}{\partial x}, g \rangle = \langle h_1, g_x \rangle + \langle h_2, g_y \rangle$$

or

$$\langle df. dg \rangle = \langle J_x \frac{\partial f}{\partial y} - J_y \frac{\partial f}{\partial x}, g \rangle + \langle h_1, g_x \rangle + \langle h_2, g_y \rangle.$$

The function in the proposition can be approximated by functions $f_n \in C_0^\infty(D, V)$ such that $f_n \to f$ in $H_{loc}^{1,p}$ and $h_1(f_n), h_2(f_n) \to 0$. Further, we have

$$J(f_n)_x \frac{\partial f_n}{\partial y} - J(f_n)_y \frac{\partial f_n}{\partial x} \to J(f)_x \frac{\partial f}{\partial y} - J(f)_y \frac{\partial f}{\partial x}.$$

but only in $L_{loc}^{p/2}$. because $J(f)_x$ contains the derivatives of f linearly. Observe that the functions in $H_{loc}^{1,p}$ are continuous and the embedding $H_{loc}^{1,p} \to C_{loc}^0$ is continuous. Thus. we get

$$\langle df, dg \rangle = \langle J_x \frac{\partial f}{\partial y} - J_y \frac{\partial f}{\partial x}, g \rangle.$$

That means by definition that f verifies

$$\triangle f = J_x \frac{\partial f}{\partial y} - J_y \frac{\partial f}{\partial x}$$

in the weak sense. This is an elliptic equation and the *regularity theorems* (see [79] or [60]) give $f \in H_{loc}^{2,p/2}$. For $p < 4$ use the Sobolev embedding ([3, Theorem 5.4]) to conclude

$$f \in H_{loc}^{1,q} \qquad q = \frac{2p}{4 - p}.$$

For $p > 2$, the exponent q is larger than p:

$$q - p = (p - 2)\frac{p}{4 - p} > p - 2 \quad \text{for} \quad 4 > p > 2.$$

so that iteratively we may assume $f \in H_{loc}^{1,3}$. We continue in the same way:

$$f \in H_{loc}^{2,3/2} \implies f \in H_{loc}^{1,6} \implies f \in H_{loc}^{2,3} \implies df \in H_{loc}^{1.3}.$$

hence the differential is continuous and we get by the *multiplication lemmas* of Sobolev functions (see [83]) that $J_x = (J \circ f)_x \in H_{loc}^{1,3}$ and $J_x \frac{\partial f}{\partial y} - J_y \frac{\partial f}{\partial x} \in H_{loc}^{1,3}$, so that f verifies the equation in the strong form and is as regular as the derivatives of J. $\qquad\qquad \square$

6.2   The Moduli Space of Holomorphic Curves

6.2.1   The setting of global analysis

For the moment the use of the regularity proposition is that we can embed the holomorphic curves in a larger set of mappings which will form a manifold in which they form a submanifold defined by a regular function. In this section we will construct this larger manifold.

Remark 6.9: Let us suppose that we are given a smooth vector bundle $p : E \to X$ over a smooth compact manifold which may have a boundary. With $H^{k,p}(X, E)$ we denote the vector space of sections, which are locally in $H^{k,p}$, i.e. if we choose a trivialization of E over an open subset $V \subset X$ the sections should appear as tupels of functions in the Sobolev space $H^{k,p}_{loc}$. This characterisation of sections of Sobolev class is clearly local: if we have a covering of X by open sets V_α. $X = \bigcup_\alpha V_\alpha$, such that over every open set V_α we can trivialize the bundle and if we are given a section which appears in all these trivialisations as a function in $H^{k,p}_{loc}$, then this section is already in $H^{k,p}(X, E)$. Because X is compact, we can define norms on $H^{k,p}(X, E)$. We choose for example a finite covering of trivializing charts such that some compact subsets in these charts still cover X. Then we take $H^{k,p}$-norms over these compact sets and take the sum. All of the norms obtained in this way are equivalent.

If we assume that we have a Sobolev imbedding into the continuous functions, $H^{k,p} \hookrightarrow C^0$, then we also get an embedding $H^{k,p}(X, E) \to C^0(X, E)$. An open set in the total space of our bundle $U \subset E$ that covers the whole manifold. $p[U] = X$, defines an open set of the continuous sections:

$$C^0(X, U) := \{ s \in C^0(X, E) \mid s(x) \in U, \, \forall x \in X \}.$$

If we have the Sobolev imbedding into the continous functions, the set

$$H^{k,p}(X, U) := H^{k,p}(X, E) \cap C^0(X, U)$$

will be open in $H^{k,p}(X, E)$.

Now let us suppose we are given a second vector bundle $F \to X$ and a map $\Phi : U \to F$ which is smooth and commutes with the projections (Φ need not be linear). This gives rise to a map on sections:

$$\Phi_* : H^{k,p}(X, U) \to H^{k,p}(X, F)$$

defined by $(\Phi_* s)(x) = \Phi(s(x))$. This map is smooth.

Let N and M be manifolds, $\dim_\mathbf{R} N = 2$ and N should be compact. A connection ∇ on the tangent bundle TM gives a diffeomorphism of a neighbourhood of the zero section of TM with a neighbourhood of the diagonal in $M \times M$:

$$\exp : TM \supset U \xrightarrow{\sim} \exp[U] \subseteq M \times M.$$

We just look at the geodesic flow defined by ∇: with $y \in M$, $v \in T_y M$ we have a unique ∇-geodesic

$$t \longmapsto \exp_y(tv),$$

which starts at y with the velocity v and which can be integrated up to $t = 1$ if v is small enough. Then $\exp(v) = (\pi(v), \exp_{\pi(v)} v)$, where we denote by $\pi : TM \to M$ the projection.

Let $f : N \to M$ be a smooth map. By f^*TM we denote the pull-back bundle and

$$f^*TM \supseteq f^*U = \{(x, v) \in f^*TM \subset N \times TM \mid v \in T_{f(x)}M \cap U\}.$$

A continuous section ξ of the bundle f^*TM which takes values in f^*U gives a variation g of the mapping f in the following way:

$$g(x) := \exp_{f(x)} \xi(x) =: (\exp_f^\nabla \xi)(x).$$

If we denote by $H^{1,p}(N, f^*TM)$ the Banach space of sections of the bundle f^*TM which are represented in a trivialisation by $H_{loc}^{1,p}$ functions, we have for $p > 2$ the Sobolev embedding

$$H^{1,p}(N, f^*TM) \subseteq C^0(N, f^*TM)$$

and $H^{1,p}(N, f^*U) = \{\xi \in H^{1,p}(N, f^*TM) \mid \xi(x) \in (f^*U)_x \, \forall x \in N\}$ is an open set in $H^{1,p}(N, f^*TM)$.

We define the set of mappings from N to M of Sobolev class $H^{1,p}$ to be the mappings, which are deformations of smooth mappings:

Definition 6.10 *The space of maps from N to M of Sobolev class $H^{1,p}$ is:*
$H^{1,p}(N, M) := \{g \in C^0(N, M) \mid \exists f \in C^\infty(N, M), \exists$ *connection* ∇ *with a set of definition* $U \subset TM$ *of the geodesic flow,* $\exists \xi \in H^{1,p}(N, f^*U)$ *with* $g = \exp_f^\nabla \xi\}$.

Proposition 6.11 $H^{1,p}(N, M)$ *is a smooth (Hausdorff) manifold modelled on the Banach spaces* $H^{1,p}(N, f^*TM)$ *with charts*

$$H^{1,p}(N, f^*U) \ni \xi \longmapsto \exp_f^\nabla \xi \in H^{1,p}(N, M).$$

indexed by $f \in C^\infty(N, M)$ *and connections* ∇ *on* TM. *The tangent space at* $f \in C^\infty(N, M)$ *is*

$$T_f H^{1,p}(N, M) = H^{1,p}(N, f^*TM).$$

Proof. The topology on $C^0(N, M)$, which is metric, induces a metric topology on the subset $H^{1,p}(N, M)$, which we will refine. For a connection ∇ we have the diffeomorphism

$$TM \supset U \xrightarrow{\sim} \exp[U] \subset M \times M,$$

such that the map

$$H^{1,p}(N, f^*U) \longrightarrow H^{1,p}(N, M)$$
$$\xi \longmapsto \exp_f^\nabla \xi$$

is a homeomorphism onto its image if we take C^0-topology on both sides. To refine the topology we have to consider the transformation of charts: We look at two connections ∇_1, ∇_2 and smooth mappings f_1. f_2 giving two parametrizations of a neighbourhood of the diagonal:

$$TM \supset U_1 \xrightarrow{\exp^{\nabla_1}} M \times M \xleftarrow{\exp^{\nabla_2}} U_2 \subset TM.$$

If we write for the open sets

$$W = \exp^{\nabla_1}(U_1) \cap \exp^{\nabla_2}(U_2) \quad \text{and} \quad W_i = (\exp^{\nabla_i})^{-1}(W),$$

we have:

$$
\begin{aligned}
H^{1,p}(N. f_i^* W_i) &= (\exp_{f_i}^{\nabla_i})^{-1}(\exp_{f_1}^{\nabla_1}[H^{1,p}(N, f_1^* U_1)] \cap \exp_{f_2}^{\nabla_2}[H^{1,p}(N, f_2^* U_2)]) \\
&\subset H^{1,p}(N, f_i^* U_i).
\end{aligned}
$$

The transformation between the charts associated to (f_1, ∇_1) and (f_2, ∇_2) is induced by the map

$$\Phi : f_1^* W_1 \subset f_1^* U_1 \to f_2^* TM$$

$$T_{f_1(x)} M \ni \xi \mapsto (\exp_{f_2(x)}^{\nabla_2})^{-1}(\exp_{f_1(x)}^{\nabla_1} \xi),$$

which is smooth and induces a smooth map

$$\Phi_* = (\exp_{f_2}^{\nabla_2})^{-1} \exp_{f_1}^{\nabla_1} : H^{1,p}(N, f_1^* W_1) \to H^{1,p}(N, f_2^* TM).$$

So the collection of maps $\exp_f^{\nabla} : H^{1,p}(N, f^* U) \to H^{1,p}(N, M)$ defines a smooth atlas. $\qquad\square$

Proposition 6.12 *Let $\{(\varphi_\alpha, U_\alpha), (\psi_\alpha, V_\alpha)\}$ be pairs of charts for N and M, respectively: $\varphi_\alpha : U_\alpha \to N$, $\psi_\alpha : V_\alpha \to M$, such that for a given map $g \in C^0(N.M)$ we have $g[\varphi_\alpha[U_\alpha]] \subseteq \psi_\alpha[V_\alpha]$. Then $g \in H^{1,p}(N, M)$ implies $\psi_\alpha^{-1} \circ g \circ \varphi_\alpha \in H_{loc}^{1,p}$. Conversely, a continuous g which in a set of charts such that $\cup_\alpha U_\alpha = N$ is represented by functions in $H_{loc}^{1,p}$, belongs to $H^{1,p}(N, M)$.*

Proof. Perhaps only the second assertion is not obvious. We just have to remember that $C^\infty(N, M)$ is dense in $C^0(N. M)$. so that we can catch g with a $f \in C^\infty(N, M)$ and a connection by $g = \exp_f^\nabla \xi$ for some $\xi \in C^0(N, f^* TM)$. If $G(x) = (f(x), g(x)) \in M \times M$, then

$$\xi(x) = (\exp^\nabla)^{-1} G(x).$$

The function G appears in coordinates as a function of type $H_{loc}^{1,p}$ and composition on the left by a smooth application produces again a $H_{loc}^{1,p}$-function, as alluded to in the remark 6.9 at the beginning. Hence $\xi \in H^{1,p}(N, f^* U)$, if $U \subset TM$ denotes the domain of definition of the connection ∇. $\qquad\square$

For a smooth vector bundle $E \to M$ we can form the pullback bundle $f^*E \to N$ for a continuous $f \in C^0(N, M)$. If $f \in H^{1,p}(N, M)$, then the transition functions of f^*E are in $H^{1,p}_{loc}$ and because L^p_{loc} is a $H^{1,p}_{loc}$-module by the multiplication lemmas for Sobolev functions (see e.g. [83]), the space $L^p(N, f^*E)$ is well defined. Thus, set-theoretically we have a bundle

$$\mathcal{E} = \bigcup_{f \in H^{1,p}(N,M)}^{\circ} L^p(N, f^*E) \longrightarrow H^{1,p}(N, M).$$

Proposition 6.13 *The bundle $\mathcal{E} \to H^{1,p}(N, M)$ has a smooth structure.*

Proof. Let ∇ be a connection on M and $f \in C^\infty(N, M)$ the base of a chart

$$H^{1,p}(N, f^*U) \overset{\exp_f^\nabla}{\longrightarrow} H^{1,p}(N, M),$$

as described above. A connection ∇^E on the bundle E gives a parallel transport operator along ∇-geodesics, τ, which is a section of the exterior tensor product $E \boxtimes E^* \to M \times M$ defined near the diagonal of $M \times M$:

$$\tau(x, y) : E_y \overset{\sim}{\longrightarrow} E_x.$$

The map

$$H^{1,p}(N, f^*U) \times L^p(N, f^*E) \longrightarrow \bigcup_{\xi \in H^{1,p}(N, f^*U)}^{\circ} L^p(N, (\exp_f \xi)^*E)$$

$$(\xi, s) \longmapsto \tau(\exp_{f(x)} \xi(x), f(x))\, s(x)$$

is a bijection and should be a trivialising chart. The transition from the chart given by $(f_1, \nabla_1, \nabla_1^E)$ to the one given by $(f_2, \nabla_2, \nabla_2^E)$ is described as follows: a section $s \in L^p(N, f_1^*E)$ is first transported parallelly along ∇_1-geodesics to $L^p(N, (\exp_{f_1}^{\nabla_1} \xi_1)^*E) \ni \tau_1(\exp_{f_1}^{\nabla_1} \xi_1(x),\, f_1(x)) s(x)$ and $\exp_{f_1}^{\nabla_1} \xi_1 = \exp_{f_2}^{\nabla_2} \xi_2$. Then this section is transported along ∇_2-geodesics to $L^p(N, f_2^*E)$. So let

$$\tau^{1,2}(x, y, z) : E_z \to E_x$$

be the parallel transport from z to y along ∇_1-geodesics with ∇_1^E and then from y to x along ∇_2-geodesics with ∇_2^E. This is defined on a neighbourhood of the diagonal in $M \times M \times M$. The change of trivialisation for the vector bundle $\mathcal{E}$ is then

$$(\xi, s) \longmapsto \left((\exp_{f_2}^{\nabla_2})^{-1}(\exp_{f_1}^{\nabla_1} \xi),\, \tau^{1,2}(f_2(x), \exp_{f_1(x)}^{\nabla_1} \xi(x), f_1(x))\, s(x) \right).$$

The application $\xi \in T_{f_1(x)}M \mapsto \tau^{1,2}(f_2(x), \exp_{f_1(x)}^{\nabla_1} \xi, f_1(x))$ is a smooth map $f_1^*TM \to \mathrm{Hom}(f_1^*E, f_2^*E)$, and with our general principle for the composition on the left, it induces a smooth map

$$H^{1,p}(N, f_1^*TM) \to H^{1,p}(\mathrm{Hom}(f_1^*E, f_2^*E)) \subset \mathcal{L}(L^p(N, f_1^*E), L^p(N, f_2^*E)).$$

Thus we got a smoothly compatible atlas of trivialisations for $\mathcal{E}$. $\square$

Proposition 6.14 *An element $g \in H^{1,p}(N, M)$ has a differential which is a section in $L^p(N, \mathrm{Hom}(TN, g^*TM))$. The function $g \mapsto dg$ defines a smooth section of the bundle*

$$\mathcal{E} = \bigcup_{g \in H^{1,p}(N,M)}^{\circ} L^p(N, \mathrm{Hom}(TN, g^*TM)) \longrightarrow H^{1,p}(N, M),$$

with a smooth structure analogously constructed as above.

Proof. In a trivialisation we have for the map $g \mapsto dg$:

$$H^{1,p}(N, f^*TM) \ni \xi \longmapsto \tau(f(x), \exp_{f(x)} \xi(x)) \circ d(\exp_f \xi)(x),$$

where f is smooth, ξ is small enough and

$$d(\exp_f \xi)(x) : T_x N \longrightarrow T_{\exp_{f(x)} \xi(x)} M$$

is the differential of the map $\exp_f \xi$. Writing out this expression in trivialisations of TM and TN shows that it is smooth in ξ. $\qquad\square$

6.2.2 The operator $\bar{\partial}_J$ and its linearization

Let us now return to our situation: $N = \Sigma$ is a Riemann surface and J an almost complex structure on M. This gives the usual decomposition into J-linear and J-antilinear maps

$$\begin{aligned}
\mathrm{Hom}(T\Sigma, f^*TM) &= \mathrm{Hom}^{1,0}(T\Sigma, f^*TM) \oplus \mathrm{Hom}^{0,1}(T\Sigma, f^*TM) \\
&= (T^{1,0}\Sigma \otimes_J f^*TM) \oplus (T^{0,1}\Sigma \otimes_J f^*TM).
\end{aligned}$$

We get the subbundle of $\mathcal{E}$ above:

$$\mathcal{E}^{0,1} = \bigcup_{f \in H^{1,p}(\Sigma,M)}^{\circ} L^p(\Sigma, T^{0,1}\Sigma \otimes_J f^*TM) \longrightarrow H^{1,p}(\Sigma, M).$$

If we choose trivialisations of $\mathcal{E}$ with connections which are compatible with J, the operation of the almost complex structure on $\mathcal{E}$ appears as an endomorphism of

$$L^p(\Sigma, \mathrm{Hom}(T\Sigma, f^*TM)).$$

The section $f \longmapsto \bar{\partial}_J f = \frac{1}{2}(df + J \circ df \circ i)$ is the projection of the section $f \mapsto df$ onto the subbundle $\mathcal{E}^{0,1}$. Consequently it is smooth. Its zero set are the holomorphic curves for J.

The set of holomorphic curves can be seen locally as the zero set of some function defined on a chart of $H^{1,p}(\Sigma, M)$ and the vector bundle construction of proposition 6.13 might seem high brow style. But it is only the canonical language to formulate a local construction, which must be done anyway.

The next task is to describe the differential of this section.

Definition 6.15 *Let $E \to \Sigma$ be a smooth complex vector bundle. A differential operator*

$$D : C^\infty(\Sigma, E) \longrightarrow C^\infty(\Sigma, T^{0,1}\Sigma \otimes E)$$

is called a $\bar{\partial}$-operator if

$$D(fs) = \bar{\partial}f \otimes s + fDs$$

for any smooth function f on Σ and any smooth section s of E.

We look again at our section $f \mapsto \bar{\partial}_J f$. On TM we choose a J-compatible connection ∇. With this connection we trivialize the bundle $\mathcal{E}^{0,1}$ in a neighbourhood of $f \in C^\infty(\Sigma, M)$. Then the section $\bar{\partial}_J$ appears in the trivialisation as

$$H^{1,p}(\Sigma, f^*TM) \supseteq W \xrightarrow{\bar{\partial}_J} W \times L^p(\Sigma, T^{0,1}\Sigma \overset{'}{\otimes} f^*TM)$$
$$\xi \longmapsto (\xi, A(\xi)).$$

Clearly, we only have to consider the differential of the operator A at the point f:

$$DA(f) : H^{1,p}(\Sigma, f^*TM) \longrightarrow L^p(\Sigma, T^{0,1}\Sigma \otimes f^*TM).$$

Lemma 6.16 *Let $T = T^\nabla$ be the torsion of the connection ∇. Let ∇^{f^*TM} be the induced connection on f^*TM. We have $f_* = df : T\Sigma \to f^*TM$. Set $\bar{\partial}^f = (\nabla^{f^*TM})^{0,1}$, the $\bar{\partial}$-operator to the connection ∇^{f^*TM}. i.e. the operator ∇^{f^*TM} composed with projection on $(0,1)$-forms. We have*

$$DA(f)\xi = \bar{\partial}^f \xi + \rho(\xi)$$

with

$$\rho(\xi)s = \frac{1}{2}\big(T(\xi, f_*s) + JT(\xi, f_*is)\big).$$

Proof. Let $\xi \in \mathbf{C}^\infty(\Sigma, f^*TM)$ and let $F : [0,1] \times \Sigma \to M$ be the map $F(t,x) = \exp^\nabla_{f(x)} t\xi(x)$, where we suppose that ξ is sufficiently small. A section s of $T\Sigma \to \Sigma$ can be lifted to a section $\tilde{s}$ of $T([0,1]\times\Sigma)$. Let $\frac{\partial}{\partial t}$ be the section of $T([0,1]\times\Sigma)$ corresponding to the parametrized interval $[0,1]$. With $f_t(x) = F(t,x)$ we have

$$F_*(\tilde{s})|_t = f_{t*}(s) = df_t(s).$$

In the trivialisation of the bundle $\mathcal{E} \to H^{1,p}(\Sigma, M)$ with ∇ we must transport $df_t(s)$ parallelly with the connection ∇^{F^*TM} along $[0,1]$, because the images of the lines $[0,1] \times \{x\}$ under F are the ∇-geodesics:

$$\tau^{\nabla^{F^*TM}}((0,x),(t,x))df_t(s) \in \mathbf{C}^\infty(\Sigma, \mathrm{Hom}(T\Sigma, f^*TM)).$$

The differential $DA(f)\xi$ applied to s is the derivation of this expression with respect to the variable t at $t = 0$ which is by definition of parallel transport equal to

$$\nabla^{F^*TM}_{\frac{\partial}{\partial t}}(F_*\tilde{s})|_{t=0}.$$

On the pull-back bundle F^*TM we still have the torsion formula: let X, Y be vector fields on $[0,1] \times \Sigma$, then

$$\nabla^{F^*TM}_{X}(F_*Y) - \nabla^{F^*TM}_{Y}(F_*X) - F_*[X,Y] = T^{\nabla}(F_*X, F_*Y)$$

(trivial, or see remark 6.17 after the proof). Using this, we get

$$\nabla^{F^*TM}_{\frac{\partial}{\partial t}}(F_*\tilde{s}) = \nabla^{F^*TM}_{\tilde{s}}(F_*\frac{\partial}{\partial t}) + T(F_*\frac{\partial}{\partial t}, F_*\tilde{s})$$

and evaluated at $t = 0$ we get as differential of the section $g \mapsto dg$ of $\mathcal{E} \to H^{1,p}(\Sigma, M)$

$$s \longmapsto \nabla^{f^*TM}_{s}\xi + T^{\nabla}(\xi, f_*s).$$

Because J is compatible with ∇, we get for $g \mapsto \bar{\partial}_J g = \frac{1}{2}(dg + Jdg\,i)$

$$(DA(f)\xi)s = \underbrace{\frac{1}{2}(\nabla^{f^*TM}_{s}\xi + J\nabla^{f^*TM}_{is}\xi)}_{\bar{\partial}^f_J\xi} + \frac{1}{2}\big(T(\xi, f_*s) + JT(\xi, f_*is)\big).$$

$\square$

Remark 6.17: Let $f : N \to M$ be a smooth map, ∇ a connection on TM. Then we have for vector fields X, Y on N the torsion formula

$$\nabla^{f^*TM}_{X}(f_*Y) - \nabla^{f^*TM}_{Y}(f_*X) - f_*[X,Y] = T^{\nabla}(f_*X, f_*Y),$$

where $f_* : TN \to f^*TM$. The formula is local. Because of derivation properties of the connection and the bracket, we only have to check the formula for the fields $\frac{\partial}{\partial x^i}$ stemming from a chart of N. So let $f : \mathbf{R}^n \supset U \to \mathbf{R}^m$. For a section s of TM one then has $\nabla^{f^*TM}_{X}(f_*s) = \nabla^{TM}_{f_*X}(s)$. Thus,

$$
\begin{aligned}
\nabla^{f^*TM}_{\frac{\partial}{\partial x^i}}(f_*\frac{\partial}{\partial x^j}) &= \nabla^{f^*TM}_{\frac{\partial}{\partial x^i}}(f^k_j \frac{\partial}{\partial y^k}) = \\
&= f^k_{ij}\frac{\partial}{\partial y^k} + f^k_j \nabla_{f_*\frac{\partial}{\partial x^i}}(\frac{\partial}{\partial y^k}) = \\
&= f^k_{ij}\frac{\partial}{\partial y^k} + f^k_j f^l_i \nabla_{\frac{\partial}{\partial y^l}}(\frac{\partial}{\partial y^k})
\end{aligned}
$$

where lower indices are derivatives with respect to the x-coordinates. So:

$$\nabla^{f^*TM}_{\frac{\partial}{\partial x^i}}(f_*\frac{\partial}{\partial x^j}) - \nabla^{f^*TM}_{\frac{\partial}{\partial x^j}}(f_*\frac{\partial}{\partial x^i}) = f^k_j f^l_i T(\frac{\partial}{\partial y^l}, \frac{\partial}{\partial y^k}).$$

Lemma 6.18 (i) *$\bar{\partial}$-operators are elliptic and all have the same symbol:*

$$\sigma_D(\xi, s_x) = \xi^{0,1} \otimes s_x$$

for a covector $\xi \in T_x^*\Sigma$ and $s_x \in E_x$. Here, $\xi^{0,1}$ is the result of the following composition:

$$T^*\Sigma \longrightarrow T^*\Sigma \otimes_{\mathbf{R}} \mathbf{C} \longrightarrow T^{1,0}\Sigma \oplus T^{0,1}\Sigma \longrightarrow T^{0,1}\Sigma.$$

(ii) *The index of a $\bar{\partial}$-operator D is given by*

$$\begin{aligned} \mathrm{ind}(D) \quad &= \quad \dim_{\mathbf{C}} \ker D - \dim_{\mathbf{C}} \mathrm{coker} D \\ &= \quad \frac{e(\Sigma)}{2} \, \mathrm{rank}_{\mathbf{C}} E + c_1(E), \end{aligned}$$

where $e(\Sigma)$ denotes the Euler number of Σ.

(iii) *For a $\bar{\partial}$-operator* unique continuation *holds: a section $s \in C^\infty(\Sigma, E)$ with $Ds = 0$ and a zero of infinite order at $x_0 \in \Sigma$ is identically zero on the whole connected component containing x_0.*

Proof. (i) The symbol is calculated by choosing f with $f(x) = 0$, $df(x) = \xi$ and a section s with $s(x) = s_x$. Then

$$\begin{aligned} \sigma_D(\xi, s_x) \quad &= \quad D(fs)(x) = \\ &= \quad (\bar{\partial} f \otimes s + f Ds)(x) \\ &= \quad \xi^{0,1} \otimes s_x. \end{aligned}$$

The application $T^*\Sigma \to T^{0,1}\Sigma$ is an antilinear isomorphism. so D is elliptic.

(ii) Let E and F be two complex vector bundles over Σ. With an isomorphism $E \to F$ we can conjugate a $\bar{\partial}$-operator on E to an operator on F which is again a $\bar{\partial}$-operator. Now any complex vector bundle E over a Riemann surface is isomorphic to a holomorphic vector bundle: if $\mathrm{rank}_{\mathbf{C}} E \geq 2$. induction over the skeleton of a triangulation of Σ yields a nonvanishing section of E, because the unit sphere bundle of E has simply connected fibres. Thus. $E \cong \mathbf{C} \oplus E'$ with $\mathrm{rank}\, E' = \mathrm{rank}\, E - 1$, and inductively $E \cong \mathbf{C}^{\mathrm{rank}E-1} \oplus L$ for some line bundle L. But on a Riemann surface any line bundle has a holomorphic structure. The isomorphism $E \cong \mathbf{C}^{\mathrm{rank}\,E-1} \oplus L = F$ conjugates our $\bar{\partial}$-operator to a $\bar{\partial}$-operator on a holomorphic bundle with the same index. On a holomorphic vector bundle F there is a canonical $\bar{\partial}$-operator: for a C^∞-function f and a holomorphic section s

$$\bar{\partial}_F(fs) := \bar{\partial} f \otimes s.$$

(This is well defined. because for an open set U which is small enough, we have $C^\infty(U, F) = C^\infty(U) \otimes_{\mathcal{O}(U)} \mathcal{O}(U, F)$ with $\mathcal{O}$ as the holomorphic sections.)

For line bundles the index of the canonical operator is calculated in books on Riemann surfaces (Gunning, Forster):

$$\mathrm{ind}(\bar{\partial}_L) = \frac{e(\Sigma)}{2} + c_1(L) \quad \text{(Riemann-Roch formula)}.$$

The difference of two $\bar{\partial}$-operators is an operator of order zero, so that they all have the same index. We get

$$\begin{aligned}
\mathrm{ind}(\bar{\partial}_F) &= (\mathrm{rank}_{\mathbf{C}} E - 1)\,\mathrm{ind}(\bar{\partial}_{\mathbf{C}}) + \mathrm{ind}(\bar{\partial}_L) \\
&= \frac{e(\Sigma)}{2}\,\mathrm{rank}_{\mathbf{C}} E + \underset{\underset{c_1(E)}{\|}}{c_1(L)}
\end{aligned}$$

(iii) This follows from Aronszajn's famous theorem: a differential operator on a (real) vector bundle whose symbol is multiplication by the norm of the cotangent vector with respect to some metric on the base gives unique continuation for the sections in its kernel. We compose the $\bar{\partial}$-operator $D : C^\infty(\Sigma, E) \to C^\infty(\Sigma, T^{0,1}\Sigma \otimes E)$ with another operator to get an operator with scalar symbol: a Riemannian metric on Σ provides a volume and $T^{0,1} \otimes T^{1,0} \cong \mathbf{C}$. Any complex bundle has a $\bar{\partial}$-operator. Just take the $(0,1)$ component of a complex connection. We then look at the following composition:

$$C^\infty(\Sigma, E) \xrightarrow{D} C^\infty(\Sigma, T^{0,1} \otimes E) \xrightarrow{c} C^\infty(\Sigma, T^{1,0} \otimes \overline{E}) \longrightarrow$$
$$\xrightarrow{P} C^\infty(\Sigma, T^{0,1} \otimes T^{1,0} \otimes \overline{E}) \longrightarrow C^\infty(\Sigma, E),$$

where P is a $\bar{\partial}$-operator on $T^{1,0} \otimes \overline{E}$ and c is a complex conjugation $T^{0,1} \xrightarrow{\sim} T^{1,0}$. The symbol is obviously $(\xi, s) \mapsto \|\xi\|^2 s$. $\qquad\square$

Remark 6.19: Integration over Σ gives the following pairings

$$C^\infty(\Sigma, E) \times C^\infty(\Sigma, T^{1,1}\Sigma \otimes E^*) \longrightarrow C^\infty(\Sigma, T^{1,1}\Sigma) \xrightarrow{\int_\Sigma} \mathbf{C}$$
$$C^\infty(\Sigma, T^{0,1} \otimes E) \times C^\infty(\Sigma, T^{1,0} \otimes E^*) \longrightarrow C^\infty(\Sigma, T^{1,1}\Sigma) \xrightarrow{\int_\Sigma} \mathbf{C}$$

With these pairings the operator $D : C^\infty(\Sigma, E) \to C^\infty(\Sigma, T^{0,1} \otimes E)$ has a transpose

$$D^t : C^\infty(\Sigma, T^{1,0} \otimes E^*) \longrightarrow C^\infty(\Sigma, T^{1,1} \otimes E^*)$$

defined by $\langle s, D^t \mathcal{O} \rangle = \langle Ds, \mathcal{O} \rangle$ with $s \in C^\infty(\Sigma, E)$, $\mathcal{O} \in C^\infty(\Sigma, T^{1,0} \otimes E^*)$. Then $-D^t$ is again a $\bar{\partial}$-operator.

6.2.3   The moduli space at regular almost complex structures

To affirm that the zero set of the section $g \mapsto \overline{\partial}_J g$ is smooth, we should know that the differential of this section is regular at the zeros. This may be false for general J. As always in such a situation we adjoin additional variables. We let J vary in a sufficiently large manifold $\mathcal{H}$, so that

$$\{(g, J) \in H^{1,p}(\Sigma, M) \times \mathcal{H} \mid \overline{\partial}_J g = 0\}$$

is smooth and then we apply the Sard-Smale theorem (cf. proposition 6.23 below) which provides regular values of the projection of this set onto $\mathcal{H}$. This $\mathcal{H}$ should be a manifold and the next lemma provides us with a sufficiently large manifold consisting of smooth ω-compatible almost complex structures.

Lemma 6.20 *Let $E \to M$ be a smooth finite dimensional vector bundle. Assume that M is compact. There exists a Banach space*

$$\mathcal{H} \subset \mathbf{C}^{\infty}(M, E) \subsetneq C^{l}(M, E)$$

which is dense in the l-times differentiable sections with their usual topology.

Proof. A sequence of C^k-norms $\|.\|_{C^k}$, $k = 0, 1, \ldots$ on E and a sequence of positive numbers $\{\varepsilon_k\}$ defines a Banach space:

$$\mathcal{H}_{\{\varepsilon_k\}} = \{f \in \mathbf{C}^{\infty}(M, E) \mid \sum_{k \geq 0} \varepsilon_k \|f\|_{C^k} < \infty\}.$$

We want to choose the sequence $\{\varepsilon_k\}$ so that $\mathcal{H}_{\{\varepsilon_k\}}$ is dense in C^l. Let $g \in C^l(M, E)$ and we may assume that g has support in a chart over which E is trivial. The C^k-norms defined by the trivialisation are compatible with the global ones. Let $\varphi \in \mathbf{C}_0^{\infty}(\mathbf{R}^n)$ with $\int_{\mathbf{R}^n} \varphi = 1$, supp $\varphi \subset B(0, 1)$ and let $\varphi_\varepsilon(x) = \varepsilon^{-n}\varphi(\frac{x}{\varepsilon})$ for $0 < \varepsilon < 1$. Convolution of g with φ_ε gives $\varphi_\varepsilon * g \in \mathbf{C}_0^{\infty}(\mathbf{R}^n)$ and $\|\varphi_\varepsilon * g - g\|_{C^l(\mathbf{R}^n)} \to 0$ as $\varepsilon \to 0$. Further,

$$\|D^{\alpha}(\varphi_\varepsilon * g)\|_{\infty} \leq \|D^{\alpha}\varphi_\varepsilon\|_1 \|g\|_{\infty}$$

for the derivatives. Obviously,

$$\|D^{\alpha}\varphi_\varepsilon\|_{L^1(\mathbf{R}^n)} = \varepsilon^{-|\alpha|} \int_{\mathbf{R}^n} |D^{\alpha}\varphi|.$$

If we have the comparison

$$\|h\|_{C^k(M, E)} \leq A_k \|h\|_{C^k(\mathbf{R}^n)},$$

then we get

$$\begin{aligned}
\varepsilon_k \|\varphi_\varepsilon * g\|_{C^k(M, E)} &\leq \varepsilon_k A_k \sum_{|\alpha| \leq k} \|D^{\alpha}\varphi\|_1 \varepsilon^{-k} \|g\|_{\infty} \\
&\leq \varepsilon_k B_k \varepsilon^{-k}
\end{aligned}$$

If we choose $\varepsilon_k \leq \frac{1}{k! B_k}$, the series $\sum_k \varepsilon_k \|\varphi_\varepsilon * g\|_{C^k}$ is convergent for all $\varepsilon > 0$. Because the base of E can be covered with a finite number of charts, we can choose the sequence so that $\mathcal{H}_{\{\varepsilon_k\}} \subseteq C^l$ is dense. $\qquad\Box$

With the assertion of corollary 6.5 in mind we get the following:

Corollary 6.21 *Let J be a ω-compatible almost complex structure on the compact symplectic manifold (M, ω) and let us denote by $h(TM, J) \subset \mathrm{End}(TM)$ the corresponding vector bundle of symmetric anti-J-linear endomorphisms of TM. Then there exists a Banach space $\mathcal{H} \subset C^\infty(M, h(TM))$ which is still dense in $C^0(M, h(TM))$.*

Recall that a section α of $h(TM, J)$ gives by $J(\alpha) = e^\alpha J e^{-\alpha}$ a ω-compatible structure. With the parameter space $\mathcal{H}$ of corollary 6.21 we can form the parametrized moduli space:

$$\mathcal{M}_{\mathcal{H}}^* := \{(f, \alpha) \in C^\infty(\Sigma, M) \times \mathcal{H} \mid \quad f \text{ is } J(\alpha)\text{-holomorphic and } \exists z \in \Sigma :$$
$$df(z) \neq 0. \; f^{-1}f(z) = \{z\}\,\}.$$

Lemma 6.22 *For a closed connected Riemann surface Σ and M, $\mathcal{H}$ as described before, the parametrized moduli space $\mathcal{M}_{\mathcal{H}}^*$ is a smooth submanifold of $H^{1,p}(\Sigma, M) \times \mathcal{H}$.*

Proof. In corollary 6.21 we constructed a Banach space $\mathcal{H}$ with a smooth injection $\mathcal{H} \subset \mathcal{J}(\omega)$. Here we have chosen a basepoint $J \in \mathcal{J}(\omega)$ and the injection is the restriction:

$$\mathcal{H} \subset C^\infty(M, h(TM, J)) \to \mathcal{J}(\omega).$$

$$\alpha \mapsto e^\alpha J e^{-\alpha}.$$

With the canonical projections Π_1 and Π_2 of $H^{1,p}(\Sigma, M) \times \mathcal{H}$ onto its factors we pull back the bundle $\mathcal{E}_J^{0,1} \to H^{1,p}(\Sigma, M)$, and we get

$$\Pi_1^* \mathcal{E}_J^{0,1} \to H^{1,p}(\Sigma, M) \times \mathcal{H}$$

with the section

$$\mathcal{P}(f, \alpha) := \frac{1}{2}(e^{-\alpha} df + J e^{-\alpha} df i).$$

Obviously, $\mathcal{P}(f, \alpha) = 0 \iff f$ is $J(\alpha)$-holomorphic. We want to calculate the differential of this section at a zero (f, α_0). We consider $\mathcal{P}$ as a section of $\Pi_1^* \mathcal{E}$ and apply the gauge transformation $\alpha \in \mathcal{H} \mapsto e^\alpha$. We get:

$$\tilde{\mathcal{P}}(f, \alpha) = \frac{1}{2}(df + J(\alpha)df i).$$

The base space is a product and we calculate each partial derivative separately. We trivialize the bundle locally with a $J(\alpha_0)$-compatible connection and consider only the vertical component of the derivative:

$$D\tilde{\mathcal{P}} : T_f H^{1,p}(\Sigma, M) \oplus T_{J(\alpha_0)}\mathcal{H} \longrightarrow L^p(\Sigma, T^{0,1}\Sigma \otimes_{J(\alpha_0)} f^*TM).$$

According to the calculation of lemma 6.16 we have for the partial derivative in the first direction

$$D_1\tilde{\mathcal{P}} : H^{1,p}(\Sigma, f^*TM) \longrightarrow L^p(\Sigma, T^{0,1}\Sigma \otimes_{J(\alpha_0)} f^*TM),$$

$$D_1\tilde{\mathcal{P}}(\xi) = \overline{\partial}^f_{J(\alpha_0)}\xi + \rho(\xi).$$

For the second partial derivative we get by the remark 6.6

$$D_2\tilde{\mathcal{P}}(\dot{\alpha}) = \frac{1}{2}[\varphi_{\alpha_0}(\dot{\alpha}), J(\alpha_0)]df\, i.$$

We want to show that $D\tilde{\mathcal{P}}$ is surjective. The operator $D_1\tilde{\mathcal{P}}(\xi)$ is an elliptic differential operator and as such it is a Fredholm operator. Thus, because $\mathrm{range}D_1\tilde{\mathcal{P}}$ has finite codimension and $\mathrm{range}D_1\tilde{\mathcal{P}} \subset \mathrm{range}D\tilde{\mathcal{P}}$, the subspace $\mathrm{range}D\tilde{\mathcal{P}}$ is closed. Let us take a vector $\eta \in L^q(\Sigma, T^{1,0}\Sigma \otimes_{J(\alpha_0)} (f^*TM)^*)$ orthogonal to the image of $\mathrm{range}D\tilde{\mathcal{P}}$. Then

$$\langle D_1\tilde{\mathcal{P}}(\xi), \eta \rangle = 0, \ \forall \xi \in H^{1,p}(\Sigma, f^*TM).$$

$$\langle [\varphi_{\alpha_0}(\dot{\alpha}), J(\alpha_0)]J(\alpha_0)df, \eta \rangle = 0, \ \forall \dot{\alpha} \in \mathcal{H}. \tag{6.1}$$

Of course $[\varphi_{\alpha_0}(\dot{\alpha}), J(\alpha_0)] = 2\varphi_{\alpha_0}(\dot{\alpha})J(\alpha_0)$. Let us look at a point $z_0 \in \Sigma$, where $df(z_0) \neq 0$, $f^{-1}f(z_0) = z_0$. Then there exists a neighbourhood W of $f(z_0)$ in M and a disc $B \subset \Sigma$ around z_0, such that $f^{-1}[W] = B$ and $f\mid_B: B \to M$ is an imbedding. Now we observe that given a nonzero vector v_0 in a symplectic vector space V with a compatible complex structure j we have $h(V, j)v_0 = V$ (to see this, it suffices to look at the problem in complex dimension 1). Thus if $\eta(z_0) \neq 0$, we find $\dot{\alpha} \in C^\infty(M, h(TM, J(\alpha_0))$ with support in W and with

$$-\langle \varphi_{\alpha_0}(\dot{\alpha})J(\alpha_0)\, df, \eta \rangle > 0.$$

We approximate the section $\dot{\alpha}$ by sections on $\mathcal{H}$ to conclude that we get a contradiction to equation (6.1) above. Thus $\eta(z_0) = 0$ and because the argument applied for z_0 works for a whole neighbourhood of z_0, the section η must be zero on an open set. But $D_1\tilde{\mathcal{P}}$ is a $\overline{\partial}$-operator and $\langle -(D_1\tilde{\mathcal{P}})^t\eta, \xi \rangle = 0$. Thus η is smooth (as a weak solution of an elliptic operator) and by Aronszajn's theorem (lemma 6.18), η vanishes everywhere. So $D\tilde{\mathcal{P}}$ is surjective at the zeroes of $\mathcal{P}$ and this section is transversal to the zero section of $\Pi_1^*\mathcal{E}_J^{0,1}$ and we may conclude that their intersection, i.e. the zero set of $\mathcal{P}$ is a smooth manifold. $\square$

We finally come to prove the proposition we aimed at in this section. To proceed in the construction of the moduli space we will use Smale's extension of Sard's theorem to Banach manifolds.

Proposition 6.23 *(i) Let $\Phi : X \to Y$ be a C^∞-map between two Banach manifolds. The map Φ is supposed to be a Fredholm map, i.e. for all $x \in X$ the differential $d\Phi(x) : T_x X \to T_{\Phi(x)} Y$ is a Fredholm operator. Then the set of critical values*

$$\{\Phi(x) \mid d\Phi(x) \text{ is not surjective}\}$$

has a dense complement in Y.

(ii) If in the situation of (i) we have further given a smooth map $g : W \to Y$ from a finite dimensional manifold W into Y and g is transversal to Φ on a closed subset $A \subset W$ (i.e. $\forall y = g(w) = \Phi(x) : g_* T_w W + \Phi_* T_x X = T_y Y$). then we can find an arbitrarily C^1-close perturbation g' to g, which is transversal to Φ and $g' \mid_A = g \mid_A$. (Of course the set A may be empty.)

A point y in Y is called a *regular value* if it is not a critical value, i.e. either $y \notin \Phi[X]$ or $y = \Phi(x)$ and $d\Phi(x)$ is surjective. In any case $\Phi^{-1}(y)$ is a closed submanifold in X, whose tangent space at $x \in \Phi^{-1}(y)$ is $\ker d\Phi(x)$.

The condition which appears in lemma 6.22 and also in proposition 6.24, namely that the holomorphic curve should be an embedding somewhere, i.e. $\exists z \in \Sigma : df(z) \neq 0$, $f^{-1}f(z) = \{z\}$, will be replaced by a homological criterion in section 6.3.1.

Proposition 6.24 *Let J be a smooth almost complex structure on a compact manifold M, $\mathcal{H} \subset \mathbf{C}^\infty(M, h(TM))$ as in corollary 6.21, Σ a closed connected Riemann surface. An element $\alpha \in \mathcal{H}$ gives the structure $J(\alpha) = e^\alpha J e^{-\alpha}$. The set*

$$\mathcal{M}^*_{J(\alpha)} = \{f \in \mathbf{C}^\infty(\Sigma, M) \mid f \text{ is } J(\alpha)\text{-holomorphic and there exists a point}$$
$$z \in \Sigma, \text{ where } df(z) \neq 0 \text{ and } f^{-1}(f(z)) = \{z\}\}$$

is a smooth manifold for a generic $\alpha \in \mathcal{H}$. Its dimension is

$$\frac{e(\Sigma)}{2} \dim_{\mathbf{R}} M + 2 \langle c_1(TM), f_*[\Sigma] \rangle.$$

Proof. For simplicity we shall write $\mathcal{P}$ instead of $\tilde{\mathcal{P}}$ from now on. In the previous lemma we arrived at

$$\mathcal{M}^*_{\mathcal{H}} = \mathcal{P}^{-1}(0) \cap \{(f, \alpha) \in C^\infty(\Sigma, M) \times \mathcal{H} \mid \exists z \in \Sigma$$
$$\text{with } df(z) \neq 0, f^{-1}(f(z)) = \{z\}\}.$$

which is a smooth manifold. Its tangent spaces are

$$T_{(f,\alpha)} \mathcal{M}^*_{\mathcal{H}} = \ker D\mathcal{P}$$
$$= \{(\xi, \dot{\alpha}) \in H^{1,p}(\Sigma, f^*TM) \times \mathcal{H} \mid D_2\mathcal{P}(\dot{\alpha}) + D_1\mathcal{P}(\xi) = 0\}.$$

(Here we noted again only the vertical components of the differential of the section $\mathcal{P}$.)

We now look at the projection

$$\tilde{\Pi}_2 : \mathcal{M}^*_{\mathcal{H}} \longrightarrow \mathcal{H}, \quad \text{restriction of } \Pi_2.$$

Its differential at a point $(f, \alpha) \in \mathcal{M}^*_{\mathcal{H}}$ is

$$T_{(f,\alpha)}\mathcal{M}^*_{\mathcal{H}} \ni (\xi, \dot{\alpha}) \longmapsto \dot{\alpha} \in \mathcal{H}.$$

So $\ker D\tilde{\Pi}_2 = \ker D_2\mathcal{P}$. Evidently, we have

$$\text{range } D\tilde{\Pi}_2 = D_2\mathcal{P}^{-1}(\text{range } D_1\mathcal{P}),$$

so that

$$\text{coker } D\tilde{\Pi}_2 = \mathcal{H}/\text{range } D\tilde{\Pi}_2 = \quad \begin{array}{c} \mathcal{H}/D_2\mathcal{P}^{-1}(\text{range } D_1\mathcal{P}) \\ \downarrow \overline{D_2\mathcal{P}} \\ L^p(\Sigma, T^{0,1} \otimes f^*TM)/\text{range } D_1\mathcal{P}, \end{array}$$

where $D_2\mathcal{P}$ passed to the quotients: $\overline{D_2\mathcal{P}}$. The operator $\overline{D_2\mathcal{P}}$ is trivially injective. Because $D\mathcal{P} = D_1\mathcal{P} \oplus D_2\mathcal{P}$ is surjective, following the proof of lemma 6.22, the application $\overline{D_2\mathcal{P}}$ is also surjective. Consequently, we have

$$\text{coker } D\tilde{\Pi}_2 \cong \text{coker } D_1\mathcal{P}.$$

Now $D_1\mathcal{P}$ is a Fredholm operator, so $\tilde{\Pi}_2$ is a Fredholm map. The Sard-Smale theorem then tells us that $\tilde{\Pi}_2$ has a dense set of regular values. So for such a regular value $\alpha \in \mathcal{H} \subset C^\infty(M, h(TM))$, the set $\mathcal{M}^*_{J(\alpha)}$ is a smooth manifold of dimension

$$\dim \ker D\tilde{\Pi}_2(f, \alpha) = \text{ind} D\tilde{\Pi}_2(f, \alpha) = \text{ind } D_1\mathcal{P}$$

$$= \frac{e(\Sigma)}{2} \dim_{\mathbf{R}} M + 2 \underbrace{\langle c_1(TM), f_*[\Sigma]\rangle}_{=c_1(f^*TM)}.$$

□

Definition 6.25 *An almost complex structure J on M is called* regular, *if for all holomorphic curves in $\mathcal{M}^*_J$ the differential operator of lemma 6.16*

$$\xi \mapsto \overline{\partial}^f_{J(\alpha_0)}\xi + \rho(\xi)$$

has vanishing cokernel.

In view of the proof of the proposition 6.24 we can formulate the conclusion of this section as a corollary:

Corollary 6.26 *There exists a dense set of smooth ω-compatible regular complex structures J (for which the moduli space $\mathcal{M}^*_J$ is a smooth manifold).*

6.2.4 Comparison of moduli spaces at different complex structures

Consider once again the assertion of lemma 6.20. We can construct a space $\mathcal{H} \subset C^\infty(M, E)$, which contains a finite number of given smooth sections. If the section s is not in $\mathcal{H}$ we form the new space $\mathcal{H}' = \mathcal{H} \oplus \mathbf{R}s$ with the norm $\|t + rs\|_{\mathcal{H}'} = \|t\|_{\mathcal{H}} + |r|$.

Proposition 6.27 *If J_0, $J_1 \in \mathcal{J}(\omega)$ are two regular structures, then there exists a smooth path $\gamma : [0, 1] \to \mathcal{J}(\omega)$, such that*

$$\mathcal{M}^*_\gamma = \{(f, t) \mid f \in \mathcal{M}^*_{\gamma(t)}\} \subset H^{1,p}(\Sigma, M) \times I$$

is a smooth manifold with boundary

$$\mathcal{M}^*_{J_0} \times \{0\} \cup \mathcal{M}^*_{J_1} \times \{1\}.$$

Proof. We choose a Banach space $\mathcal{H} \subset \mathcal{J}(\omega)$ which contains J_0 and J_1 as indicated before. Then as in lemma 6.22 we have the following smooth submanifold of $H^{1,p}(\Sigma, M) \times \mathcal{H}$:

$$\mathcal{M}^*_{\mathcal{H}} = \{(f, J) \mid f \text{ is J-holomorphic and} \exists z \in \Sigma : df(z) \neq 0, \, f^{-1}f(z) = \{z\}\}.$$

The index of the projection

$$\Pi : \mathcal{M}^*_{\mathcal{H}} \to \mathcal{H}$$

is again equal to the index of the $\bar{\partial}$-operator corresponding to these maps (proof of proposition 6.24). Thus J_0, J_1 are regular values for Π. We join J_0 to J_1 by an arbitrary smooth path in $\mathcal{H}$. With the second part of the Smale-Sard theorem we can perturb the path to get a path γ which is transversal to the projection Π and this perturbation can be done such that the endpoints of γ are J_0, J_1 respectively. Then we look at the diagram:

$$
\begin{array}{ccc}
\mathcal{M}^*_\gamma = \mathcal{M}^*_{\mathcal{H}} \times_{\mathcal{H}} I & \xrightarrow{\;id \times \gamma\;} & \mathcal{M}^*_{\mathcal{H}} \\
\downarrow & & \downarrow \Pi \\
I & \xrightarrow{\;\gamma\;} & \mathcal{H}
\end{array}
$$

and remember that the fiber product of two transversal mappings is smooth to see that $\mathcal{M}^*_\gamma$ is smooth with boundary as stated. $\qquad\square$

6.3 Compactness of the Moduli Space

In this section we consider some aspects of the global structure of the moduli space of holomorphic curves. The aim of the first section 6.3.1 is corollary 6.34, which replaces the somewhat technical condition of nondegeneracy of the curves appearing in proposition 6.22 by a condition on the homology class represented by the curves. The sections 6.3.2 up to 6.3.4 construct tools to force the moduli space to be compact. This will be used in section 6.3.5, which treats a question of existence for holomorphic curves by some sort of continuity method as announced in section 6.1.2.

6.3.1   A homological criterion for injectivity

We want to consider a holomorphic curve $f : \Sigma \to M$ from a Riemann surface into an almost complex manifold, for which there does not exist a point $z_0 \in \Sigma$ such that $df(z_0) \neq 0$ and $f^{-1}f(z_0) = \{z_0\}$. Certainly we expect the set of critical points of f to be discrete in Σ, so that essentially the map f must cover its image several times. So we will desingularise the image of f to get a Riemann surface of which Σ will be a branched cover and a factorisation of f, such that the homology class of f will be a multiple of the class of this desingularisation. If we prescribe that the homology class of f is indivisible, there must be a point $z_0 \in \Sigma$ such that $df(z_0) \neq 0$ and $f^{-1}f(z_0) = \{z_0\}$.

Remark 6.28: Let us denote by $D = \{z \in \mathbf{C} \mid |z| < 1\}$ the standard open disc. The theorem of Aronszajn has the following version, which is stronger than the one cited in 6.18. We call a map $f : D \to \mathbf{R}^{2n}$ with $f(0) = 0$ *flat at the origin* $0 \in D$. if we have for arbitrarily high $N > 0$ an estimation

$$|f(z)| \leq c(N)|z|^N,$$

for z in some neighbourhood of the origin. The theorem of Aronszajn asserts that a smooth $f : D \to \mathbf{R}^{2n}$, flat at the origin, which satisfies the differential inequality

$$|\triangle f(z)| \leq c|f(z)| + |df(z)|\}$$

with some constant $c > 0$, must be equal to zero on a whole neighbourhood of the origin.

In this section we can forget that the almost complex structure on M stems from a symplectic one.

Corollary 6.29 *Let M and Σ be equipped with some auxiliary metric. Let $f : \Sigma \to M$ be a holomorphic curve which is flat at a point $z_0 \in \Sigma$, i.e. for all $N > 0$ there exists a $c(N) > 0$ with*

$$\operatorname{dist}_M(f(z), f(z_0)) \leq c(N)\operatorname{dist}_\Sigma(z, z_0)^N.$$

for z in some neighbourhood of z_0. Then f is constant on the whole connected component of z_0.

Proof. If f is flat at z_0 then all the derivatives of f with respect to some choice of coordinate systems around z_0 and $f(z_0)$ must vanish. Thus the set of flat points of a smooth map is closed. Relative to some charts we can express f by a function $g : D(z_0) \to \mathbf{R}^{2n}$ defined on some disc centered at z_0. Then as in the proof of proposition 6.8 the function g satisfies a differential equation of the form

$$\triangle g = q(g, dg, dg).$$

Here q is a bilinear form in the last two arguments with coefficients which are smooth functions in the first. Thus on a compact neighbourhood of z_0 we get

$$|\triangle g(z)| \le c|dg(z)|^2 \le c'|dg(z)|.$$

and with the theorem cited in the previous remark 6.28, we can conclude that $g = 0$ on a neighbourhood of z_0. Thus the set of flat points is also open. $\quad\square$

Corollary 6.30 *For a smooth holomorphic curve $f : \Sigma \to M$ which is nonconstant on every connected component, the set of critical points*

$$C = \{z \in \Sigma \mid df(z) = 0\}$$

is discrete.

Proof. Let $z_0 \in \Sigma$ be a critical point. Without loss of generality we may assume that $\Sigma = D$. $z_0 = 0$, $M = \mathbf{R}^{2n}$ and $f(0) = 0$. Let J denote the almost complex structure on $\mathbf{R}^{2n}$. The value $J(0) = J_0$ gives to the vector space $\mathbf{R}^{2n}$ the structure of a complex vector space. If f were flat at the origin, the map f would be constant on the component of z_0 (corollary 6.29). Let T_l be the lowest order term in the Taylor expansion at the origin:

$$f(z) = T_l(z) + O(|z|^{l+1}).$$

T_l is a homogenous polynomial of degree l in x, y where $z = x + iy$. Thus we get for the almost complex structure on $\mathbf{R}^{2n}$

$$J(f(z)) = J_0 + O(|z|^l).$$

The comparison of terms of equal order in the equation $df\, i = (J \circ f)\, df$ leads to:

$$dT_l \circ i = J_0 \circ dT_l.$$

Thus dT_l must be a J_0-holomorphic polynomial: $T_l(z) = az^l$, with $a \in \mathbf{R}^{2n}\setminus\{0\}$ and

$$f(z) = az^l + O(|z|^{l+1}).$$

This shows that the zero set of the differential

$$df(z) = laz^{l-1} + O(|z|^l)$$

contains zero as an isolated point. $\quad\square$

The implicit function theorem for holomorphic mappings permits to bring an honest holomorphic curve to a standard form near a regular point. The same is true for pseudoholomorphic curves.

Lemma 6.31 *Let $f : D \to M$ be a holomorphic curve defined on the disc and which is regular at the origin: $df(0) \neq 0$. Then there exists a diffeomorphism $\Phi : V \to \mathbf{R}^{2n}$ defined on a neighbourhood of $f(0)$ with the following properties:*

(i) $(\Phi \circ f)(x + iy) = (x, y, 0, \ldots, 0)$ on $f^{-1}V$.

(ii) $\Phi_* J$ is standard along $\mathbf{R}^2 \times \{0\} \subseteq \mathbf{R}^{2n}$.

Proof. If we restrict f to a small neighbourhood of the origin we may assume that it is an imbedding. Along f we choose vector fields $a_1, b_1, \ldots a_n, b_n$ which form at each point a base of the corresponding tangent space to M in the following way:

$$a_1 = f_* \frac{\partial}{\partial x}, \ b_1 = f_* \frac{\partial}{\partial y},$$

where x, y are the standard coordinates on the disc and the other fields are chosen transversal to f such that

$$J a_i = b_i, \ J b_i = -a_i.$$

We pick a connection ∇ on M which is compatible with J and by integrating ∇-geodesics, we define the following map:

$$\mathbf{R}^2 \times \mathbf{R}^{2n-2} \longrightarrow M$$

$$(x, y, x_2 \cdot y_2, \ldots, x_n, y_n) \mapsto \exp^{\nabla}_{f(x+iy)}(x_2 a_2 + \ldots + y_n b_n).$$

The differential of this map is an isomorphism in a neighbourhood of the origin and it is also (J_0, J)-linear if we denote again by J_0 the standard almost complex structure on $\mathbf{R}^{2n}$. The inverse of this local diffeomorphism gives Φ. $\square$

With the aid of this lemma we can show that the intersection of two holomorphic curves cannot be too wild.

Proposition 6.32 *Consider two holomorphic curves $f, g : D \to M$ defined on the disc, such that there exist two sequences $\{z_i\}_{i\in\mathbf{N}}, \{w_i\}_{i\in\mathbf{N}}$ in D with the properties:*

(i) $f(z_i) = g(w_i). \forall i \in \mathbf{N}$

(ii) $\lim_{i\to\infty} z_i = \lim_{i\to\infty} w_i = 0$

(iii) $z_i \neq 0, w_i \neq 0. \forall i \in \mathbf{N}$.

If f is regular at the origin then there exists a holomorphic map $\psi : W \to D$ defined on an open neighbourhood of the origin in D such that

$$g = f \circ \psi.$$

Proof. Because the property in question is a local one. we may again assume that $M = \mathbf{R}^{2n}$ with a variable almost complex structure. By lemma 6.31 we may assume that f has the following form:

$$f : D \subset \mathbf{R}^2 \hookrightarrow \mathbf{R}^{2n} = \mathbf{R}^2 \times \mathbf{R}^{2n-2} =: V_1 \oplus \tilde{V},$$

such that the almost complex structure J is standard along $V_1 : J\,|_{V_1} = J_0$. Corresponding to this decomposition the map g can be written as $g = (g_1, \tilde{g})$. Let us first assume that $\tilde{g}$ is not flat at the origin. Then we may proceed as in the proof of lemma 6.31 and write

$$\tilde{g}(z) = T_l(z) + O(|z|^{l+1})$$

$$g_1(z) = p(z) + O(|z|^{l+1}).$$

where T_l is a nonvanishing homogenous polynomial in $x. y$ of degree l and p is also a polynomial. Because $J\,|_{V_1} = J_0$ we have

$$J(v_1, \tilde{v}) = J_0 + O(|\tilde{v}|).$$

if we let $v = (v_1, \tilde{v})$ vary in a compact neighbourhood of the origin. Thus,

$$J(g(z)) = J_0 + O(|z|^l)$$

and comparing terms of equal order in the equation $dg\,i = (J \circ g)dg$ shows that

$$\tilde{g}(z) = az^l + O(|z|^{l+1}),$$

where $a \in \mathbf{R}^{2n} \setminus \{0\}$. Thus the intersection points of f and g could not accumulate at the origin.

Thus $\tilde{g}$ is flat at the origin and we will show that it satisfies the necessary differential inequality needed to apply Aronszajn's theorem. The function g satisfies:

$$\triangle g(z) = \frac{\partial J(g(z))}{\partial x} \frac{\partial g}{\partial y}(z) - \frac{\partial J(g(z))}{\partial y} \frac{\partial g}{\partial x}(z).$$

The endomorphism J can be decomposed according to the decomposition $\mathbf{R}^{2n} = V_1 \oplus \tilde{V}$: if π_1, π_2 denote the canonical projections of $V_1 \oplus \tilde{V}$ we write

$$J_1 = \pi_2 J|_{V_1} : V_1 \to \tilde{V}.$$

$$\tilde{J} = \pi_2 J|_{\tilde{V}} : \tilde{V} \to \tilde{V}.$$

Thus we get:

$$\triangle \tilde{g}(z) = \frac{\partial \tilde{J}(g(z))}{\partial y} \frac{\partial \tilde{g}}{\partial x}(z) - \frac{\partial \tilde{J}(g(z))}{\partial x} \frac{\partial \tilde{g}}{\partial y}(z) +$$
$$+ \frac{\partial J_1(g(z))}{\partial y} \frac{\partial g_1}{\partial x}(z) - \frac{\partial J_1(g(z))}{\partial x} \frac{\partial g_1}{\partial y}(z).$$

The first two terms on the right may be estimated by $|d\tilde{g}(z)|$ for z varying in a compact set. For the second pair we split the differential

$$d(J_1 \circ g) = \tilde{d}J_1 \circ d\tilde{g} + d^1 J_1 \circ dg_1 \, .$$

where $dJ_1 = \tilde{d}J_1 + d^1 J_1$ is the decomposition of the differential according to $\mathbf{R}^{2n} = V_1 \oplus \tilde{V}$. Because $d^1 J = d^1 J_1 = 0$ along V_1, we have $|d^1 J_1(g(z))| \leq const|\tilde{g}(z)|$. Finally all this adds up to the desired estimation:

$$| \triangle \tilde{g}(z)| \leq c\{|\tilde{g}(z)| + |d\tilde{g}(z)|\}$$

and we conclude that $\tilde{g}$ vanishes in a neighbourhood of the origin. We finish by taking $\psi = g_1$. $\qquad\qquad\qquad\qquad\qquad\qquad\qquad\qquad\qquad\qquad\qquad\qquad\quad\square$

After these preliminaries we can prove the proposition we aimed at. We call a homology class $\alpha \in H_*(M, \mathbf{Z})$ *divisible* if there exists a $\beta \in H_*(M, \mathbf{Z})$ and a $d \in \mathbf{Z}$, $d \neq \pm 1$, such that $\alpha = d\beta$.

Proposition 6.33 *Let Σ be a closed connected Riemann surface and M an almost complex manifold. Let $f : \Sigma \to M$ be a holomorphic curve, such that the homology class $f_*[\Sigma] \in H_2(M, \mathbf{Z})$ is not divisible. Then there exists a point $z_0 \in \Sigma$, such that*

$$df(z_0) \neq 0, \ f^{-1}f(z_0) = \{z_0\}.$$

Proof. Let us assume that there exists no point with the required property. We will construct a desingularisation Σ' of the image of f, which will be a Riemann surface and we will get a factorization

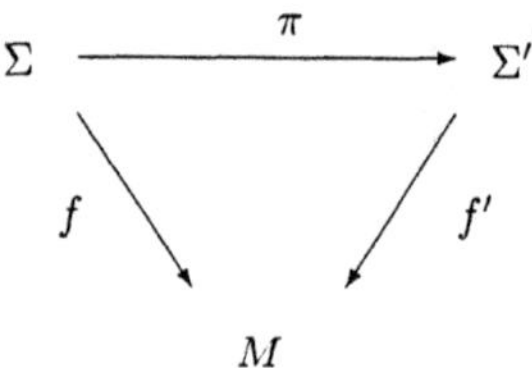

with a ramified covering π of degree $d > 1$. Then $f_*[\Sigma] = f'_*\pi_*[\Sigma] = d\,f'_*[\Sigma']$ and the class of f is divisible.

The critical points of f form a finite set $C \subset \Sigma$. All the fibers of $f\mid_{\Sigma \setminus C}$ must have more than one point. Let the set $B \subseteq \Sigma \setminus C$ be defined by $z \in B \Leftrightarrow$

$$\exists z' \in f^{-1}f(z) : \forall \text{ neighbourhoods } V, V' \text{ of } z, z' \text{ resp., one has } f[V] \neq f[V'].$$

Then, by proposition 6.32 the points of B can only accumulate on C. Then $B \subset \Sigma \setminus C$ is discrete and $\Gamma = f[\Sigma \setminus (C \cup B)] \subset M$ is a submanifold, i.e a Riemann surface. We have the factorization:

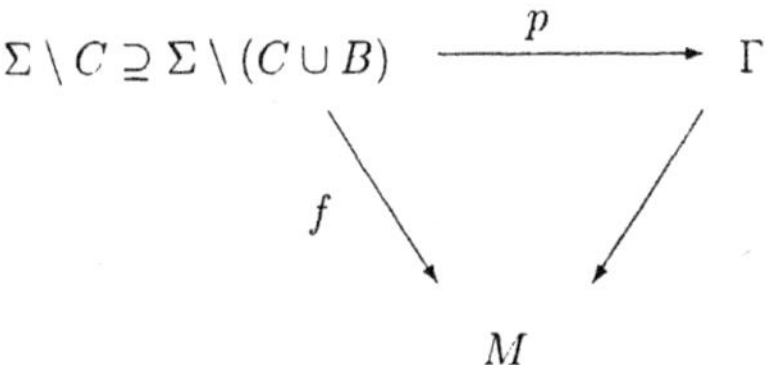

where p is a covering of degree $d > 1$. In a first step we will extend p over the points of B.

For a point $b \in B$ we can find a disc $D(b) \subseteq \Sigma \setminus C$ centered at b with $D(b) \cap B = \{b\}$ and such that $f\,|_{D(b)}$ is an imbedding. Thus the punctured disc $D(b)^* = D(b) - \{b\}$ gets imbedded into Γ by f. We can glue holomorphically to Γ a full disc via $f\,|_{D(b)^*}$. Another point b' in the fiber $f^{-1}f(b)$ has either a neighbourhood V' for which there exists a neighbourhood V of b, such that $f[V'] = f[V]$. or for all such pairs of neighbourhoods which are small enough one has $f[V] \cap f[V'] = f(b) = f(b')$ (see proposition 6.32). Thus we can unambigously complete the surface Γ to a surface Γ', such that p extends to a (unramified) covering $p' : \Sigma \setminus C \to \Gamma'$ and such that we still have a factorization:

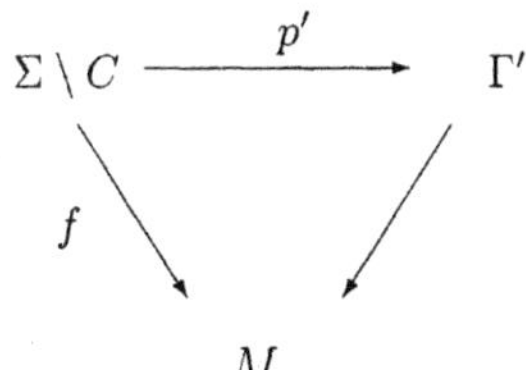

Now we have to extend this diagram over the critical points. The surface $\Sigma \setminus C$ is *of finite topological type*, by what we mean that its fundamental group $\pi_1(\Sigma \setminus C)$ is finitely generated. $\pi_1(\Gamma')$ contains $\pi_1(\Sigma \setminus C)$ as a subgroup of index $\leq d$. Thus Γ' is also of finite topological type. Consequently, there exists a closed surface Σ', such that Γ' is the complement of a finite set of points of Σ'. Thus we find inside Γ' a compact set K, such that its complement $\Gamma' \setminus K$ is a finite union of disjoint smooth punctured discs. Holomorphically these punctured discs are annuli:

$$\Gamma' \setminus K \cong \bigcup_{i=1}^{m} A(r_i).$$

where $A(r) = \{z \in \mathbf{C} \mid r < |z| < 1\}$ and $r < 1$.

Now $p'^{-1}(K) \subseteq \Sigma$ is closed. Let $\Sigma \setminus p'^{-1}(K) = \bigcup_{j=1}^{m'} U_j$ be the decomposition in connected components. Then $C \subset \Sigma \setminus p'^{-1}(K)$ and

$$p' \mid U_j \setminus (U_j \cap C) \to A(r_i)$$

should be a covering for suitable i, j. Then $U_j \cap C = \{c_j\}$ and $U_j \setminus \{c_j\} \cong A(\tilde{r}_j)$, where $\tilde{r}_j$ is some root of r_i. Then $\tilde{r}_j = 0$ and $r_i = 0$ for all i. Now that we know that the ends of Γ' are punctured holomorphic discs, the holomorphic structure of Γ' determines unambigously such a structure on Σ' and we can extend p' to $\pi : \Sigma \to \Sigma'$ by Riemann's theorem of removability of singularities. $\qquad\square$

For a homology class $\alpha \in H_2(M; \mathbf{Z})$ we define

$$\mathcal{M}_J(\alpha) := \{f : \Sigma \to M \mid f \text{ is } J\text{-holomorphic and } f_*[\Sigma] = \alpha\}$$

and more generally by adding the hypothesis on the homology class represented by the curves we define $\mathcal{M}_\mathcal{H}(\alpha)$ and $\mathcal{M}_\gamma(\alpha)$.

Obviously we have versions of corollary 6.26 and proposition 6.27 with this additional restriction on the homology class. Thus we get the following assertion:

Corollary 6.34 *Let $\alpha \in H_2(M; \mathbf{Z})$ be an indivisible homology class. Then there is a dense set of regular compatible almost complex structures $J \in \mathcal{J}(\omega)$, such that $\mathcal{M}_J(\alpha)$ is smooth.*

For two regular structures J_1, J_2 there exists a smooth path $\gamma : I \to \mathcal{J}(\omega)$, such that $\mathcal{M}_\gamma(\alpha)$ is a smooth cobordism from $\mathcal{M}_{J_1}(\alpha)$ to $\mathcal{M}_{J_2}(\alpha)$.

6.3.2 An apriori inequality

With this section we begin to study in which way a sequence of holomorphic curves can diverge in the moduli space. If the Riemann surface has a noncompact automorphism group, there are obviously sequences in the moduli space which have no convergent subsequence. But there might also be extrinsic reasons for a lack of convergence. Let us consider the following example.

We blow up $\mathbf{C}^2$ at the origin to get the space

$$\tilde{\mathbf{C}}^2 = \{(x, y, [t_0, t_1]) \mid xt_1 - yt_0 = 0\} \subset \mathbf{C}^2 \times \mathbf{P}^1.$$

with its canonical projection $\pi : \tilde{\mathbf{C}}^2 \to \mathbf{C}^2$, which restricts to an isomorphism on the complement of the exceptional divisor $\pi^{-1}(0, 0)$. There are two affine charts

$$A_1 : (u, v) \mapsto (v, uv, [1, u])$$

$$A_2 : (u, v) \mapsto (uv, v, [u, 1]).$$

In these charts the exceptional divisor appears as the set $\{v = 0\}$. For $\varepsilon \neq 0$ we define the holomorphic curve $f_\varepsilon(t) = \pi^{-1}(\varepsilon.t)$. In the charts A_1, A_2 this line appears as:

$$f_\varepsilon^1(t) = (\frac{t}{\varepsilon}, \varepsilon) \quad \text{and} \quad f_\varepsilon^2(t) = (\frac{\varepsilon}{t}.t).$$

As long as $t \neq 0$, these maps converge locally uniformly for $\varepsilon \to 0$ to $f(t) = \pi^{-1}(0,t) = (0,t,[0,1])$. To recognize what happens at $t = 0$ we rescale to a new parametrisation $t' = \varepsilon^{-1}t$ and get in the first chart $t' \mapsto (t',\varepsilon)$, which converges for $\varepsilon \to 0$ to a parametrisation of the exceptional divisor omitting the point $(0,0,[0.1]) \in \pi^{-1}(0,0)$. But the same point is also missing on the curve f. We conclude that we must consider the singular object $\pi^{-1}(\{0\} \times \mathbf{C}) \cup \pi^{-1}(0.0)$, a line with a sphere attached at the origin, as some sort of limit of the family f_ε.

In the following we will derive an apriori inequality which permits us to control this appearance of additional spheres.

Remark 6.35: In this section we consider again a symplectic manifold (M,ω) with a compatible almost complex structure J. On the surface Σ we choose a metric μ in the conformal class of the complex structure, and on M we choose a metric g which is not necessarily compatible with $(\omega.J)$. The differential of a smooth map $f : \Sigma \to M$ is pointwise a homomorphism $df(z) : T_z\Sigma \to T_{f(z)}M$ between euclidean vector spaces and as such it has a norm

$$\|df\|^2 = \mathrm{Tr}((df)^* \circ df).$$

The metric on Σ determines a volume form dv_μ and the *energy* of f is defined by

$$E(f) = \int_\Sigma \|df\|^2 dv_\mu.$$

We define the *area* of f to be

$$a(f) = \int_\Sigma f^*\omega.$$

For these two quantities we have the following relations:

(i) Even if g is not compatible with ω, the form $\|df\|^2 dv_\mu$ only depends on the conformal class of the metric μ,

(ii) if $(\omega.J,g)$ are compatible and f is holomorphic. then

$$f^*\omega = \frac{1}{2}\|df\|^2 dv_\mu.$$

To see this, let us consider the differential $df : T\Sigma \to f^*TM$ as a f^*TM-valued 1-form. The metric on TM provides a pairing of f^*TM-valued 1-forms: evaluated on two tangent vectors v, w the paired forms should give

$$g(\alpha . \beta)(v, w) = g(\alpha(v), \beta(w)) - g(\alpha(w). \beta(v)).$$

The metric on Σ gives the *Hodge star operator* $* : \Lambda^q T^*\Sigma \to \Lambda^{2-q} T^*\Sigma$, defined by $\alpha \wedge *\beta = \mu(\alpha, \beta) dv_\mu$, for scalar valued forms. This extends to an operation on f^*TM-valued forms and then we have

$$\|df\|^2 dv_\mu = g(df, *df).$$

(Take an orthonormal basis e_1, e_2 of a tangent space and ω^1, ω^2 dual 1-forms. Then $*df = *(e_1(f)\omega^1 + e_2(f)\omega^2) = e_1(f)\omega^2 - e_2(f)\omega^1$ such that

$$\|df\|^2 dv_\mu(e_1. e_2) = \|df(e_1)\|^2 + \|df(e_2)\|^2 = g(df. *df)(e_1. e_2).)$$

But in the middle dimension the Hodge star is always conformally invariant. As for the second assertion, let v, w be two tangent vectors on Σ. For a 1-form the Hodge star is $*\alpha(v) = -\alpha(iv)$. Thus

$$
\begin{aligned}
g(df, *df)(v, w) &= -g(df(v), df(iw)) + g(df(w). df(iv)) \\
&= -g(df(v), Jdf(w)) + g(df(w). Jdf(v)) \\
&= -2g(df(v), Jdf(w)) \\
&= 2\omega(df(v), df(w)).
\end{aligned}
$$

We formulate our conclusion as a lemma:

Lemma 6.36 *Let $f : \Sigma \to M$ be a holomorphic curve, let μ be a metric on Σ in the conformal class of i and let g be a metric on M compatible with (ω, J). Then $f^*\omega \geq 0$ and if Σ is compact, the energy is a topological invariant:*

$$\frac{1}{2} E(f) = \int_\Sigma f^*\omega = \langle [\omega], f_*[\Sigma] \rangle,$$

where $[\omega]$ denotes the de Rham class of the symplectic structure.

Remark 6.37: By the way, the same sort of calculations as in remark 6.35 shows another important property of holomorphic curves. Let $f : \Sigma \to M$ be a holomorphic curve and let (ω, J, g) be compatible. Then the metric induced by g on Σ, f^*g, has the volume form

$$dv_{f^*g} = f^*\omega.$$

To see this. let $x \in T_z\Sigma$ be a tangent vector. Then x and ix are an orthogonal basis of $T_z\Sigma$ and

$$
\begin{aligned}
f^*\omega(x,ix) &= \omega(df(x), df(ix)) = \omega(df(x). Jdf(x)) \\
&= f^*g(x,x) = dv_{f^*g}(x.ix).
\end{aligned}
$$

On the other hand, if we look at a variation f_t, $t \in (-1.1)$. $f_0 = f$ of f such that $f_t(z) = f(z)$ outside a compact set $K \subseteq \Sigma$, we get for two vectors $e_1, e_2 \in T_z\Sigma$ with $f_t^*g(e_1.e_2) = 0$:

$$
\begin{aligned}
|f_t^*\omega(e_1,e_2)| &= |\omega(df_t(e_1). df_t(e_2))| \\
&= |g(Jdf_t(e_1). df_t(e_2))| \\
&\leq \|df_t(e_1)\| \, \|df_t(e_2)\| \\
&= dv_{f_t^*g}(e_1.e_2).
\end{aligned}
$$

Thus

$$
\int_K dv_{f_t^*g} \geq \int_K f_t^*\omega = \int_K f^*\omega = \int_K dv_{f^*g}.
$$

We see that a holomorphic curve is a *minimal surface* with respect to the compatible metrics.

Remark 6.38: To prove the a priori inequality of proposition 6.42 we shall use a connection on the tangent bundle with a special torsion tensor. In general we cannot expect to have a connection compatible with the almost complex structure and without torsion (because the existence of such a connection is equivalent to the integrability of the almost complex structure).

We first consider the *decomposition of forms* with values in a complex vector bundle. On an almost complex manifold M we can decompose the **R**-linear differential forms with values in a complex vector bundle E according to type:

$$
\begin{aligned}
\mathrm{Hom}_{\mathbf{R}}(\Lambda_{\mathbf{R}}^r TM, E) &= (\mathrm{Hom}_{\mathbf{R}}(\Lambda_{\mathbf{R}}^r TM. \mathbf{R}) \otimes_{\mathbf{R}} \mathbf{C}) \otimes_{\mathbf{C}} E \\
&= \mathrm{Hom}_{\mathbf{C}}(\Lambda_{\mathbf{C}}^r (TM \otimes_{\mathbf{R}} \mathbf{C}), \mathbf{C}) \otimes_{\mathbf{C}} E \\
&= \Lambda_{\mathbf{C}}^r (T^{1,0}M \oplus T^{0.1}M) \otimes_{\mathbf{C}} E \\
&= \bigoplus_{p+q=r} (\Lambda^p T^{1,0}M \otimes_{\mathbf{C}} \Lambda^q T^{0,1}M) \otimes_{\mathbf{C}} E \\
&= \bigoplus_{p+q=r} \Lambda^{p,q} TM \otimes_{\mathbf{C}} E
\end{aligned}
$$

where $\mathrm{Hom}_{\mathbf{C}}(TM \otimes \mathbf{C}, \mathbf{C}) = T^{1,0}M \oplus T^{0,1}M$ according to the decomposition $TM \otimes_{\mathbf{R}} \mathbf{C} = TM \oplus \overline{TM}$ with respect to the eigenvalues $\pm i$ of the almost complex structure.

A connection ∇ on the tangent bundle has a *torsion* $T : TM \otimes TM \to TM$ defined by

$$T(X,Y) = \nabla_X Y - \nabla_Y X - [X,Y]$$

for two vector fields X, Y. This tensor can be considered as a 2-form with values in TM. Thus it may be decomposed according to type as described before. We observe that if the torsion tensor has no $(1,1)$-component, its restriction to a complex line in a tangent space will be zero.

Lemma 6.39 *On an almost complex manifold with a hermitian metric there exists a unique connection which is compatible with the almost complex structure and with the metric such that its torsion tensor has vanishing $(1,1)$-component.*

Proof. We recall what we mean by *compatibility*. The connection ∇ is compatible with the metric g if for all tripels of vector fields $X, Y. Z$ it is true that

$$X(g(Y,Z)) = g(\nabla_X Y, Z) + g(Y, \nabla_X Z).$$

and ∇ is compatible with the almost complex structure J if $\nabla_X(JY) = J\nabla_X Y$. By linearity we may extend ∇ to operate on sections of the complex tangent bundle $T_{\mathbf{C}}M = TM \otimes_{\mathbf{R}} \mathbf{C}$ and by sesquilinearity we can also extend the hermitian metric. We have the splitting $TM \otimes_{\mathbf{R}} \mathbf{C} = TM \oplus \overline{TM}$ and we call the sections of the first summand of type $(1,0)$ and of the second of type $(0,1)$. These two subbundles are mutually orthogonal. Because ∇ is compatible with J, it preserves type: for X any section and Z a section of type $(1,0)$ the derivative $\nabla_X Z$ is again of type $(1,0)$ etc.

We choose an orthonormal complex frame of type $(1,0)$: $Z_1, \ldots, Z_n$. Then we have for all i

$$\nabla_X Z_i = \sum_j \Theta_i^j(X) Z_j \tag{6.2}$$

with some 1-forms Θ_i^j. The compatibility with g gives $\Theta_i^j = -\bar{\Theta}_j^i$. Let us again by T denote the torsion tensor as a $T_{\mathbf{C}}$-valued 2-form (extended by $\mathbf{C}$-linearity to $T_{\mathbf{C}}M$). Let $\alpha^1, \ldots \alpha^n$ be the $(1,0)$-forms dual to the Z_i's. Then the definition of the torsion reads

$$d\alpha^j = -\sum_i \Theta_i^j \wedge \alpha^i + \alpha^j \circ T. \tag{6.3}$$

Here $\alpha^j \circ T(X,Y) = \alpha^j(T(X,Y))$. On the other hand, an arbitrary antihermitian matrix of 1-forms defines a connection via (6.2) whose torsion can be read off from equation (6.3). The tensor T has no $(1,1)$-component exactly if $\alpha^j \circ T$ has no $(1,1)$-component for all j.

We pick an arbitrary connection ∇ which is compatible with J and g. Locally we try to adapt the connection given by $\tilde{\Theta}_j^i = \Theta_j^i + \gamma_j^i$ to the imposed conditions:

$$da^i = -\sum_j (\Theta^i_j + \gamma^i_j) \wedge \alpha^j + \underbrace{\sum_j \gamma^i_j \wedge \alpha^j + \alpha^i \circ T}_{\alpha^i \circ \tilde{T}}$$

Thus the conditions are

$$\gamma^i_j = -\bar{\gamma}^j_i,$$

$$\sum_j \gamma^i_j \wedge \alpha^j + \alpha^i \circ T \text{ has no (1,1)-component.}$$

Expanding with respect to α^i we get

$$\alpha^i \circ T = \sum_{k,l} (A^i_{k,l}\alpha^k \wedge \alpha^l + A^i_{k,\bar{l}}\alpha^k \wedge \bar{\alpha}^l + A^i_{\bar{k},\bar{l}}\bar{\alpha}^k \wedge \bar{\alpha}^l)$$

$$\gamma^i_j = \sum_k (B^i_{j,k}\alpha^k + B^i_{j,\bar{k}}\bar{\alpha}^k).$$

and we see that the conditions have a unique solution for the $B^i_{j,k}$ and $B^i_{j,\bar{k}}$. Thus locally there exists a unique connection satisfying the conditions. Because of local uniqueness, the connections corresponding to overlapping patches must coincide on the intersection. Thus we get global existence and uniqueness. $\square$

Remark 6.40: We recall some generalities from differential calculus. Let $E \to X$ be a smooth vector bundle and let $\Omega^p_X(E)$ be $C^\infty(X. \Lambda^p T^* X \otimes E)$. Then a connection, which is a differential operator $\nabla : \Omega^0_X(E) \to \Omega^1_X(E)$, extends to a covariant exterior differential

$$d^\nabla : \Omega^p_X(E) \to \Omega^{p+1}_X(E)$$

by the definition

$$d^\nabla(\alpha \otimes s) = d\alpha \otimes s + (-1)^{\deg\alpha}\alpha \wedge \nabla s.$$

With metrics on X and on E we may form the formally adjoint operators

$$(d^\nabla)^* : \Omega^p_X(E) \to \Omega^{p-1}_X(E)$$

and the covariant *Hodge-Laplace operator*:

$$\triangle = (d^\nabla)^* d^\nabla + d^\nabla(d^\nabla)^*.$$

The metric on X gives the Levi-Civita connection on TX and on $\Lambda^p T^* X$. Thus we can differentiate covariantly the sections in $\Omega^p_X(E)$ and we get a differential operator

$$\nabla : \Omega^p_X(E) \to C^\infty(X, T^* X \otimes \Lambda^p T^* X \otimes E)$$

which again has a formally adjoint operator and we can form $\nabla^*\nabla$. Although the operators $\triangle$ and $\nabla^*\nabla$ are both of second order, their difference is of order zero. The *Weitzenböck formulae* describe this difference for low degrees, and in particular for $p = 1$ we have

$$\triangle s = \nabla^*\nabla s + s \circ \text{Ricc} + K^E(s).$$

Here we consider the Ricci tensor of the metric on X, which by its usual definition is a symmetric tensor of type $(0,2)$, as a symmetric endomorphism of the tangent bundle. The operator K^E is an endomorphism which is pointwise a linear function of the curvature tensor of the connection on E.

Remark 6.41: In the proof of the following proposition 6.42 we shall make use of *Morrey's mean value inequality* (for a proof see [48, Thm 9.20]).

Let $D(1,0) = \{z \in \mathbf{C} : |z| < 1\}$ be the standard disc with its standard metric and let u be a positive smooth function on it, which satisfies the following differential inequality

$$\triangle u = d^*du \leq au,$$

with a positive number a. Then u satisfies also

$$u(0) \leq c(a) \int_{D(1,0)} u$$

with a constant c independent of u.

Proposition 6.42 *Let (M,ω) be a closed symplectic manifold with a compatible almost complex structure J. Let Σ be a compact Riemann surface and $U \subseteq \Sigma$ an open subset with $U \cap \partial\Sigma = \emptyset$. The manifolds M, Σ are equipped with the hermitian metrics g and μ and the energy of a map $f : \Sigma \to M$ is defined with respect to them (see remark 6.35),*

$$e(f)dv_\mu = g(df, *df) = \text{Tr}((df)^t \circ df)dv_\mu.$$

In this situation there exist constants $C, \varepsilon, r_0 > 0$ such that for every holomorphic map $f : \Sigma \to M$ with

$$\int_{D(r,z)} f^*\omega < \varepsilon$$

on a geodesic disc centered at $z \in U$ with radius $r \leq r_0$ the following estimate holds:

$$\sup_{D(\frac{r}{2},z)} e(f) \leq \frac{C}{r^2} \int_{D(r,z)} f^*\omega.$$

Proof. (1) The differential of f, $df : T\Sigma \to f^*TM$, is a 1-form with values in f^*TM. On M we choose a connection with torsion tensor T and we pull it back to f^*TM where we note it by ∇. Then we have the relation of remark 6.17 which can be written in the style of remark 6.40 as

$$d^\nabla df = f^*T.$$

By the hypothesis on f the subspaces $f_*T_z\Sigma \subset T_{f(z)}M$ are complex lines. If we choose on M the canonical connection of lemma 6.39 we have

$$d^\nabla df = 0.$$

The adjoint operator $(d^\nabla)^*$ can be expressed by the Hodge star operator: $(d^\nabla)^* = - * d^\nabla *$. But on a 1-form α on Σ the Hodge star operates as the complex structure: $(*\alpha)(v) = -\alpha(iv)$ (because the metric μ is in the conformal class of i). Thus we get

$$
\begin{aligned}
(d^\nabla)^* df &= - * d^\nabla * df \\
&= *d^\nabla (df \circ i) \\
&= *d^\nabla (J df) \\
&= *J d^\nabla df = 0.
\end{aligned}
$$

We conclude that df is a harmonic section, $\triangle df = 0$. We put this immediately into the Weitzenböck formula of remark 6.40:

$$0 = \nabla^*\nabla df + df \circ \mathrm{Ricc} + K^{f^*TM}(df).$$

With this equation we estimate the ordinary Laplacian of the energy function: $\triangle e(f) = d^* de(f)$. Let us also look at the scalar product on 1-forms

$$\langle \alpha, \beta \rangle dv_\mu = g(\alpha, *\beta).$$

Then $e(f) = \langle df, df \rangle$ and

$$
\begin{aligned}
\frac{1}{2}\triangle e(f) &= \langle \nabla^*\nabla df, df \rangle - \langle \nabla df, \nabla df \rangle \\
&\leq \langle \nabla^*\nabla df, df \rangle \\
&\leq |\langle df \circ \mathrm{Ricc}, df \rangle| + |\langle K^{f^*TM}(df), df \rangle| \\
&\leq \frac{1}{2} c_1 (e(f) + e(f)^2).
\end{aligned}
$$

(2) Because Σ is compact, we can cover it by charts which are holomorphically discs and such that the standard metric of the discs and the metric on Σ are compatible with global comparison constants. Because the energy functions relative to these metrics are also comparable, we may assume that

$\Sigma = D(r_0, 0) = \{z \in \mathbf{C} : |z| < r_0\}$ for some small r_0. We may also assume that the disc $D(r, z)$ of the assertion is the disc $D(r, 0)$ in standard metric and centered at the origin.

We first derive an inequality of the form

$$\sup_{D(\frac{3r}{4}, 0)} e(f) \leq \frac{C'}{r^2}$$

with a constant C' independent of f and of $r \leq r_0$. Let us define the function

$$s(t) = t^2 \sup_{D(r-t, 0)} e(f)$$

for $t \in [0, r]$. This function attains a maximum at $t_0 \in (0, r]$. If $s(t_0) \leq 1$ we are done by taking $t = r/4$.

Thus let us assume that $s(t_0) > 1$ and let $z_0 \in D(r - t_0, 0)$ be a point at which $e(f)(z_0) = \sup_{D(r-t_0, 0)} e(f) = e_0$. We rescale the map f in a neighbourhood of z_0 by a factor

$$\delta = \frac{1}{2e_0^{1/2}} \leq \frac{t_0}{2}.$$

We take the map

$$D(1.0) \xrightarrow{\ \sigma\ } D(\delta, z_0) \subseteq D(\frac{t_0}{2}, z_0), \ y \mapsto \delta y + z_0.$$

The rescaled map $f^\sigma(y) = f(\sigma(y))$ then has the energy function

$$e(f^\sigma)(y) = \delta^2 e(f)(\sigma(y)).$$

But $D(t_0/2, z_0) \subseteq D(r - t_0/2, 0)$ so that

$$e(f)(z) \leq \frac{4}{t_0^2} s(\frac{t_0}{2}) \leq \frac{4}{t_0^2} s(t_0) = 4e_0,$$

for $z = \sigma(y)$, and we get

$$e(f^\sigma)(y) \leq 4\delta^2 e_0 = 1.$$

The differential inequality of (1) then leads to

$$\triangle e(f^\sigma) \leq c_1 e(f^\sigma)(1 + e(f^\sigma)) \leq 2c_1 e(f^\sigma),$$

and the Morrey inequality cited in remark 6.41 may be applied to get

$$e(f^\sigma)(0) \leq c_2 \int_{D(1,0)} e(f^\sigma) dv_\mu \leq c_2 \int_{D(r,0)} f^* \omega.$$

But $e(f^\sigma)(0) = \delta^2 e(f)(z_0) = 1/4$ which leads to a contradiction if $\int_{D(r,0)} f^*\omega$ is too small compared with the universal constant c_2.

(3) Let us now consider a point $z_0 \in D(r/2.0)$. Then $D(r/4, z_0) \subseteq D(3r/4.0)$. We rescale with the map

$$D(1,0) \xrightarrow{\varphi} D(\tfrac{r}{4}, z_0)$$

$$y \mapsto \frac{r}{4}y + z_0,$$

to get as before

$$\begin{aligned}
e(f^\varphi)(y) &= \frac{r^2}{16} e(f)(\varphi(y)) \\
&\leq \frac{r^2}{16} \sup_{D(\frac{3r}{4},0)} e(f) \\
&\leq \frac{C'}{16}.
\end{aligned}$$

Thus the differential inequality of (1) and the Morrey inequality can again be applied as in (2) to get

$$\frac{r^2}{16} e(f)(z_0) = e(f^\varphi)(0) \leq c_3 \int_{D(1,0)} e(f^\varphi) dv_\mu \leq c_3 \int_{D(r,0)} f^*\omega.$$

$\square$

6.3.3 Weak compactness

This and the following section will translate the basic a priori inequality of proposition 6.42 into more tangible properties of holomorphic curves.

Corollary 6.43 *Let (M,ω) be a closed symplectic manifold with a compatible almost complex structure J and let Σ be a closed connected Riemann surface. Then there exists a constant $c > 0$, such that a holomorphic curve $f : \Sigma \to M$ with $\int_\Sigma f^*\omega \leq c$ is constant.*

Proof. By proposition 6.42 there exists a constant c'. such that

$$\sup_\Sigma e(f) \leq c' \int_\Sigma f^*\omega.$$

as soon as the area of f is small enough. But if $\gamma : [0.1] \to \Sigma$ is a smooth path on Σ, we have a path $f \circ \gamma$ joining $f(\gamma(1))$ and $f(\gamma(0))$ with a length $l(f \circ \gamma)$ estimated by

$$l(f \circ \gamma) \leq \int_0^1 \|df(\gamma(t))\| \, \|\dot\gamma\| dt \leq \left(\sup_\Sigma e(f)\right)^{\frac{1}{2}} l(\gamma).$$

Thus there is a constant $c > 0$, such that for $\int_\Sigma f^*\omega \leq c$ the whole image of f lies in a geodesic ball of M. Consequently the homology class $f_*[\Sigma]$ is trivial and $\int_\Sigma f^*\omega = 0$. Thus $e(f) = 0$ and f is constant. $\square$

Lemma 6.44 *Let Σ, M and $U \subseteq \Sigma$ be as in the statement of proposition 6.42. There exists a constant $c > 0$, such that any sequence of holomorphic curves $f_n : U \to M$, $n \in \mathbf{N}$, with $\int_\Sigma f_n^*\omega \leq c$ has a subsequence converging in C^∞-topology.*

Proof. By proposition 6.42 we know: If c is small enough, then the differentials $\{df_n\}$ are locally uniformly bounded in sup-norm and so the family $\{f_n\}$ is equicontinuous. The manifold M is assumed to be compact, so that there exists a subsequence converging locally uniformly. We call it again $\{f_n\}$. Now we may look at this sequence in some coordinate charts where the f_n's satisfy the equation

$$\triangle f_n = q(f_n, df_n, df_n)$$

(cf. p. 103). The sequence $\{f_n\}$ is bounded in $H^{1,p}_{loc}$ for any p. Thus $q(f_n, df_n, df_n)$ is bounded in L^p_{loc} and elliptic regularity for the Laplace operator implies that $\{f_n\}$ is bounded in $H^{2,p}_{loc}$ for any p. Arguing inductively, we see that the sequence is bounded in $H^{k,p}_{loc}$ and by the Sobolev imbeddings the sequence is bounded in C^∞-topology. Then there exists a converging subsequence. $\square$

Now we come to the proposition we aimed at in this section and which describes the possible behaviour of sequences of holomorphic curves.

Proposition 6.45 *(Weak compactness) Let Σ and M be closed, J a compatible almost complex structure on M, and let $f_n : \Sigma \to M$, $n \in \mathbf{N}$, be a sequence of J-holomorphic curves with bounded area:*

$$a(f, \Sigma) = \int_\Sigma f^*\omega \leq A.$$

Then there exist finitely many points $\{x^1, \ldots, x^l\}$ in Σ and a subsequence of $\{f_n\}$ which converges in C^∞-topology on $\Sigma \setminus \{x^1, \ldots, x^l\}$.

Proof. We can find a sequence of finite open coverings of Σ

$$\mathbf{N} \ni k \mapsto \{V_i^k\}_{i \in I_k}, \; |I_k| < \infty,$$

such that

(i)

$$\sup_{i \in I_k} \mathrm{diam}(V_i^k) \to 0 \quad (k \to \infty),$$

(ii) there exists a number B, such that for each k the intersection of more than B elements of $\{V_i^k\}_{i \in I_k}$ is empty.

(If you do not believe this, see remark 6.46). Let $c > 0$ be the constant given in lemma 6.44 as bound on the area. With the second property we can estimate the number of sets in $\{V_i^k\}_{i \in I_k}$ satisfying

$$\int_{V_i^k} f_n^* \omega > c$$

(the "bad" sets). We have for all n

$$BA \geq c|\{i \in I_k \mid \int_{V_k^i} f_n^* \omega > c\}|$$

so that the number of bad sets is bounded by a number independent of k and n, say by l. We construct l sequences of points in Σ, indexed by $\mathbf{N} \times \mathbf{N}$: For $(n, k) \in \mathbf{N} \times \mathbf{N}$ we choose points

$$\{x_{(n,k)}^1, \ldots, x_{(n,k)}^l\}$$

in the bad sets V_i^k, and if we do not have enough bad sets to get l points we add arbitrary points to complete. We number these points in arbitrary manner and receive l sequences. Of these we choose subsequences, i.e. subsequences of $\{f_n\}$ and of $k \mapsto \{V_i^k\}_{i \in I_k}$, such that we may assume that all the sequences $\{x_{(n,k)}^i\}$ converge to x^i, $i = 1, \ldots, l$ respectively.

Now let $K \subseteq \Sigma \setminus \{x^1, \ldots, x^l\}$ be a compact set. Because of the first property of the sequence of coverings, we can assume for $n.k$ large enough that K intersects only sets V_i^k with $\int_{V_i^k} f_n^* \omega \leq c$. Then by lemma 6.44 we can extract a subsequence of $\{f_n\}$ converging in C^∞-topology on K.

Now we choose a sequence $\{K_m\}_{m \in \mathbf{N}}$ of compact sets in $\Sigma \setminus \{x^1, \ldots, x^l\}$ with

$$\bigcup_{m \in \mathbf{N}} K_m = \Sigma \setminus \{x^1, \ldots, x^l\}.$$

For each m we have a subsequence converging on K_m. The diagonal sequence then is what we were after. $\qquad\square$

Remark 6.46: The sequence of coverings we needed in the proof of proposition 6.45 can be obtained in the following manner: We choose first a smooth triangulation of Σ. Then we refine this triangulation by dividing in the following way

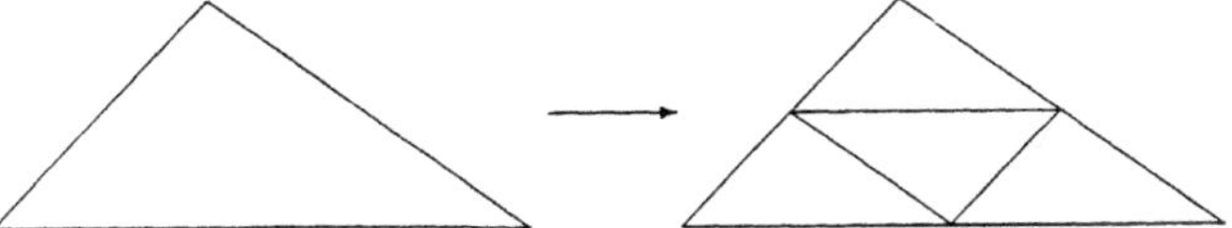

Thus the number of edges touching a vertex is kept bounded as we go on refining. To each vertex we take the open star i.e. the (simplicial) interior of all the simplices touching the vertex. This procedure yields the required open covering.

Remark 6.47: Because the compatible metric is a smooth function of the compatible almost complex structure, the basic a priori estimate of proposition 6.42 is uniform in the almost complex structure J. Thus we have the following extension of proposition 6.45 about weak compactness.

Corollary 6.48 *Let* $\{J_n\}_{n\in\mathbf{N}}$, $J_n \in \mathcal{J}(\omega)$, *be a convergent sequence of compatible almost complex structures on the closed symplectic manifold* (M,ω) *and let* $f_n : \Sigma \to M$ *be a sequence of* J_n*-holomorphic maps from a closed Riemann surface. If* $\int_\Sigma f_n^*\omega \leq A < \infty$ *then there exists a finite set* $C \subset \Sigma$ *and a subsequence of* $\{f_n\}$ *converging in* C^∞*-topology on* $\Sigma \setminus C$ *to a curve holomorphic for* $\lim_{n\to\infty} J_n$.

6.3.4   Removing singularities

This section is devoted to prove the following proposition which allows to fill in missing points on holomorphic curves as for example the points appearing for nonconverging sequences (cf. proposition 6.45 and corollary 6.48).

Proposition 6.49 *Let* $D^* = \{z \in \mathbf{C} \mid 0 < |z| < 1\}$ *be the disc without its center and let* (M,ω) *be a closed symplectic manifold with a compatible almost complex structure* J *(and also the corresponding hermitian metric). A smooth* J*-holomorphic curve* $f : D^* \to M$ *with finite area*

$$a(D^*, f) = \int_{D^*} f^*\omega < \infty$$

has a smooth J*-holomorphic extension to all of* $D = D(1,0)$.

Proof. We want to apply proposition 6.8 on regularity of weak holomorphic curves. But for the moment we do not know if f is regular enough near the origin, i.e. the differential should be p-integrable with $p > 2$ and f should be at least continuous at the origin. We note in the following $a(U, f) = \int_U f^*\omega$.

(1) There exists a constant $c_1 > 0$, such that for $0 < r < 1/2$ we have

$$l(r) = \int_0^{2\pi} \|df(re^{i\varphi})\| r\, d\varphi \leq 2\pi c_1 a(D(2r,0), f)^{1/2}.$$

We remark that $a(D^*, f) < \infty$ implies that $\lim_{r \to 0} l(r) = 0$, i.e. the image of D^* is at worst a thin tube. The apriori estimate of proposition 6.42 gives a constant c' such that

$$\|df(z)\|^2 \leq \frac{c'}{(\frac{|z|}{2})^2} a(D(\frac{|z|}{2}, z), f) \leq \frac{c_1^2}{|z|^2} a(D(2|z|, 0), f). \tag{6.4}$$

Integration yields the estimate above.

(2) There exists a constant $c_2 > 0$ such that for r sufficiently small:

$$a(D(r, 0), f) \leq c_2 l(r)^2.$$

To see this, we inspect what happens to an annulus

$$A(r, r') = \{z \in \mathbf{C} \mid r' \leq |z| \leq r\} \subseteq D^*$$

for $0 < r' < r$. If r is sufficiently small, $l(r)$ will be smaller than the injectivity radius of M and we can construct the following map

$$\varphi : D \times [r', r] \longrightarrow M,$$

which fills the tube $f[A(r, r')]$: We fiber the disc D by a linear projection along a fixed direction. At time $t \in [r', r]$ the fiber with endpoints $e^{i\alpha}$, $e^{i\alpha'}$ will be mapped linearly onto the unique geodesic from $f(te^{i\alpha})$ to $f(te^{i\alpha'})$, where we identified $S^1 \times [r', r]$ with $A(r, r')$ in a standard way. Thus we get by Stokes:

$$0 = \int_{\partial(D \times [r', r])} \varphi^* \omega = \int_{A(r, r')} f^* \omega - \int_{D \times \{r'\}} \varphi^* \omega + \int_{D \times \{r\}} \varphi^* \omega.$$

To estimate the last two terms we observe that $\varphi[S^1 \times \{r\}]$ lies in a geodesic ball whose radius is estimated by $l(r)$. By integrating radially in this ball, we find a primitive λ for the closed form ω whose magnitude can also be estimated by $l(r)$. Thus the two terms are estimated by $l(r)^2 + l(r')^2$. The limit $r' \to 0$ gives the desired inequality.

(3) For r_0 small enough there exist constants $c_3, \alpha > 0$ such that for $|z| < r_0$ it is true that

$$\|df(z)\| \leq \frac{c_3}{|z|^{1-\alpha}}.$$

To see this, we look at the function

$$a(r) = a(D(r, 0), f) = \int_{D(r, 0)} f^* \omega = \frac{1}{2} \int_{D(r, 0)} e(f) dv_\mu.$$

Then we have by the inequality of (2) and with the Cauchy inequality

$$a(r) \ \leq \ c_2 \Big[\int_0^{2\pi} \| df(re^{i\varphi}) \| r d\varphi \Big]^2$$

$$\leq \ 2\pi r c_2 \int_0^{2\pi} \| df(re^{i\varphi}) \|^2 r d\varphi$$

$$= \ 4\pi r c_2 a'(r).$$

Thus

$$\frac{a'(r)}{a(r)} \geq \frac{1}{4\pi c_2 r}$$

and integrated from r to $r_0 > r$

$$a(r) \leq a(r_0) \Big(\frac{r}{r_0} \Big)^{2\alpha}, \ \alpha = \frac{1}{8\pi c_2} > 0.$$

We put this into the estimate (6.4) of (1) to get the inequality we wanted.

(4) Because of the estimate in (3),

$$\int_0^{r_0} \| df(re^{i\varphi}) \| dr < \infty$$

and f is Lipschitz with a global Lipschitz constant. Thus it can be extended to a continuous function $\tilde{f} : D \rightarrow M$. Also (3) shows that $df \in L^p$ for a $p < \frac{2}{1-\alpha} > 2$, such that $\tilde{f} \in H^{1,p}(D)$ for a $p > 2$. A simple verification with the help of cut-off functions shows that $\tilde{f}$ is a weak holomorphic curve and it must finaly be smooth by propositon 6.8. $\qquad\qquad \square$

Now we can look at the result of proposition 6.46 with the help of proposition 6.49. A holomorphic map $f : \Sigma \rightarrow M$ defines a positive measure

$$C(\Sigma) \ni h \mapsto a_f(h) = \int_\Sigma h f^* \omega.$$

Because the space of bounded measures is weakly compact, a sequence $f_k : \Sigma \rightarrow M$ of holomorphic curves with bounded area has a subsequence (again called f_k) which converges uniformly outside a finite set $C \subset \Sigma$ to a map $f'_\infty : \Sigma \backslash C \rightarrow M$ and such that the measures a_{f_k} converge weakly to some measure λ. The map f'_∞ also has bounded area and can be extended by proposition 6.49 to a map $f_\infty : \Sigma \rightarrow M$. For a point $z \in \Sigma$ we define

$$m(z) = \lim_{\varepsilon \rightarrow 0} \lim_{k \rightarrow \infty} a(D(\varepsilon, z), f_k), \qquad\qquad (6.5)$$

where we have made a choice of a holomorphic chart around z. Obviously this number is well defined and independent of the choice of the chart. For a point

$z \notin C$ we have $\lim_{k \to \infty} a(D(\varepsilon, z), f_k) = a(D(\varepsilon, z), f_\infty)$ and thus $m(z) = 0$. If we denote by δ_z the Dirac measure situated at z, then we have

$$\lambda = a_{f_\infty} + \sum_{z \in C} m(z)\delta_z. \tag{6.6}$$

It suffices to look at open sets $U \subseteq \Sigma$, for which we have

$$
\begin{aligned}
\lambda(U) &= \lim_{k \to \infty} a(U, f_k) \\
&= \lim_{k \to \infty} a\Big(U \setminus \bigcup_{z \in C \cap U} D(\varepsilon, z), f_k\Big) + \sum_{z \in C \cap U} \lim_{k \to \infty} a(D(\varepsilon, z), f_k) \\
&= \lim_{\varepsilon \to 0} a\Big(U \setminus \bigcup_{z \in C \cap U} D(\varepsilon, z), f_\infty\Big) + \sum_{z \in C \cap U} \lim_{\varepsilon \to 0} \lim_{k \to \infty} a(D(\varepsilon, z), f_k) \\
&= a_{f_\infty}(U) + \sum_{z \in C} m(z)\delta_z(U).
\end{aligned}
$$

If we remember the model of failure of uniform convergence of page 126 we see that at a singular point a nonconstant holomorphic sphere splits off and we expect that by this process the area of f_∞ may be strictly smaller than the limit of $a(\Sigma, f_k)$. The missing area should belong to some holomorphic sphere. By corollary 6.43 we have a positive gap

$$\rho = \min\{a(S^2, g) \mid g : S^2 \to M \text{ holomorphic and nonconstant}\}.$$

Proposition 6.50 *Let $f_k : \Sigma \to M$ be a sequence of holomorphic curves with bounded area prepared as before (i.e. the corresponding measures converge and outside $C \subset \Sigma$ the sequence converges locally uniformly to f_∞). Let the set C be minimal. Then*

(i) $z \in C \Rightarrow m(z) \geq \rho$,

(ii) $a(\Sigma, f_\infty) + \sum_{z \in C} m(z) = \lim_{k \to \infty} a(\Sigma, f_k)$.

Proof. (i) The set C being minimal, uniform convergence must fail at each point $z \in C$. Thus there exists a sequence $\{z_k\}$ with

- $z_k \to z$

- $e_k = e(f_k)(z_k) = \sup_{D(1,z)} e(f_k) \to \infty$.

We renormalize to $g_k(y) = f_k(e_k^{-1/2}y + z_k)$ which is defined on the disc $D(\tfrac{1}{2}e_k^{1/2}, 0)$ for sufficiently large k. But:

$$|dg_k(0)| = 1, \qquad \sup_{D(\frac{1}{2}e_k^{1/2}, 0)} |dg_k| \leq 1.$$

Thus $\{g_k\}$ has a subsequence, again denoted $\{g_k\}$, which converges locally uniformly to a holomorphic map $g_\infty : \mathbf{C} \to M$. The area of g_∞ is bounded. Thus it may be extended to a holomorphic sphere $g_\infty : S^2 \to M$ which is not constant. We get

$$
\begin{aligned}
\rho \;\;\leq\;\; & a(\mathbf{C}, g_\infty) = \lim_{N \to \infty} a(D(N, 0), g_\infty) \\
=\;\; & \lim_{N \to \infty} \lim_{k \to \infty} a(D(N, 0), g_k) \\
=\;\; & \lim_{N \to \infty} \lim_{k \to \infty} a(D(N e_k^{-1/2}, z_k), f_k) \\
\leq\;\; & \lim_{\varepsilon \to 0} \lim_{k \to \infty} a(D(\varepsilon, z), f_k) \\
=\;\; & m(z).
\end{aligned}
$$

(ii) follows from (6.6). $\qquad\qquad\qquad\qquad\qquad\qquad\qquad\qquad\qquad\qquad\square$

6.3.5 A compactness and an existence result

In this section we specialize to $\Sigma = S^2$. The Riemann sphere has a large automorphism group which tends to inhibit compactness of the moduli space. Thus we have to fix three distinct points in S^2 and avoid that they approach in the image. If the energy of the spheres is too small, it is impossible to split off a further sphere. Thus we expect the following proposition to hold.

Proposition 6.51 *Let S^2 be the Riemann sphere with three distinct points s_1, s_2, s_3. Let $\alpha \in H_2(M, \mathbf{Z})$ be an indivisible class with $\langle [\omega], \alpha \rangle < 2\mathcal{R}$, where the constant $\mathcal{R}$ should be a lower bound for the energy of all nonconstant holomorphic spheres for all ω-compatible almost complex structures on M. For a positive δ let*

$$
\mathcal{M}_\omega^\delta(\alpha) = \{(f, J) \in \mathcal{M}_{\mathcal{J}(\omega)}(\alpha) \mid \mathrm{dist}(f(s_i), f(s_j)) \geq \delta \text{ for } i \neq j\}
$$

(The distance is measured with some fixed auxiliary metric). Then the projection

$$
\Pi : \mathcal{M}_{\mathcal{J}(\omega)}(\alpha) \longrightarrow \mathcal{J}(\omega)
$$

gets proper when restricted to $\mathcal{M}_\omega^\delta(\alpha)$.

Proof. Given a converging sequence $\{J_k\}$ in $\mathcal{J}(\omega)$ and a sequence $\{(f_k, J_k)\}$ in $\mathcal{M}_\omega^\delta(\alpha)$ we may assume thet there are finitely many points $\{z^1, \ldots, z^l\} \subset S^2$ such that $\{f_k\}$ converges uniformly on compact sets in $S^2 \setminus \{z^1, \ldots, z^l\}$ to a map $f_\infty : S^2 \to M$ and such that we have the relation

$$
a(S^2, f_\infty) + \sum_{i=1}^l m(z^i) < 2\mathcal{R}
$$

with $m(z^i) \geq \rho_J \geq \mathcal{R}$ (see proposition 6.50). Thus $l \leq 1$ and at least two of the points s_1, s_2, s_3 are not in $\{z^1, \ldots, z^l\}$. Thus f_∞ is not constant and $a(S^2, f_\infty) \geq \mathcal{R}$. Hence $l = 0$. $\qquad\Box$

Let us consider the following situation. By X we denote a closed symplectic manifold with symplectic structure ω_X. Let S^2 be the Riemann sphere with its standard symplectic structure ω_{S^2} with volume 4π. On the product manifold $M = S^2 \times X$ we have the symplectic structure $\omega_0 = \omega_{S^2} \oplus \omega_X$. The sum of the standard almost complex structure of S^2 and a ω_X-compatible structure J_X gives a ω_0-compatible structure $J_0 = J_{S^2} \oplus J_X$. Let $\alpha = [S^2] \times 1 \in H_2(M, \mathbf{Z})$. Then the set of J_0-holomorphic spheres in the class α are the compositions

$$S^2 \longrightarrow S^2 \longrightarrow S^2 \times \{x\} \subseteq M$$

with an arbitrary point $x \in X$. (The projection of a J_0-holomorphic sphere on the second factor is again a holomorphic sphere. but with trivial homology class, so that it must be constant). Thus we have $\dim \mathcal{M}_{J_0}(\alpha) = \dim X + 6$. On the other hand the index of the relevant $\bar{\partial}$-operator is

$$2\left(\frac{e(S^2)}{2}\operatorname{rank}_{\mathbf{C}}(TS^2 \oplus TX \mid_{S^2 \times \{x\}}) + c_1(TS^2 \oplus TX \mid_{S^2 \times \{x\}})\right) = 6 + \dim X.$$

Consequently, the structure J_0 is regular (see page 117).

If we choose a point $s_1 \in S^2$ we have the smooth evaluation map

$$e_{J_0} : \mathcal{M}_{J_0}(\alpha) \longrightarrow M$$

$$f \mapsto f(s_1).$$

This map is surjective.

Proposition 6.52 *In the situation we have described. let J be an arbitrary ω_0-compatible almost complex structure on $M = S^2 \times X$. If we assume that the integral of ω_X vanishes on spherical classes, then the evaluation map $e_J : \mathcal{M}_J(\alpha) \to M$ is surjective.*

Proof. Let us first assume that J is a regular structure. Then we have by proposition 6.27 a path $\gamma : I \to \mathcal{J}(\omega)$ such that

$$\mathcal{M}_\gamma(\alpha) = \{(f, t) \mid f_*[S^2] = \alpha, \ f \text{ is } \gamma(t)\text{-holomorphic}\}$$

is a manifold with boundary

$$\mathcal{M}_{J_0}(\alpha) \amalg \mathcal{M}_J(\alpha).$$

Let us choose two further points $s_2, s_3 \in S^2$ and let

$$L_i = \{s_i\} \times X \xrightarrow{j_i} M \quad (i = 1, 2, 3)$$

be the injections. Then we can find arbitrarily small C^1-perturbations $\tilde{j}_i$ which are transversal to the evaluation map $e_\gamma : \mathcal{M}_\gamma(\alpha) \to M$. Thus

$$\tilde{\mathcal{M}}_\gamma(\alpha) = \{(f,t) \in \mathcal{M}_\gamma(\alpha) \mid f(s_i) \in L_i, \ i = 1,2,3\}$$

is a manifold with boundary

$$\tilde{\mathcal{M}}_{J_0}(\alpha) \amalg \tilde{\mathcal{M}}_J(\alpha).$$

Now we have

$$\dim \tilde{\mathcal{M}}_{J_0}(\alpha) = \dim \tilde{\mathcal{M}}_J(\alpha) = \dim \mathcal{M}_{J_0}(\alpha) - 3\dim S^2 = \dim X.$$

If the perturbations of the j_i's are small enough there exists a $\delta > 0$ such that for all $(f,t) \in \tilde{\mathcal{M}}_\gamma(\alpha)$

$$\operatorname{dist}(f(s_i), f(s_j)) \geq \delta \text{ for } i \neq j.$$

Now we observe that we can take $\mathcal{R} = 4\pi$ and with proposition 6.51 the manifold $\tilde{\mathcal{M}}_\gamma(\alpha)$ is compact and $\tilde{\mathcal{M}}_{J_0}(\alpha), \tilde{\mathcal{M}}_J(\alpha)$ are (nonoriented) cobordant. The evaluation map e_{J_0} has nontrivial degree, hence also e_J and it must also be surjective (if you do not believe this, look at remark 6.53).

Now let J be any compatible structure on M and let $p \in M$ be an arbitrary point. Then we may approximate J by regular structures $J_k \to J$. For $\delta > 0$

$$\Pi : \mathcal{M}_\omega^\delta(\alpha) \longrightarrow \mathcal{J}(\omega)$$

is proper and over each J_k there is a f_k such that $f_k(s_1) = p$. Then there exists a subsequence f_{k_m} converging in $\mathcal{M}_\omega^\delta(\alpha)$ to (f,J) and obviously $f(s_1) = p$. $\square$

Remark 6.53: A closed manifold $Y^{(n)}$ has a fundamental class $[Y]$ which is the generator of $H_n(Y; \mathbf{Z}_2) = \mathbf{Z}_2$, and the localization map $H_n(Y; \mathbf{Z}_2) \to H_n(Y, Y \setminus \{p\}; \mathbf{Z}_2)$ is an isomorphism for each point p. A map $f : Z^{(n)} \to Y^{(n)}$ between closed manifolds of equal dimension such that the point p is not in the image: $f[Z] \subset Y \setminus \{p\}$ has $f_*[Z] \mid_{(Y,Y-p)} = 0$. Thus also $f_*[Z] = 0 \in H_n(Y; \mathbf{Z}_2)$. This shows that a map of degree 1 must be surjective.

A compact manifold with boundary $(Z^{(n+1)}, \partial Z)$ has a fundamental class in $H_{n+1}(Z, \partial Z; \mathbf{Z}_2) = \mathbf{Z}_2$. Via the boundary operator $\partial : H_{n+1}(Z, \partial Z) \xrightarrow{\sim} H_n(\partial Z)$, this yields the fundamental class of the boundary manifold. A map $f : (Z, \partial Z) \to Y$ satisfies $f_*[\partial Z] = 0 \in H_n(Y; \mathbf{Z}_2)$, because one has the factorization

$$
\begin{array}{ccc}
H_{n+1}(Z, \partial Z) & \xrightarrow{\sim} & H_n(\partial Z) \\
\downarrow & & \downarrow \\
H_{n+1}(Y, Y) & \xrightarrow{\partial} & H_n(Y).
\end{array}
$$

This shows that the two boundary components in the proof of proposition 6.52 give the same homology class under the evaluation map. The standard reference for these kinds of things is of course chap. 6.3 of [97].

6.3.6 An application

In this section we shall prove a theorem of Gromov which states that a ball in $\mathbf{R}^{2n}$ cannot be squeezed in certain directions by a symplectic diffeomorphism. (For another proof using symplectic capacities see section 4.2.) Let $\mathbf{R}^{2n}$ be equipped with the standard symplectic structure ω which is described relative to the standard basis by $\omega(e_{2i-1}, e_{2j}) = \delta_{ij}$.

Proposition 6.54 *We consider an open set $U \subseteq D(R,0) \times \mathbf{R}^{2n-2} \subseteq \mathbf{R}^{2n}$ and a symplectic diffeomorphism $U \xrightarrow{\sim} B_r$ with an open ball in $\mathbf{R}^{2n}$. Then $R \geq r$.*

Proof. First we remark that we can embed (by an elementary formula) the disc $D(R,0)$ symplectically as $S^2 \setminus \{\infty\}$ into the sphere with the homogeneous symplectic form of volume πR^2 on S^2. If we denote by $\varphi : B_r \xrightarrow{\sim} U$ the symplectic diffeomorphism of the hypothesis, for every $\delta > 0$ the ball $B_{r-\delta}$ is mapped onto a bounded set in $D(R,0) \times \mathbf{R}^{2n-2}$. Hence we can embed $V_\delta = \varphi[B_{r-\delta}(0)]$ into a symplectic manifold of the form

$$M = S^2 \times T,$$

where T is a compact standard torus with its symplectic form. On V_δ we have the almost complex structure coming from the standard structure on $B_r \subseteq \mathbf{R}^{2n}$. It is obviously compatible with the symplectic form on M. But remember that a compatible almost complex structure can be identified with a smooth section of a vector bundle (see corollary 6.5). Thus we can extend the almost complex structure given on V_δ to the whole of M without changing it on $V_{2\delta}$. Now we can apply proposition 6.52 because $\pi_2(T) = 0$ and we get a holomorphic sphere passing through the image of the center of B_r and whose homology class is $[S^2] \times 1$ such that its area must be πR^2. The part of this sphere inside of $V_{2\delta}$ can be pulled back with φ, and we get a holomorphic curve inside $B_{r-2\delta}$ passing through the center and of area $\leq \pi R^2$. But by an inequality on complex subvarieties of dimension 1 in a ball, its volume must be larger than $\pi(r-2\delta)^2$. Thus $r - 2\delta \leq R$ for all $\delta > 0$. $\qquad\square$

This result immediately implies that a product of two open discs $D(r_1,0) \times D(r_1',0)$ can be diffeomorphic to another such product $D(r_2,0) \times D(r_2',0)$ by a symplectic map if and only if $\{r_1, r_1'\} = \{r_2, r_2'\}$.
First the two volumes must be equal, i.e. $r_1 r_1' = r_2 r_2'$. Then if $r_1 \leq r_1'$ we have the ball $B_{r_1} \subset D(r_1,0) \times D(r_1',0)$ such that the last proposition implies that $r_1 \leq \min(r_2, r_2') = r_2$, say. By symmetry $r_1 = r_2$ and we are done.

This is of course only the beginning of the whole story. But here we abandon the patient reader and hope that he returns with more courage to the work of Gromov [50] or to more recent applications as in [73] and [74].

7 Gromov's Compactness Theorem from a Geometrical Point of View

As a complement to the chapter on pseudoholomorphic curves we draw now our attention to the original proof of Gromov's compactness theorem [50]. Pansu was the first to shed light on the ideas of this proof in his preprint [85]. Recently all the details have been worked out carefully by Hummel [61], whose "Diplomarbeit" is the main reference for this chapter.

The real advantage of this approach is its geometrical intuition and its analogy to the holomorphic theory which occurs on different occasions: in fact, pseudoholomorphic versions of the Schwarz Lemma, of the Weierstrass Theorem and of a well known monotonicity property will prove to be extremely important tools. One further example of a classical result generalized for pseudoholomorphic maps is a theorem on removing singularities similar to proposition 6.49. On the other hand, Hummel investigates a very delicate notion of convergence, which is closely related to convergence of hyperbolic structures on Riemann surfaces, using elementary coordinates on the Riemann moduli space, the Fenchel-Nielsen parameters. So the geometrical aspects of pseudoholomorphic curves and of Gromov's compactness result are indeed illuminated by analogy with classical facts.

It is not our intention to reproduce all the details of Pansu's and Hummel's work. On the contrary, most of the tools mentioned above, important as they are, are just quoted without proofs. However, as an alternative to some of the results stated in Chapter 6, we try at least to present the main ideas and steps in Gromov's original approach. Moreover the interested reader will find an extensive presentation of the subject in [61] and [85].

Chapter 7 is divided into three sections: in the first one we recall some definitions and formulate a number of important properties of pseudoholomorphic curves. The second section treats Riemann surfaces and their deformation and introduces the notion of cusp-curves. The third section is entirely devoted to Gromov's compactness theorem and its proof.

7.1 Gromov-Schwarz Lemma, Monotonicity, Removing Singularities

Definition 7.1 *Let (N, j) and (M, J) be almost complex manifolds. A C^∞-map $f : (N, j) \to (M, J)$ is called (j, J)-holomorphic (or simply* pseudoholomorphic*), if it is complex linear, i.e.*

$$df \circ j = J \circ df.$$

For the special case where f is a pseudoholomorphic map defined on a closed Riemann surface (S, j) we introduce the term J-holomorphic (or pseudoholomorphic*) curve. In this case j denotes the usual complex structure on the Rie-*

mann surface S. By analogy a J-holomorphic curve with boundary is given by a pseudoholomorphic map defined on a Riemann surface with boundary.

Remember that a Hermitian metric μ on an almost complex manifold (M, J) is defined as a Riemannian metric which is J-invariant:

$$\mu(u, v) = \mu(Ju, Jv) \qquad (\forall u, v \in T_p M; p \in M).$$

(M, J, μ) then denotes a Hermitian manifold. A first crucial result due to Gromov is a generalization of the classical Schwarz lemma:

Lemma 7.2 *(Gromov-Schwarz)* [61, p. 13] *Let (M, J, μ) be a compact Hermitian manifold. Then there exist constants $\varepsilon_0 > 0$ and $c > 0$ with the following properties: If g is a J-holomorphic map $g : D \to M$ defined on the open unit disc $D \subseteq \mathbf{C}$ such that the image $g(D)$ is contained in some ε_0-ball $B_{\varepsilon_0} \subseteq M$, then the norm of the differential of g at the origin is bounded by c: $\|dg_0\| < c$.*

A proof of this lemma is presented in all the details in [61, I.4]; the ideas go back to Gromov [50] and Pansu [85].

We should mention that D, as an open subset of $\mathbf{C}$, inherits the standard complex structure j and the Hermitian metric ν from the complex plane. The (j, J)-holomorphic map $g : D \to M$ is conformal. Indeed, if $v \in TS$, then $\nu(v, jv) = 0 = \mu(dg(v), Jdg(v)) = g^*\mu(v, jv)$ and $g^*\mu(v, v) = g^*\mu(jv, jv)$. Therefore $g^*\mu = \|dg\|^2 \nu$. Here $\|dg\|$ denotes the norm of dg with respect to the Hermitian metrics ν and μ.

The conformality of g is an important property used in the proof of the Gromov-Schwarz lemma. Another fact which enters into the proof is the area minimizing property of pseudoholomorphic curves. Let us be more precise: The area of an immersion $f : S \to (M, J, \mu)$ is defined by

$$\mathrm{Area}(f) := \int_S \sigma_{f^*\mu}.$$

$\sigma_{f^*\mu}$ is the volume element on S with respect to $f^*\mu$. If f is pseudoholomorphic, then, given a Hermitian metric h on the Riemann surface S, the area coincides with the energy

$$\mathrm{E}(f) := \int_S \|df\|^2 \sigma_h,$$

(where σ_h is the volume form with respect to h).

The following estimate

$$\mathrm{Area}(f) \geq \int_S f^*\omega,$$

an immediate consequence of the Wirtinger inequality formulated in the second chapter[1], is true for every differentiable map $f : (S, h) \to (M, J, \mu)$, provided

---

[1] In the special case in question, the Wirtinger inequality follows easily from the fact that $\omega(v, w) \leq 1$ for any two orthonormal $v, w \in TM$, with equality iff $w = Jv$.

$\omega(\cdot,\cdot) := \mu(J\cdot,\cdot)$ defines a compatible symplectic structure (remember definition 6.1) on (M,J,μ). Equality holds if and only if f is J-holomorphic. So a pseudoholomorphic curve is area minimizing in its homology class in $H_2(M,\mathbf{R})$. In addition. if $U \subset M$ is an open subset with the property that $\omega|_U$ is exact. then a J-holomorphic curve $f : S \to U$ with boundary in U is area minimizing among all the differentiable maps $\varphi : S \to U$ with $\varphi|_{\partial S} = f|_{\partial S}$.

The idea of the proof of the Gromov-Schwarz lemma now is first to derive in several steps a sort of isoperimetrical relation in analogy with the Euclidean case, where area minimizing maps of surfaces with boundary satisfy such an isoperimetric inequality. By the way, as a corollary of one of these steps one gets the monotonicity of pseudoholomorphic curves formulated in the lemma below. As regards the proof of the Gromov-Schwarz lemma, it is not hard to bring it to an end: from the isoperimetric inequality the boundedness of the norm of dg_0 is easily derived.

Lemma 7.3 *(Monotonicity)* [61, p. 19] *Let $(M. J, \mu)$ be a compact Hermitian manifold. Then there exist constants $\varepsilon_0, C_{ML} > 0$ with the following properties: If $f : (S. j) \to (M, J, \mu)$ is a J-holomorphic curve with boundary, $s_0 \in S \setminus \partial S$ and $r \in (0. \varepsilon_0)$ such that $f(\partial S)$ lies outside of a closed r-ball $\overline{B_r}(f(s_0)) \subseteq M$, then the following inequality holds*

$$\mathrm{Area}(f(S) \cap \overline{B_r}(f(s_0))) \geq C_{ML} \cdot r^2.$$

We often use the notation $\mathrm{Area}(f(S) \cap U)$ instead of $\mathrm{Area}(f|_{f^{-1}(U)})$, where U is an open subset of M.

As mentioned above, this result already appears in the proof of the Gromov-Schwarz lemma and it is of course not by accident that similar properties show up in the theory of minimal surfaces. In the domain of pseudoholomorphic curves the monotonicity in fact turns out to be an efficient result, as we will see in the third section, when we prove Gromov's compactness theorem.

We proceed with a theorem on removing singularities and with a convergence result. In both cases the reader will recognize the similarity to well known facts in classical function theory.

Theorem 7.4 *(Removing Singularities)* [61, p. 30] *Let S be a Riemann surface and (M, J, μ) a Hermitian manifold. Given a J-holomorphic map $f : S \setminus \{a\} \to M$ (where $a \in S$ is an inner point of S), whose image is relatively compact, suppose that there exists a neighbourhood G of a in S such that $f|_{G \setminus \{a\}}$ has finite area. Then f extends to a J-holomorphic map defined on the whole surface S.*

This is a variant of Proposition 6.49. The proof suggested by [61] and [85] uses the fact that, roughly speaking, the differentials of pseudoholomorphic maps are again pseudoholomorphic. The same argument is needed to prove a generalized version of the Weierstrass Theorem.

Theorem 7.5 *(Generalized Weierstrass Theorem)* [61, p. 45] *Let $(j_n)_{n\geq 1}$ be a sequence of complex structures on a compact connected surface S which converges in the C^∞-topology to a complex structure j on S and let $(f_n)_{n\geq 1}$ be a sequence of (j_n, J)-holomorphic maps $f_n : S \to (M, J, \mu)$. Then the following holds: If $(f_n)_{n\geq 1}$ converges to a map $f : S \to (M, J, \mu)$ in the C^0-topology, then it convergences in the C^∞-topology and moreover f is (j, J)-holomorphic.*

By C^k-topology we mean the weak C^k-topology, in other words, the topology of locally uniform C^k-convergence $(0 \leq k < \infty)$.

7.2 Deformation of Surfaces and Convergence of Hyperbolic Structures

We begin this section with a summary of some definitions and classical facts about Riemann surfaces. Let S be a Riemann surface. By the uniformization theorem, its universal covering surface $\tilde{S}$ is conformally equivalent to $\mathbf{C}$, $\hat{\mathbf{C}}$ or $\mathbf{H} := \{z \in \mathbf{C} : \operatorname{Im} z > 0\}$. The covering group Γ is a discrete group of Möbius transformations acting freely on $\tilde{S}$. S is then conformally equivalent to $\tilde{S}/\Gamma$. Note that if the universal covering is the upper halfplane $\mathbf{H}$, then a complete Hermitian metric with constant curvature -1 is induced on S, the Poincaré metric.

We are mainly interested in hyperbolic surfaces. These are connected oriented Riemann surfaces which are equipped with the Poincaré metric h and which have finite volume. Moreover the boundary components are supposed to be simple closed geodesics (with respect to h). The *signature* (g, m, k) of such a hyperbolic surface S is defined by its genus g, by the number m of boundary components and by the number k of punctures. If the Euler characteristic $2g - 2 + m + k$ is positive, then the volume of S is $2\pi(2g - 2 + m + k)$.

The Riemann moduli space $\mathfrak{R}_{(g,m,k)}$ is the space of equivalence classes of conformally equivalent Riemann surfaces with signature (g, m, k). If $2g-2+m+k > 0$, then a parametrization of $\mathfrak{R}_{(g,m,k)}$ (which by the way is the quotient of the Teichmüller space $\mathcal{T}_{(g,m,k)}$ with the corresponding modular group) is given by the Fenchel-Nielsen coordinates. These parameters are deduced from the pants decomposition of the given hyperbolic surface.

Pants Decomposition

How does this decomposition look like? A pair of pants is a hyperbolic surface with signature $(0, 3, 0)$. Let us point out that the conformal equivalence class of a pair of pants Y is uniquely determined by the lengths $(l(\gamma_1), l(\gamma_2), l(\gamma_3))$ of its boundary components $\gamma_1, \gamma_2, \gamma_3$. If one or more boundary components degenerate to a puncture, then we get a hyperbolic surface of signature $(0, 2, 1), (0, 1, 2)$ or $(0, 0, 3)$. In such a case we use the term "degenerate pair of pants". We set $l(\gamma) := 0$, if γ is a degenerate boundary component, i.e. a puncture. Observe that on every non-degenerate boundary component γ_i $(i = 1, 2, 3)$ there is a distinguished point, namely the initial point of the shortest geodesic which connects γ_i with γ_{i-1} (set $\gamma_0 = \gamma_3$).

Now, given a hyperbolic surface S with signature (g, m, k), there are $r :=
3g - 3 + m + k$ simple closed and pairwise disjoint geodesics $\gamma_1, \ldots, \gamma_r$ such that $S \setminus (\gamma_1 \cup \ldots \cup \gamma_r)$ consists of $N := 2g - 2 + m + k$ pairs of pants. Taking into account a *marking* of the surface S (this is a minimal ordered set of elements generating the fundamental group $\pi_1(S)$ of S), the r geodesics and the resulting pants decomposition of S are uniquely determined. (Note that pairs of pants are themselves marked by the numeration of their boundary components.)

The Fenchel-Nielsen parameters of S are now given by

$$
\begin{aligned}
(l(\gamma_1), \ldots, l(\gamma_r)) &\in \mathbf{R}_+^r \\
(l(\gamma_{r+1}), \ldots, l(\gamma_{r+m})) &\in \mathbf{R}_+^m \\
(\alpha(\gamma_1), \ldots, \alpha(\gamma_r)) &\in [0, 1)^r,
\end{aligned}
$$

where $\gamma_{r+1}, \ldots, \gamma_{r+m}$ are the geodesic boundary components of S and $\alpha(\gamma_i)$ is the twist parameter at γ_i. By the latter we mean the normalized signed length modulo $\mathbf{Z}$ along γ_i measured between the two distinguished points on γ_i. Note that there are indeed two distinguished points on γ_i: one from each of the two pairs of pants which are separated by γ_i. Obviously a point in $\mathbf{R}_+^{r+m} \times [0, 1)^r$ uniquely determines a marked hyperbolic surface S: S can be constructed by glueing together $2g - 2 + m + k$ pairs of pants; the glueing data are provided by the Fenchel-Nielsen parameters.

In summary, there is a one-one relation between $\mathfrak{R}_{(g,m,k)}$ and the $(6g - 6 + 3m + 2k)$-dimensional space $\mathbf{R}_+^{r+m} \times [0, 1)^r$ given by the Fenchel-Nielsen coordinates.

Convergence of Hyperbolic Structures

We deal now with the convergence problem of hyperbolic structures on a surface S with a given signature, in other words: we are interested in convergence within the Riemann moduli space and especially at its boundary, as we will see. More about these questions can be found in the standard literature of Teichmüller theory ([1], for instance).

Let us consider a sequence $(S_n^*)_{n \geq 1}$ of marked hyperbolic surfaces, all of the same signature (g, m, k). We assume that the lengths of all the closed geodesics on S_n^* which are not boundary components are bounded from below, independently of n, by $c > 0$. Moreover the lengths of the boundary components of S_n^* should be bounded from above, again independently of n.

In view of the marking, there is a uniquely determined decomposition of S_n^* into pairs of pants $Y_n^1, \ldots, Y_n^N$ $(N = 2g - 2 + m + k)$ for all $n \geq 1$. By a theorem of Bers (see [1, p. 98]) we can assume without loss of generality that the length of all the boundary components of all the pairs of pants are bounded from above by a constant independent of n.

Fenchel-Nielsen coordinates are now available for all the hyperbolic surfaces S_n^*, namely

$$
\begin{aligned}
(l_{n1}, \ldots, l_{nr}) &\in \mathbf{R}_+^r \\
(L_{n1}, \ldots, L_{nm}) &\in \mathbf{R}_+^m \\
(\alpha_{n1}, \ldots, \alpha_{nr}) &\in [0,1)^r
\end{aligned}
$$

with $r = 3g - 3 + m + k$. The boundedness of all the parameters enables us to take a subsequence of $(S_n^*)_{n \geq 1}$, again denoted by $(S_n^*)_{n \geq 1}$, such that the parameters converge for $n \to \infty$, i.e.

$$
l_{ni} \to l_i > 0, \quad L_{nj} \to L_j, \quad \alpha_{ni} \to \alpha_i
$$

for $i = 1, \ldots, r$ and $j = 1, \ldots, m$. Let k' be the number of L_j which are equal to zero. $(l_i)_{i=1,\ldots,r}$, $(L_j)_{j=1,\ldots,m}$ and $(\alpha_i)_{i=1,\ldots,r}$ are the Fenchel-Nielsen coordinates of a uniquely determined hyperbolic surface S^* with signature $(g, m - k', k + k')$.

Let S_n and S be the one-point compactifications at the punctures of S_n^* and S^*, respectively. After some reordering we can use the following notation:

- $\gamma_n^1, \ldots, \gamma_n^m$ for the non-degenerate boundary components of S_n^*

- $\gamma_n^{m+1}, \ldots, \gamma_n^{m+k}$ for the degenerate boundary components of S_n^*

- $\gamma^{m+1}, \ldots, \gamma^{m+k}$ and, in addition, $\gamma^1, \ldots, \gamma^{k'}$ for the degenerate boundary components of S^*

- $\gamma^{k'+1}, \ldots, \gamma^m$ for the non-degenerate boundary components of S^* .

Proposition 7.6 ([61, III.3.1]) *There is a sequence of continuous maps* $\varphi_n : S_n \to S$ *with the following properties:*

1. *$\varphi_n(\gamma_n^i) = \gamma^i$ for $i = 1, \ldots, m + k$*

2. *$\varphi_n|_{S_n \setminus \bigcup \{\gamma_n^i : i = 1, \ldots, k'\}}$ is a diffeomorphism onto its image. Its inverse we denote by $\psi_n : S \setminus \bigcup \{\gamma^i : i = 1, \ldots k'\} \to S_n \setminus \bigcup \{\gamma_n^i : i = 1, \ldots, k'\}$*

3. *ψ_n enables us to pull back the hyperbolic structure h_n of S_n^* to a Riemannian metric $\psi_n^* h_n$ on S^*. Then $(\psi_n^* h_n)_{n \geq 1}$ converges to the hyperbolic structure h on S^* in the C^∞-topology.*

4. *For the complex structure j_n of S_n we have a similar result: $(\psi_n^* j_n)_{n \geq 1}$ converges in the C^∞-topology to the complex structure j on S restricted to $S \setminus \bigcup \{\gamma^i : i = 1, \ldots, k'\}$.*

For the proof we refer to [61]. The idea is to reduce the statements to pairs of pants. Then one has to solve some convergence problems for hexagons in the universal covering space using hyperbolic geometry. Similar techniques occur for example in [37].

Deformations and Cusp-Curves

We are going to consider deformations of surfaces and we will introduce the notion of cusp-curves, in other words we arrive at another central point of this section.

Let S be a closed surface and $(\gamma^i)_{i \in I}$ a finite family of simple, closed and pairwise disjoint curves in S. Then $S \backslash \bigcup \{\gamma^i : i \in I\}$ consists of a finite number of components, each being homeomorphic to a closed surface from which a certain finite number of points have been deleted. $\hat{S}$ is now defined as the surface we get from $S \backslash \bigcup \{\gamma^i : i \in I\}$ by one-point compactification. Corresponding to every $\gamma^i \in S$, there are two points s_i' and s_i'' in $\hat{S}$, possibly lying in different components of $\hat{S}$. We now identify s_i' and s_i'' $(\forall i \in I)$ and get a new topological space $\bar{S}$ in this way. Intuitively, we get $\bar{S}$ from S by shrinking each curve γ_i to a point. Let $\alpha : \hat{S} \to \bar{S}$ denote the canonical projection. $\bar{s}_i := \alpha(s_i') = \alpha(s_i'')$ are called singular points of the singular surface $\bar{S}$,

$$si(\bar{S}) := \{ \text{ singular points of } \bar{S} \}.$$

Note that $\alpha|_{\hat{S} \backslash \{s_i', s_i'' : i \in I\}}$ is a diffeomorphism onto its image. Given a closed surface S, we shall always denote the corresponding singular surface by $\bar{S}$. $\hat{S}$ is the disjoint union of the components of $\bar{S}$.

Definition 7.7 *A* deformation *of a surface S into a singular surface $\bar{S}$ is a continuous surjective map $\varphi : S \to \bar{S}$ with the following properties:*

1. *The preimage $\varphi^{-1}(\{\bar{s}\})$ of every singular point $\bar{s} \in si(\bar{S})$ is a simple closed curve in S.*

2. *$\varphi|_{S \backslash \varphi^{-1}(si(\bar{S}))}$ is a diffeomorphism onto $\bar{S} \backslash si(\bar{S})$. Its inverse we denote by $\psi : \bar{S} \backslash si(\bar{S}) \to S \backslash \varphi^{-1}(si(\bar{S}))$.*

It is time to present the definition of cusp-curves. a generalization of pseudoholomorphic curves. (Every J-holomorphic curve is a cusp-curve.)

Definition 7.8 *A complex structure $\bar{j}$ on a singular surface $\bar{S}$ is a complex structure on $\hat{S}$. We call $(\bar{S}, \bar{j})$ a singular Riemann surface or a Riemann surface with singular points. A continous map $\bar{f} : (\bar{S}, \bar{j}) \to (M, J)$ (where (M, J) is an almost complex manifold) is called J-holomorphic, if $\bar{f} \circ \alpha$ is J-holomorphic. Then $\bar{f}$ is called a* cusp-curve *in M. $\mathrm{Area}(\bar{f}) := \mathrm{Area}(\bar{f} \circ \alpha)$ is the area of the cusp-curve $\bar{f}$.*

We will see in section 7.3 that under certain circumstances there is a compactification of the space of pseudoholomorphic curves by means of cusp-curves. We close this section with some remarks about the

Thick-Thin Decomposition of Hyperbolic Surfaces

Let us first have a look at annuli. It is a well known fact that each annulus A is conformally equivalent

a) either to the elliptic cylinder $\mathbf{C}/\mathbf{Z}$ $(\simeq S^1 \times \mathbf{R} \simeq \mathbf{C} \setminus \{0\})$

b) or to the parabolic cylinder $\mathbf{H}/\langle P \rangle$ $(\simeq S^1 \times (0,\infty) \simeq \mathbf{D} \setminus \{0\})$, where P is the standard parabolic transformation in $PSL(2,\mathbf{R})$; $P : z \mapsto z+1$. $\langle P \rangle$ is the cyclic group generated by P.

c) or to exactly one of the hyperbolic cylinders $\mathbf{H}/\langle T_l \rangle$ with $l > 0$, where T_l is a standard hyperbolic transformation in $PSL(2,\mathbf{R})$. $T_l : z \mapsto e^l z$ $(l > 0)$. We have $\mathbf{H}/\langle T_l \rangle \simeq S^1 \times (0,r)$ with $r = \frac{2\pi^2}{l} > 0$.

In the two hyperbolic cases b and c the modulus of the annulus A is defined by $\mathrm{Mod}A := \infty$ and $\mathrm{Mod}A := r$, respectively. (For details consult [16]).

Definition 7.9 *Given a hyperbolic surface* $S = \mathbf{H}/\Gamma$, *the injectivity radius of* S *at a point* $z \in S$ *is defined by* $\mathrm{injrad}(S,z) := \frac{1}{2}\inf_{\gamma \in \Gamma \setminus \{id\}} d(z.\gamma(z))$, *where* d *is the hyperbolic metric.*

The thick-thin decomposition of a hyperbolic surface describes the (thin) components with small injectivity radius. The thick components are those with large injectivity radius. For details we recommend [16].

Theorem 7.10 *(Thick-Thin Decomposition)   Let* S *be a hyperbolic surface and* $U \subseteq S$ *a connected component of* $\{s \in S : \mathrm{injrad}(S,s) < 1\}$. *Then one of the following statements is true:*

> 1. *either* U *contains a simple closed geodesic* γ *of* S *with length* $l(\gamma) < 2\,\mathrm{arsinh}1$ *and is isometric to*

$$\{w \in \mathbf{H}/\langle T_l \rangle : \mathrm{injrad}(\mathbf{H}/\langle T_l \rangle, w) < \mathrm{arsinh}1\}$$

> 2. *or* U *is isometric to*

$$\{w \in \mathbf{H}/\langle P \rangle : \mathrm{injrad}(\mathbf{H}/\langle P \rangle, w) < \mathrm{arsinh}1\}$$

The thick-thin decomposition will prove to be a useful tool in the proof of Gromov's compactness theorem.

7.3 Gromov's Compactness Theorem

We start with the central notion of convergence.

Definition 7.11 *Let (M, J, μ) be a Hermitian manifold, S a closed surface and $(j_n)_{n \geq 1}$ a sequence of complex structures on S. Suppose that $f_n : (S, j_n) \to (M, J, \mu)$ is a sequence of (j_n, J)-holomorphic curves in M. We say that $(f_n)_{n \geq 1}$ converges to a cusp-curve $\bar{f} : (\bar{S}, \bar{j}) \to (M, J, \mu)$ if the following conditions hold:*

1. *For each $n \geq 1$ there exists a deformation $\varphi_n : S \to \bar{S}$ such that $(\varphi_n^{-1})^* j_n$ converges to $\bar{j}|_{\bar{S} \setminus si(\bar{S})}$ in the C^∞-topology.*

2. *$(f_n \circ \varphi_n^{-1})_{n \geq 1}$ converges to $\bar{f}|_{\bar{S} \setminus si(\bar{S})}$ in the C^∞-topology.*

3. *$\lim_{n \to \infty} \mathrm{Area}(f_n) = \mathrm{Area}(\bar{f})$.*

Remark: If f_n converges to a cusp-curve $\bar{f}$ as described above, then $\bar{f}$ is independent of the parametrization. To be more precise: if $\chi_n : (S, j_n) \to (S, j_n)$ are conformal maps (for $n \geq 1$), then $(f_n \circ \chi_n)_{n \geq 1}$ converges to the same $\bar{f}$ as $(f_n)_{n \geq 1}$. The parametrization of $\bar{f}$ is not unique either. But one can show that if $\bar{f}$ and $\tilde{f}$ are two different limits of the same sequence, then their images in (M, J, μ) agree.

We are now ready to state Gromov's compactness theorem [50] for pseudoholomorphic curves.

Theorem 7.12 *(Gromov's Compactness Theorem)* [61, p. 66] *Let (M, J, μ) be a compact Hermitian manifold, S a closed surface and $(j_n)_{n \geq 1}$ a sequence of complex structures on S. Assume that $f_n : (S, j_n) \to (M, J, \mu)$ is a sequence of (j_n, J)-holomorphic curves in (M, J, μ) with $\mathrm{Area}(f_n) \leq C$ for a constant C independent of n. Then there is a subsequence of $(f_n)_{n \geq 1}$, which converges to a cusp-curve $\bar{f} : (\bar{S}, \bar{j}) \to (M, J, \mu)$.*

Remark: If in addition (M, J, μ) is endowed with a symplectic structure ω such that J is compatible with ω (see definition 6.1), then the condition $\mathrm{Area}(f_n) \leq C$ ($\forall n \geq 1$) is fulfilled provided that $(f_n)_{n \geq 1}$ is a sequence of J-holomorphic curves $f_n : (S, j_n) \to (M, J, \mu)$ in a fixed homology class.

Proof of the Compactness Theorem

We reproduce the proof given by [61], which works out the ideas of [85]. Given a compact Hermitian manifold (M, J, μ), we can choose $\varepsilon_0 \in (0, \mathrm{injrad}(M))$ such that the conditions of the Gromov-Schwarz lemma as well as the conditions of the monotonicity lemma hold. Let us start with the investigation of J-holomorphic curves $f : (S, j) \to (M, J, \mu)$ whose area is bounded by a constant: $\mathrm{Area}(f) \leq C$.

Lemma 7.13 *(Rescaling)* [61, p. 67] *If $f : S \to M$ is a J-holomorphic curve in M with $\mathrm{Area}(f) \leq C$, then there exist a universally bounded number of points $s_1, \ldots, s_k$ and a universal constant $\rho_0 > 0$ such that, with respect to the Poincaré metric h on $S^* := S \setminus \{s_1, \ldots, s_k\}$, every ρ_0-ball in (S^*, h) is mapped into an appropriate ε_0-ball in M.*

Remark: By a universal constant we mean in this section a constant which depends at most on (M, J, μ), ε_0 and C. Universally bounded stands for bounded by a universal constant.

Proof of Lemma 7.13. Without loss of generality we may assume that $f : S \to M$ is not a constant map. Given $s \in S$ we define $U(s)$ as the connected component of $f^{-1}(B_{\varepsilon_0/18}(f(s)))$ in S which contains s.

First step: An estimation for $\mathrm{Area}(f|_{\overline{U(s)}})$. One has to check that $\partial U(s) \neq \emptyset$ for the topological boundary $\partial U(s) = \overline{U(s)} \setminus U(s)$. This is indeed true, because we can assume without loss of generality ([61, lemma I.6.1]) that there is no J-holomorphic curve in M whose image is completely contained in a single ε_0-ball in M. For topological reasons we have then

$$f(\partial U(s)) \subseteq \partial B_{\varepsilon_0/18}(f(s)).$$

Apply now the monotonicity lemma to $f|_{\overline{U(s)}} : \overline{U(s)} \to M$:

$$\mathrm{Area}(f|_{\overline{U(s)}}) \geq \mathrm{Area}\left(f(\overline{U(s)}) \cap \overline{B_{\frac{1}{2}\varepsilon_0/18}(f(s))}\right) \geq C_{ML}\left(\frac{1}{2} \cdot \frac{\varepsilon_0}{18}\right)^2.$$

Note that $f(\partial U(s)) \cap \overline{B_{\frac{1}{2}\varepsilon_0/18}(f(s))} = \emptyset$.

Second step: We have to choose a discrete set of points $F \subset S$ with the property that $U(s_1) \cap U(s_2) = \emptyset$ for all $s_1, s_2 \in F$. Because of $\mathrm{Area}(f|_{\overline{U(s)}}) \geq C_{ML}\left(\frac{1}{2}\frac{\varepsilon_0}{18}\right)^2$ the number of points in F is universally bounded:

$$\#F \leq \frac{\mathrm{Area}(f)}{\mathrm{Area}(f|_{\overline{U(s)}})} \leq \frac{\mathrm{Area}(f)}{C_{ML}} \cdot \left(\frac{36}{\varepsilon_0}\right)^2.$$

Therefore let us choose the set F with maximal number of points. The consequence is: To each $s \in S$ there exists a point $s' \in F$ with $U(s) \cap U(s') \neq \emptyset$.

Third step: Note that $\#F$ is certainly greater than 2. The reason is again that the image of f is not contained in a single ε_0-ball. So it is possible to endow $S^* := S \setminus F$ with the Poincaré metric h.

Fourth step: Let $A \subseteq S^*$ be an embedded closed annulus in S^*. Given $s \in A$, there are two possibilities:

1. $U(s) \not\subseteq A \setminus \partial A$. Then it is obvious that $d(f(s), f(\partial A)) < \frac{\varepsilon_0}{18} < \frac{\varepsilon_0}{6}$.

2. $U(s) \subseteq A \setminus \partial A$. Because of the maximality of F it is possible to choose some $s_0 \in F$ such that $U(s_0) \cap U(s) \neq \emptyset$. $s_0 \notin A$ as $A \subset S^* = S \setminus F$. Therefore we find some $s' \in U(s_0) \cap \partial A$ with

$$
\begin{aligned}
d(f(s), f(\partial A)) &\leq d(f(s), f(s')) \\
&\leq d(f(s), f(s_0)) + d(f(s_0), f(s')) \\
&< \frac{\varepsilon_0}{9} + \frac{\varepsilon_0}{18} = \frac{\varepsilon_0}{6}.
\end{aligned}
$$

From 1 and 2 we conclude

$$
d(f(s), f(\partial A)) < \frac{\varepsilon_0}{6} \qquad (\forall s \in A)
$$

and

$$
d(f(\partial_0 A), f(\partial_1 A)) < \frac{\varepsilon_0}{3}.
$$

where $\partial_0 A$ and $\partial_1 A$ are the two boundary components of the annulus A.

Last step: We finish the proof of Lemma 7.13 by means of the following result:

Lemma 7.14 [61, p. 69]   *There is a universal constant $\rho_0 > 0$ such that every ρ_0-ball in (S^*, h) is contained in an annulus A embedded in S^* with the property: $l(f|_{\partial_i A}) \leq \frac{\varepsilon_0}{6}$ for both boundary components $\partial_0 A$ and $\partial_1 A$ of A.*

This lemma guarantees in our situation the existence of a universal constant $\rho_0 > 0$ with the following property: Given a ρ_0-ball $B_{\rho_0}(s^*) \subset (S^*, h)$. there is a embedded closed annulus $A \subset S^*$ with $l(f|_{\partial_i A}) \leq \frac{\varepsilon_0}{6}$ $(i = 0, 1)$. Then for any $s, s' \in A$ we have

$$
\begin{aligned}
d(f(s), f(s')) &\leq d(f(s), f(\partial A)) + l(f|_{\partial_0 A}) + l(f|_{\partial_1 A}) \\
&\quad + d(f(\partial_0 A), f(\partial_1 A)) + d(f(\partial A), f(s')) \\
&\leq \frac{\varepsilon_0}{6} + \frac{\varepsilon_0}{6} + \frac{\varepsilon_0}{6} + \frac{\varepsilon_0}{3} + \frac{\varepsilon_0}{6} = \varepsilon_0.
\end{aligned}
$$

Immediately we conclude that the diameter of $f(A)$ is less than ε_0, and the diameter of $f(B_{\rho_0}(s^*))$ is even smaller. $\square$

Proof of Lemma 7.14. We begin with some preliminaries: First, if $\varphi : S \times (0, T) \to M$ is a J-holomorphic map of the hyperbolic annulus $S^1 \times (0, T)$ with modulus $T < \infty$ into M, then we are able to estimate the area of φ in the usual way

$$
\text{Area}(\varphi) \geq \frac{1}{2\pi} \int_0^T l^2(\varphi|_{S^1 \times \{t\}}) \, dt
$$

because φ is a conformal map. So there is a $t_0 \in (0, T)$ with

$$l^2(\varphi|_{S^1 \times \{t_0\}}) \leq \frac{2\pi \operatorname{Area}(\varphi)}{T}. \tag{7.1}$$

Second, let us choose $\rho_0 > 0$ small enough such that the following conditions are fulfilled:

(a) $3\rho_0 < \sqrt{\rho_0}$

(b) $\log(\tanh \frac{\sqrt{\rho_0}}{2}) - \log(\tanh \frac{3\rho_0}{2}) \geq 36 \cdot \frac{2\pi C}{\varepsilon_0^2}$

(c) $r := 2\sqrt{\rho_0} + \rho_0 < \operatorname{arsinh} 1$ and $\frac{1}{\sinh r} - 1 \geq 36 \cdot \frac{2C}{\varepsilon_0^2}$. As a consequence we have: if $\operatorname{injrad}(S^*, s) < 2\sqrt{\rho_0}$, then $\operatorname{injrad}(S^*, \cdot)|_{B_{\rho_0}(s)} < r < \operatorname{arsinh} 1$.

Note that ρ_0 only depends on (M, J, μ), ε_0 and C. Given any ρ_0-ball $B_{\rho_0}(s)$ in S^*, the main idea of the proof is to construct three annuli A_0, A', A_1 embedded in (S^*, h) with moduli $\operatorname{Mod} A_0, \operatorname{Mod} A', \operatorname{Mod} A_1$ and the following properties:

1. $B_{\rho_0}(s) \subseteq A'$

2. A' is closed. A_0 and A_1 are half-open

3. The three annuli are lying one in another, i.e. there are conformal transformations

$$\begin{aligned}
\varphi_0 &: \quad S^1 \times (0, \operatorname{Mod} A_0] \to A_0 \\
\varphi' &: \quad S^1 \times [\operatorname{Mod} A_0, \operatorname{Mod} A_0 + \operatorname{Mod} A'] \to A' \\
\varphi_1 &: \quad S^1 \times [\operatorname{Mod} A_0 + \operatorname{Mod} A', \operatorname{Mod} A_0 + \operatorname{Mod} A' + \operatorname{Mod} A_1) \to A_1
\end{aligned}$$

which can be joined to an embedding of a single annulus

$$\varphi : S^1 \times (0, \operatorname{Mod} A_0 + \operatorname{Mod} A' + \operatorname{Mod} A_1) \to S^*$$

4. $\operatorname{Mod} A_i \geq 36 \cdot \frac{2\pi C}{\varepsilon_0^2}$ (for $i = 0, 1$).

Once such annuli have been constructed, the proof is easily finished: Indeed, using (7.1) we find $t_0 \in (0, \operatorname{Mod} A_0]$ and $t_1 \in [\operatorname{Mod} A' + \operatorname{Mod} A_0, \operatorname{Mod} A' + \operatorname{Mod} A_0 + \operatorname{Mod} A_1)$ such that

$$l^2(f \circ \varphi_i|_{S^1 \times \{t_i\}}) \leq \frac{2\pi \operatorname{Area}(f)}{\operatorname{Mod} A_i} \leq \frac{\varepsilon_0^2}{36} \qquad (i = 0, 1)$$

(for the second inequality we used 4 and the condition $\operatorname{Area}(f) \leq C$). Set $A := \varphi(S^1 \times [t_0, t_1])$. Then $B_{\rho_0}(s) \subseteq A' \subseteq A$ and A is the annulus of lemma 7.14.

But now the question is: How do we construct the three annuli A_0, A', A_1? In this situation the thick-thin decomposition proves to be very helpful. Let us distinguish whether $s \in S^*$ (the center of the given ρ_0-ball) lies in a thick or in a thin component of (S^*, h).

Case 1: $\mathrm{injrad}(S^*, s) \geq 2\sqrt{\rho_0}$. Then we choose some $s' \in S^*$ with $d(s, s') = 2\rho_0$ and set

$$
\begin{aligned}
A_0 &:= B_{\sqrt{\rho_0}}(s') \setminus B_{3\rho_0}(s') \\
A' &:= \overline{B_{3\rho_0}}(s') \setminus B_{\rho_0}(s') \\
A_1 &:= \overline{B_{\rho_0}}(s') \setminus \{s'\}.
\end{aligned}
$$

The conditions 1,2 and 3 are fulfilled and the moduli of A_0 and A_1 are calculated to be

$$
\begin{aligned}
\mathrm{Mod} A_1 &= \infty \\
\mathrm{Mod} A_0 &= \log(\tanh \frac{\sqrt{\rho_0}}{2}) - \log(\tanh \frac{3\rho_0}{2}) \geq 36 \cdot \frac{2\pi C}{\varepsilon_0^2}
\end{aligned}
$$

(the last inequality holds because of condition (b)). Therefore 4 holds, too.

Case 2: $\mathrm{injrad}(S^*, s) < 2\sqrt{\rho_0}$. Condition (c) and the theorem about the thick-thin decomposition enables us to find a neighbourhood U of the ρ_0-ball in $B_{\rho_0}(s)$ in S^* such that

- either U is isometric to
  $\{w \in \mathbf{H}/\langle T_l \rangle : \mathrm{injrad}(\mathbf{H}/\langle T_l \rangle, w) < \mathrm{arsinh} 1\}$ (for some $l < 2r$),
- or U is isometric to $\{w \in \mathbf{H}/\langle P \rangle : \mathrm{injrad}(\mathbf{H}/\langle P \rangle, w) < \mathrm{arsinh} 1\}$.

So. U can be identified with the corresponding subset of the hyperbolic cylinder $\mathbf{H}/\langle T_l \rangle$ or the parabolic cylinder $\mathbf{H}/\langle P \rangle$.

In the hyperbolic case let us define A_0 and A_1 to be the two connected components of the subset

$$
\{w \in \mathbf{H}/\langle T_l \rangle : r \leq \mathrm{injrad}(\mathbf{H}/\langle T_l \rangle, w) < \mathrm{arsinh} 1\} \subseteq U
$$

and A' the subset lying between them

$$
A' = \{w \in \mathbf{H}/\langle T_l \rangle : \mathrm{injrad}(\mathbf{H}/\langle T_l \rangle, w) \leq r\} \subseteq U.
$$

Because of condition (c) the ρ_0-ball $B_{\rho_0}(s)$ lies in A' and for the moduli of A_0 and A_1 elementary estimations yield

$$
\mathrm{Mod} A_0 = \mathrm{Mod} A_1 \geq \pi(\frac{1}{\sinh r} - \frac{1}{\sinh \mathrm{arsinh} 1}).
$$

Therefore, using property (c),

$$
\mathrm{Mod} A_0 = \mathrm{Mod} A_1 \geq 36 \cdot \frac{2\pi C}{\varepsilon_0^2}.
$$

Therefore A_0, A', A_1 can be determined in the hyperbolic case.

In the parabolic situation appropriate annuli are defined by

$$A_0 \ := \ \{w \in \mathbf{H}/\langle P\rangle : r \le \mathrm{injrad}(\mathbf{H}/\langle P\rangle, w) < \mathrm{arsinh}1\} \subseteq U$$
$$A' \ := \ \{w \in \mathbf{H}/\langle P\rangle : r' \le \mathrm{injrad}(\mathbf{H}/\langle P\rangle, w) < r\} \subseteq U$$
$$A_1 \ := \ \{w \in \mathbf{H}/\langle P\rangle : \mathrm{injrad}(\mathbf{H}/\langle P\rangle, w) < r'\} \subseteq U$$

where $r' \in (0.r)$ is chosen in such a way that $B_{\rho_0}(s) \subseteq A'$. In fact the moduli of A_0 and A_1 are, by elementary calculations, $\mathrm{Mod}A_1 = \infty$ and

$$\mathrm{Mod}A_0 = \pi\left(\frac{1}{\sinh r} - \frac{1}{\sinh \mathrm{arsinh}1}\right) \ge 36\frac{2\pi C}{\varepsilon_0^2}.$$

This finishes the proof of lemma 7.14. $\square$

Corollary 7.15 *For a J-holomorphic curve $f : S \to M$ with $\mathrm{Area}(f) \le C$ the differential df on S^* is universally bounded with respect to the Poincaré metric.*

Proof. Given $s \in S^*$. Let $p : (\mathbf{D}, \lambda) \to S^*$ be the universal covering of S^* with $p(0) = s$. Here $\mathbf{D}$ is the unit disc with Poincaré metric λ. By lemma 7.13 the image of $B_{\rho_0}(0) \subseteq (\mathbf{D}, \lambda)$ under $f \circ p$ is contained in an appropriate ε_0-ball in (M, J, μ). Using the Gromov-Schwarz lemma the norm of $d(f \circ p)$ at the point 0 can be estimated by a universal constant. Therefore $\|df_s\|$ is universally bounded as well. $\square$

We start now with the proof of the compactness theorem:

Choice of a Subsequence

Assume that S is a closed Riemann surface of genus g and $(j_n)_{n\ge 1}$ a sequence of complex structures on S. Let S_n denote the Riemann surface (S, j_n) and let $f_n : S_n \to M$ $(n \ge 1)$ be a sequence of (j_n, J)-holomorphic curves in the compact Hermitian manifold (M, J, μ) with uniformly bounded area:

$$\mathrm{Area}(f_n) \le C \qquad (\forall n \ge 1).$$

A remark concerning the notation: we shall always omit the subindices of subsequences; so, for example, a subsequence of $(S_n)_{n\ge 1}$ will again be denoted by $(S_n)_{n\ge 1}$.

By means of lemma 7.13 we first choose a finite subset $F_n \subseteq S_n$ for each $n \ge 1$ such that $f_n|_{S_n^*}$ is universally Lipschitz bounded on $S_n^* := S_n \setminus F_n$ with respect to the Poincaré metric h_n. The number k_n of points in F_n is universally bounded. Taking a subsequence of $(f_n)_{n\ge 1}$ we can assume that $k_n =: k$ does not depend on n, and consequently the hyperbolic surfaces are all of the same signature $(g, 0, k)$.

For each $n \geq 1$ let us denote the geodesics on $(S_n^*. h_n)$ with length less than $2\operatorname{arsinh}1$ by γ_n^i $(i = 1, 2, \ldots, l_n)$. These geodesics are pairwise disjoint (see for example [16], collar theorem) and their number is bounded by $3g - 3 + k$ for every $n \geq 1$. Again we can take a subsequence such that $l_n =: l'$ does not depend on n.

In a further step a marking of S should be taken into account. In other words: we wish to have a diffeomorphism $\chi_n : S_1^* \to S_n^*$ such that $\chi_n(\gamma_1^i) = \gamma_n^i$ for $i = 1, \ldots. l'$. This can indeed be realized if we take once again a subsequence of $(f_n)_{n \geq 1}$.

All those i for which the lengths $l(\gamma_n^i)$ do not converge to 0 for $n \to \infty$ are not interesting for us and the corresponding geodesics γ_n^i are no longer considered. So for every n there are l_0 geodesics left: $\gamma_n^1, \ldots, \gamma_n^{l_0}$. Moreover $\chi(\gamma_1^i) = \gamma_n^i$ for all $i \in \{1, \ldots, l_0\} =: I$ (if $l_0 = 0$. then $I = \emptyset$). Let $S_n^1, \ldots, S_n^{m_0}$ denote the components of $S_n^* \setminus \bigcup \{\gamma_n^i : i \in I\}$. $\chi_n(S_1^\nu) = S_n^\nu$ for $\nu = 1, \ldots, m_0$.

Shrinking the curves γ_1^i for $i \in I$ we get, as in section 7.2, a singular surface $\bar{S}$. At this point the convergence of hyperbolic structures enters:

Proposition 7.16 ([61, IV.3.4]) *There exist a subsequence of $(S_n)_{n \geq 1}$. a complex structure $\bar{j}$ on $\bar{S}$ and a finite subset $F \subseteq \bar{S} \setminus si(\bar{S})$ such that for every $n \geq 1$ a deformation $\varphi_n : S_n \to \bar{S}$ exists with the following properties:*

1. *φ_n maps F_n bijectively onto F.*

2. *$(\varphi_n^{-1})^* h_n$ converges to the Poincaré metric h on $\bar{S} \setminus (F \cup si(\bar{S}))$ in the C^∞-topology for $n \to \infty$.*

3. *$(\varphi_n^{-1})^* j_n$ converges in the C^∞-topology to $\bar{j}|_{\bar{S} \setminus si(\bar{S})}$ for $n \to \infty$.*

How do we find the subsequence, the deformations and the complex structure $\bar{j}$? We successively apply the Proposition of section 7.2 (describing convergence of hyperbolic structures) to $(S_n^1)_{n \geq 1}, (S_n^2)_{n \geq 1}, \ldots, (S_n^{m_0})_{n \geq 1}$. In this way we get the subsequence $f_n : S_n \to M$, the deformations φ_n and the singular Riemann surface $(\bar{S}, \bar{j})$.

The corollary formulated above as well as the convergence of $(\varphi_n^{-1})^* h_n$ to h on $\bar{S} \setminus (F \cup si(\bar{S}))$ (with respect to the C^∞-topology) imply the equicontinuity and Lipschitz boundedness of $f_n \circ \varphi_n^{-1}|_{\bar{S} \setminus (F \cup si(\bar{S}))}$ on compact subsets. The theorem of Arzela-Ascoli therefore guarantees the existence of a subsequence of $f_n \circ \varphi_n^{-1}$ which, restricted to $\bar{S} \setminus (F \cup si(\bar{S}))$, converges to a continuous map

$$\bar{f} : \bar{S} \setminus (F \cup si(\bar{S})) \to (M. J, \mu)$$

in the C^0-topology. By the generalized Weierstrass theorem and because of the convergence of $(\varphi_n^{-1})^* j_n$ in the C^∞-topology. we can conclude that the $((\varphi_n^{-1})^* j_n. J)$-holomorphic maps $f_n \circ \varphi_n^{-1}$ converge to $\bar{f}$ on $\bar{S} \setminus (F \cup si(\bar{S}))$ even in the C^∞-topology, and in addition $\bar{f}$ is $(\bar{j}, J)$-holomorphic.

The Convergence of Area

First we note that

$$\int_{\bar{S}\backslash(F\cup si(\bar{S}))}\sigma_h = \int_{\bar{S}\backslash(F\cup si(\bar{S}))}\sigma_{(\varphi_n^{-1})^*h_n} = 2\pi(2g-2+k) < \infty$$

for all n. Also, because of the uniform convergence $(\varphi_n^{-1})^*h_n \to h$ on any compact subset $K \subset \bar{S}\setminus(F\cup si(\bar{S}))$, we observe that

$$\lim_{n\to\infty}\int_K \sigma_{(\varphi_n^{-1})^*h_n} = \int_K \sigma_h.$$

This enables us to find for any $\varepsilon > 0$ an open neighbourhood U of $\bar{S}\setminus(F\cup si(\bar{S}))$ in $\bar{S}$ with the property

$$\int_{U\backslash(F\cup si(\bar{S}))}\sigma_h \le \frac{\varepsilon}{2}.$$

Together with the fact that $|\int_{U\backslash(F\cup si(\bar{S}))}\sigma_{(\varphi_n^{-1})^*h_n} - \int_{U\backslash(F\cup si(\bar{S}))}\sigma_h| \le \frac{\varepsilon}{2}$ for n large enough, this immediately implies

$$\int_{U\backslash(F\cup si(\bar{S}))}\sigma_{(\varphi_n^{-1})^*h_n} \le \varepsilon \qquad (n \text{ large enough}).$$

The estimate

$$\mathrm{Area}(f_n\circ\varphi_n^{-1}|_{U\backslash si(\bar{S})}) = \int_{\varphi_n^{-1}(U\backslash si(\bar{S}))}\|df_n\|^2\sigma_{h_n} \le \mathrm{const}\cdot\varepsilon \qquad (7.2)$$

is then easily derived from the fact that $\|df_n\|$ is universally bounded on S_n^*. On the other hand the uniform C^1-convergence of $(f_n\circ\varphi_n^{-1})$ on $\bar{S}\setminus U$ implies

$$\lim_{n\to\infty}\mathrm{Area}(f_n\circ\varphi_n^{-1}|_{\bar{S}\setminus U}) = \mathrm{Area}(\bar{f}|_{\bar{S}\setminus U}).$$

We conclude that

$$\limsup_{n\to\infty}\mathrm{Area}(f_n\circ\varphi_n^{-1}) \le \mathrm{Area}(\bar{f}).$$

The C^1-convergence itself yields

$$\liminf_{n\to\infty}\mathrm{Area}(f_n\circ\varphi_n^{-1}) \ge \mathrm{Area}(\bar{f}).$$

Thus we have proved

$$\lim_{n\to\infty}\mathrm{Area}(f_n\circ\varphi_n^{-1}) = \mathrm{Area}(\bar{f}).$$

The Continuation of $\bar{f}$ to $\bar{S}$

So far the $(\bar{j}, J)$-holomorphic map $\bar{f}$ is only defined on $\bar{S} \setminus (F \cup si(\bar{S}))$. So $\hat{f} := f \circ \alpha$ is defined on $\alpha^{-1}(\bar{S} \setminus (F \cup si(\bar{S})))$. The fact that $\hat{f}$ has finite area enables us to apply the theorem on removing singularities. $\hat{f}$ then extends to a J-holomorphic map on $\hat{S}$ ($\hat{S}$ and $\alpha : \hat{S} \to \bar{S}$ have been introduced in section 7.2).

In order to finish the proof of the compactness theorem two properties should be verified:

(a) $\bar{f}$ is compatible with α. This provides the $(\bar{j}, J)$-holomorphic cusp-curve $\bar{f} : \bar{S} \to M$, now defined on the whole surface $\bar{S}$.

(b) $(f_n \circ \varphi_n^{-1})$ converges to $\bar{f}|_{\bar{S} \setminus si(\bar{S})}$ in the C^{∞}-topology.

Property (a) is verified in the following way: assume that there exists a point $\bar{s} \in si(\bar{S})$ such that for the preimages $s'. s'' \in \hat{S}$ of $\bar{s}$ under α we have $\hat{f}(s') \neq \hat{f}(s'')$. Let $\rho := \min\{\varepsilon_0, \frac{1}{4}d(\hat{f}(s'), \hat{f}(s''))\}$. Then in $\hat{S}$ we can choose compact connected neighbourhoods U' and U'' of s' and s'', respectively, such that $\hat{f}(U') \subseteq B_\rho(\hat{f}(s'))$ and $\hat{f}(U'') \subseteq B_\rho(\hat{f}(s''))$. Taking U' and U'' small enough we get, because of (7.2)

$$\mathrm{Area}(f_n|_{\varphi_n^{-1}\alpha(U' \cup U'')}) < C_{ML} \cdot \rho^2 \qquad (n \text{ large enough}).$$

We now apply the monotonicity lemma to

$$f_n|_{\varphi_n^{-1}\alpha(U' \cup U'')} : \varphi_n^{-1}\alpha(U' \cup U'') \to M.$$

Observe that the uniform convergence of $(f_n \circ \varphi_n^{-1})$ implies

$$f_n(\partial(\varphi_n^{-1}\alpha(U' \cup U''))) \subseteq B_\rho(\hat{f}(s')) \cup B_\rho(\hat{f}(s'')).$$

Moreover there is a point $s_0 \in \varphi_n^{-1}\alpha(U' \cup U'')$ such that

$$\overline{B_\rho}(\hat{f}_n(s_0)) \cap [B_\rho(\hat{f}_n(s')) \cup B_\rho(\hat{f}_n(s''))] = \emptyset.$$

So the monotonicity lemma gives

$$C_{ML} \cdot \rho^2 \leq \mathrm{Area}\big(f_n(\varphi_n^{-1}\alpha(U' \cup U'')) \cap \overline{B_\rho}(f_n(s_0))\big) \leq \mathrm{Area}(f_n|_{\varphi_n^{-1}\alpha(U' \cup U'')}).$$

But this contradicts the estimate above: $\mathrm{Area}(f_n|_{\varphi_n^{-1}\alpha(U' \cup U'')}) < C_{ML} \cdot \rho^2$.

Property (b): $(f_n \circ \varphi_n^{-1})$ converges on $\bar{S} \setminus si(\bar{S})$ to $\bar{f}$ in the C_0-topology. Otherwise there would be a point $s_0 \in F$, a number $\varepsilon > 0$, a sequence $(s_n)_{n \geq 1}$ of points in $\bar{S} \setminus si(\bar{S})$ and a subsequence of $(f_n \circ \varphi_n^{-1})_{n \geq 1}$ such that

$$d(f_n \circ \varphi_n^{-1}(s_n), \bar{f}(s_n)) > \varepsilon \qquad (\forall n \geq 1).$$

This and the fact that $d(\bar{f}(s_n), \bar{f}(s_0)) < \frac{\varepsilon}{2}$ for large n induces

$$f_n \circ \tilde{r}_n^{-1}(s_n) \notin B_{\varepsilon/2}(\bar{f}(s_0)) \qquad (n \text{ large enough}).$$

(Use the triangle inequality).

In a further step. using the continuity of $\bar{f}$, we are able to choose a compact connected neighbourhood $U \subseteq \bar{S} \backslash si(\bar{S})$ of s_0 such that

$$\text{Area}(\bar{f}|_U) \;\; < \;\; C_{ML} \cdot \left(\frac{\varepsilon}{4}\right)^2 \quad \text{and}$$
$$\bar{f}(U) \;\; \subseteq \;\; B_{\varepsilon/4}(\bar{f}(s_0)).$$

Moreover, without loss of generality $\partial U \cap F = \emptyset$ and because of the C^0-convergence of $(f_n \circ \varphi_n^{-1})$ outside of F, we deduce

$$f_n \circ \varphi_n^{-1}(\partial U) \subseteq B_{\varepsilon/4}(\bar{f}(s_0)).$$

Now. $f_n \circ \varphi_n^{-1}(s_n) \notin B_{\varepsilon/2}(\bar{f}(s_0))$ and $f_n \circ \varphi_n^{-1}(\partial U) \subseteq B_{\varepsilon/4}(\bar{f}(s_0))$ immediately imply

$$f_n \circ \varphi_n^{-1}(\partial U) \text{ lies outside of } B_{\varepsilon/4}(f_n \circ \varphi_n^{-1}(s_n)).$$

This allows the application of the monotonicity lemma to $f_n|_{\varphi_n^{-1}(U)} : \varphi_n^{-1}(U) \to M$ as follows:

$$\text{Area}\left(f_n \circ \varphi_n^{-1}(U) \cap B_{\varepsilon/4}(f_n \circ \varphi_n^{-1}(s_n))\right) \geq C_{ML}\left(\frac{\varepsilon}{4}\right)^2.$$

Therefore $\text{Area}(f_n \circ \varphi_n^{-1}(U)) \geq C_{ML}(\frac{\varepsilon}{4})^2$ yields a contradiction to the choice of U. So the C^0-convergence is proved. C^∞-convergence follows from the generalized Weierstrass theorem.

This finally is the end of the proof of Gromov's compactness theorem. $\square$

We close this chapter with some remarks on bubbles. Under certain circumstances the limits of convergent sequences of pseudoholomorphic curves are special cusp-curves. Roughly speaking, the tied up surfaces on which these cusp-curves are defined, look like the original S to which a finite number of spheres (so called bubbles) have been attached. In fact, this occurs if all the pseudoholomorphic curves f_n of the convergent sequence are parametrized by the same Riemann surface (S, j).

Theorem 7.17 *Let $f_n : (S, j) \to (M, J, \mu)$ (where $n \geq 1$) be a sequence of (j, J)-holomorphic curves with $\text{Area}(f_n) < C$ which converges to the cusp-curve $\bar{f} : (\bar{S}, \bar{j}) \to (M, J, \mu)$ in M. Then $(\hat{S}, \bar{j})$ is conformally equivalent to the disjoint union of S with at most finitely many 2-spheres.*

It is not too difficult to show that $(\hat{S}, \bar{j})$ is diffeomorphic to S and a certain number of disjoint spheres. One has to prove the existence of a constant

$c > 0$ (depending only on (S, j)) with the following properties: If F is a finite set of points in S and $S \setminus F$ is endowed with the Poincaré metric, then each simple closed geodesic on $S \setminus F$ of length less than c is homotopic to zero in S. These null homotopic geodesics are responsible for the tied up 2-spheres. The conformal equivalence stated in the theorem above is somewhat more involved (consult [61, Thm 4.3]).

We remark that intuitively the bubbles are pseudoholomorphic curves $S^2 \to M$ catching the area "lost" at a finite number of points $\bar{s} \in (\bar{S}, \bar{j})$ in the limit process.

8 Contact structures

Contact manifolds are odd-dimensional manifolds $M = M^{2n+1}$ which carry a completely nonintegrable field ξ of hyperplanes. Locally such a field is determined by a one-form α, $\xi = \{\alpha = 0\}$ and the nonintegrability condition can be stated as $\alpha \wedge (d\alpha)^n \neq 0$. Contact manifolds are intimately related to symplectic manifolds. They can naturally be embedded as surfaces of contact type in their symplectification, which is a symplectic manifold of dimension $2n + 2$ naturally associated to M (see section 8.2).

We consider general properties of contact manifolds in the first section. Among the examples, the surfaces of contact type in a symplectic manifold are particularly important, because they arise as integral surfaces $M = \{H = 0\}$ of Hamiltonian systems. They have proved useful in connection with the problem of finding periodic solutions of Hamiltonian systems. In 1979 Weinstein had conjectured that such periodic solutions exist on surfaces on contact type. In 1987 Viterbo gave an affirmative solution to the conjecture and also showed that the vanishing of the first homology class (as it was originally required) was an irrelevant condition in this context.

In complex analysis contact manifolds M also arise as boundaries $M = \partial\Omega$ of strictly pseudoconvex domains Ω in complex manifolds. The field of hyperplanes is the field ξ of maximal complex invariant subspaces of the tangent space. The pseudoconvexity condition guarantees the complete nonintegrability of ξ.

In sections 8.4 to 8.6 we study contact structures on 3-manifolds. The chapter ends with a section on Eliashberg's classification results. The contact structures on 3-manifolds are quite peculiar. They can be divided up into tight and overtwisted structures (this terminology is due to Eliashberg). The standard structure on the sphere S^3 is tight (Bennequin's theorem), but it is the only (positively oriented) tight contact structure on S^3. Overtwisted structures are characterised by the property that the characteristic foliations of embedded 2-dimensional surfaces may contain closed leaves. The overtwisted structures are quite flexible, the tight structures on the contrary are rather rigid.

Index theorems for embedded surfaces and invariants for transversal and Legendrian curves determine the analysis of contact 3-manifolds. Auxiliary complex structures play a critical role in the proof that certain contact structures are tight. Some of our efforts are indeed spent on proving Gromov's theorem, which says that fillable contact structures are tight (see section 8.6.3).

8.1 Contact manifolds

Contact manifolds are differentiable manifolds S of odd dimension $2n + 1$ with a completely nonintegrable distribution ξ of hyperplanes in the tangent space

TS. Locally, the hyperplanes can be described as the kernel of a one-form α

$$\xi = \{X \in TS : \alpha(X) = 0\}.$$

The nonintegrability condition takes the form

$$d\alpha|_\xi \text{ is nondegenerate}$$

(i.e. $d\alpha$ induces a symplectic structure in each contact plane) and this can be reformulated as

$$(d\alpha)^n|_\xi \neq 0.$$

The n-th exterior product of $d\alpha$ is thus a nonvanishing volume element on ξ.

The conditions above are independent of the choice of the one-form α. If $\tilde\alpha = f\alpha$ is a different one-form describing locally the same distribution ξ, then $f \neq 0$ and

$$\begin{aligned}
d\tilde\alpha &= f\,d\alpha + df \wedge \alpha, \\
d\tilde\alpha|_\xi &= f\,d\alpha|_\xi.
\end{aligned}$$

The forms $d\tilde\alpha$ and $d\alpha$ restricted to ξ are therefore simultaneously nondegenerate.

If n is odd, then the sign of $\alpha \wedge (d\alpha)^n$ is independent of the sign of α. The contact manifold is oriented. If n is even, then the sign of $(d\alpha)^n|_\xi$ is independent of the sign of α. This means that the hyperplanes from the contact distribution ξ are oriented.

A *contact transformation* $f : M \to N$ between contact manifolds (M, ξ) and (N, η) is a diffeomorphism such that the tangent mapping f_* maps ξ onto η. If the contact structures are described locally by one-forms

$$\xi = \{\alpha = 0\}, \qquad \eta = \{\beta = 0\},$$

then $f^*\beta = \lambda\alpha$ with λ a nonvanishing scalar function.

There are no local invariants for contact structures.

Theorem 8.1 (Darboux) *Locally any contact structure is equivalent to the standard structure $\{\alpha_0 = 0\}$ in $\mathbf{R}^{2n+1}$,*

$$\alpha_0 = \sum_{j=1}^{n}(-y_j\,dx_j + x_j\,dy_j) + dz$$

(the coordinates of points in $\mathbf{R}^{2n+1}$ are denoted by $x_1, \ldots, x_n, y_1, \ldots, y_n, z$).

Proof. If a given contact structure is locally described by the one-form β defined in a neighbourhood of the origin in $\mathbf{R}^{2n+1}$, then there is a linear mapping $l : \mathbf{R}^{2n+1} \to \mathbf{R}^{2n+1}$ such that at the origin

$$\begin{aligned}
l^*\beta &= \alpha_0, \\
l^*d\beta &= d\alpha_0.
\end{aligned}$$

First map the contact planes at the origin $\{\alpha_0 = 0\}$ and $\{\beta = 0\}$ onto each other by a linear mapping which is symplectic with respect to the forms $d\alpha_0$ and $d\beta$. In the second step extend this to a linear map of the whole tangent space at the origin, by mapping the vector X which is determined by the equations

$$\alpha_0(X) = 1 \quad \text{and} \quad X\lrcorner d\alpha_0 = 0$$

onto the vector Y determined by the corresponding equations for β:

$$\beta(Y) = 1 \quad \text{and} \quad Y\lrcorner d\beta = 0.$$

The pullback of the form β by the linear mapping l is denoted by α_1

$$\alpha_1 = l^*\beta$$

and we define the homotopy

$$\alpha_t = (1 - t)\alpha_0 + t\alpha_1.$$

At the origin $\alpha_0 = \alpha_t = \alpha_1$ and $d\alpha_0 = d\alpha_t = d\alpha_1$. There exists therefore a neighbourhood U of the origin such that for all $t \in [0, 1]$ the form $d\alpha_t$ restricted to $\{\alpha_t = 0\}$ is nondegenerate. In the proof of Gray's theorem we will construct an isotopy of mappings φ_t, defined in a neighbourhood of the origin, such that

$$\varphi_t^*\alpha_t = \alpha_0 \qquad t \in [0, 1].$$

It then follows that $\alpha_0 = \varphi_1^*\alpha_1 = \varphi_1^* l^*\beta$. This will complete the proof of Darboux' theorem. $\qquad\square$

Theorem 8.2 (Gray) *Given a smooth family ξ_t, $t \in [0, 1]$ of contact structures on a compact manifold M without boundary, there exists a 1-parameter family φ_t of diffeomorphisms $\varphi_t : M \to M$ such that for all t*

$$\varphi_{t*}\xi_0 = \{\varphi_{t*}X : X \in \xi_0\} = \xi_t.$$

Proof. The proof once again is based on Moser's method. We start off with the fundamental formula (see e.g. [51])

$$\frac{d}{dt}(\varphi_t^*\vartheta_t) = \varphi_t^*\{\frac{d\vartheta_t}{dt} + v\lrcorner d\vartheta_t + d(v\lrcorner\vartheta_t)\},$$

where φ_t is the flow of the time dependent vector field v and the differential form ϑ is allowed to depend on t.

Choose differential forms ϑ_t which locally represent ξ_t and depend smoothly on t. The vector field v will have to be determined in such a way that

$$\varphi_t^*\vartheta_t = \lambda_t\vartheta_0$$

with some nonvanishing scalar function λ_t. This will then mean

$$\frac{d}{dt}(\varphi_t^*\vartheta_t) = \frac{d\lambda_t}{dt}\vartheta_0 = \left(\frac{d}{dt}\log\lambda_t\right)\varphi_t^*\vartheta_t$$
$$= \varphi_t^*\left((\frac{d}{dt}\log\lambda_t)\circ\varphi_t^{-1}\vartheta_t\right).$$

Combined with the fundamental formula this leads to the equation

$$\frac{d\vartheta_t}{dt} + v\lrcorner d\vartheta_t + d(v\lrcorner\vartheta_t) = \mu\vartheta_t \tag{8.1}$$

which has to be solved for v (v will depend on t). The function μ is taking the place of $(\frac{d}{dt}\log\lambda_t)\circ\varphi_t^{-1}$.

The characteristic vector field T_t (or Reeb vector field for the form ϑ_t) is determined by the equations

$$\vartheta_t(T_t) = 1 \quad \text{and} \quad T_t\lrcorner d\vartheta_t = 0.$$

If $T = T_t$ is inserted in the preceding equation, then μ can be expressed in dependence of v and $\vartheta = \vartheta_t$:

$$\frac{d\vartheta}{dt}(T) + d(v\lrcorner\vartheta)(T) = \mu\vartheta(T) = \mu.$$

There exists now a unique vector field $v \in \xi_t$ which solves (8.1). For $v \in \xi_t$ the equation (8.1) reduces to

$$v\lrcorner d\vartheta = \mu\vartheta - \frac{d\vartheta}{dt}$$

The function μ is determined such that the form $\gamma = \mu\vartheta - \frac{d\vartheta}{dt}$ satisfies $\gamma(T) = 0$. Since $d\vartheta|\xi_t$ is non-degenerate, there exists then a unique $v \in \xi_t$ with $v\lrcorner d\vartheta = \gamma$.

Next, observe that the vector field $v \in \xi_t$ is canonically defined. It is independent of the choice of the differential forms ϑ_t which represent ξ_t. If $\tilde\vartheta_t = a_t\vartheta_t$, $a_t \neq 0$ is a different form representing ξ_t and if $w \in \xi_t$ is the solution of

$$\frac{d}{dt}(a_t\vartheta_t) + w\lrcorner d(a_t\vartheta_t) = \tilde\mu\vartheta_t$$

(with some different $\tilde\mu$), then

$$a_t\frac{d\vartheta_t}{dt} + w\lrcorner(da_t\wedge\vartheta_t) + w\lrcorner(a_td\vartheta_t) = \tilde\mu\vartheta_t - \frac{da_t}{dt}\vartheta_t.$$

Using

$$w\lrcorner(da_t\wedge\vartheta_t) = (w\lrcorner da_t)\vartheta_t - (w\lrcorner\vartheta_t)da_t = w(a_t)\vartheta_t$$

we get

$$a_t(\frac{d\vartheta_t}{dt} + w\lrcorner d\vartheta_t) = (\tilde\mu - \frac{da_t}{dt} - w(a_t))\vartheta_t.$$

Applying the last equation to T_t yields

$$a_t \mu = \tilde{\mu} - \frac{da_t}{dt} - w(a_t)$$

and v and w are solutions of the same equation, hence $v = w$.

The vector field v can thus be defined globally and it follows that

$$\frac{d}{dt}(\varphi_t^* \vartheta_t) = \varphi_t^*(\mu \vartheta_t) = (\mu \circ \varphi_t)\varphi_t^* \vartheta_t$$

$$\frac{d}{dt}[\log(\varphi_t^* \vartheta_t)(x)] = \mu \circ \varphi_t$$

$$\varphi_t^* \vartheta_t(x) = \vartheta_0(x) \exp \int_0^t (\mu \circ \varphi_t) dt$$

$\square$

The proof shows that the deformation φ_t is uniquely determined by the requirement that the generating vector field v be "horizontal" i.e. $v \in \xi_t$.

To complete the proof of Darboux' theorem we show that $v = 0$ at the origin (for all t). If v and φ_t are determined from the differential forms α_t: $\varphi_t^* \alpha_t = \alpha_0$. then $\lambda_t \equiv 0$, hence $\mu \equiv 0$, and one has at the origin

$$\frac{d}{dt}\alpha_t = 0.$$

The equation for v is $v \lrcorner d\alpha_t = -\frac{d}{dt}\alpha_t$. Together with the nondegeneracy of $d\alpha_t$ this shows that $v = 0$ at the origin.

A curve γ in a contact manifold (M, ξ) is *transversal*, if at all points p on the curve the contact plane ξ is transversal to the tangent vector $\frac{d\gamma}{dt}$ to the curve.

With the above methods we can classify (following Lutz [68] and Martinet [69]) the germs of contact structures transversal to a circle. For this, consider the manifold $S^1 \times \mathbf{R}^{2n}$ with the standard contact form

$$\alpha_0 = d\vartheta + \sum_{j=1}^n (-y_j dx_j + x_j dy_j)$$

and the circle $\gamma = S^1 \times \{0\}$ parametrized by $\vartheta \in [0, 2\pi)$.

Theorem 8.3 (Lutz, Martinet) *Assume that ξ is a contact structure defined in a neighbourhood of γ and transversal to γ. Then there exists a diffeomorphism φ defined in a neighbourhood of γ such that $\varphi|_\gamma = id$ and $\varphi_* \xi = \xi_0$ (with ξ_0 the standard contact structure).*

Proof. In a neighbourhood of γ, the contact structure ξ can be defined with a single differential form α and $\alpha(\frac{\partial}{\partial\vartheta}) \neq 0$ on γ by transversality. Without loss of generality it can be assumed that $\alpha(\frac{\partial}{\partial\vartheta}) = 1$. There exists then a function f defined near γ such that $f|_\gamma \equiv 1$ and

$$\frac{\partial}{\partial\vartheta}\lrcorner\, d(f\alpha) = 0 \quad \text{on } \gamma.$$

In fact, since $\frac{\partial}{\partial\vartheta}f = 0$ and $\alpha(\frac{\partial}{\partial\vartheta}) = 1$ along γ, the equation

$$\begin{aligned}
0 &= \frac{\partial}{\partial\vartheta}\lrcorner\, d(f\alpha) = \frac{\partial}{\partial\vartheta}\lrcorner\, f\, d\alpha + \frac{\partial}{\partial\vartheta}\lrcorner\,(df \wedge \alpha) \\
&= f(\frac{\partial}{\partial\vartheta}\lrcorner\, d\alpha) + (\frac{\partial f}{\partial\vartheta})\alpha - df\alpha(\frac{\partial}{\partial\vartheta})
\end{aligned}$$

reduces to

$$df = \frac{\partial}{\partial\vartheta}\lrcorner\, d\alpha \quad \text{on } \gamma.$$

Since $(\frac{\partial}{\partial\vartheta}\lrcorner\, d\alpha)(\frac{\partial}{\partial\vartheta}) = 0$, such a function clearly exists.

We conclude that near γ the contact structure ξ can be represented by a differential form α with the property that at all points of γ

$$\frac{\partial}{\partial\vartheta}\lrcorner\, d\alpha = 0 \quad \text{and} \quad \alpha(\frac{\partial}{\partial\vartheta}) = 1.$$

The $\mathbf{R}^{2n}$ fiber bundles $\xi_0 \to S^1$ and $\xi \to S^1$ are symplectic vector bundles with symplectic forms given by $d\alpha_0|_{\xi_0}$ and $d\alpha|_\xi$. At each point $(\vartheta.0)$ of γ there is a linear map

$$h_\vartheta : (\xi_0)_{(\vartheta,0)} \to \xi_{(\vartheta,0)}$$

such that $h_\vartheta^* d\alpha = d\alpha_0$. This map is unique up to composition with elements of $Sp(n,\mathbf{R})$. But because the group $Sp(n,\mathbf{R})$ is connected, the mappings h_ϑ can be adjusted along γ to depend smoothly on ϑ. There results a bundle map $h : \xi_0 \to \xi$ between the bundles $\xi_0 \to S^1$ and $\xi \to S^1$ such that

$$h^* d\alpha = d\alpha_0.$$

Consider an arbitrary Riemannian metric on $S^1 \times \mathbf{R}^{2n}$. The associated exponential map establishes a diffeomorphism

$$\exp : \xi \to S^1 \times \mathbf{R}^{2n}$$

between a neighbourhood of the zero section of the vector bundle $\xi \to S^1$ and a neighbourhood of the circle γ in $S^1 \times \mathbf{R}^{2n}$ (and a similar diffeomorphism $\exp_0$ for ξ_0).

The composed mapping $\psi = \exp \circ h \circ (\exp_0)^{-1}$ is a diffeomorphism in a neighbourhood of γ which fixes γ pointwise. The tangent map at a point $(\vartheta, 0) \in \gamma$ is given by

$$\psi_* |_{\xi_0} = h,$$
$$\psi_* \frac{\partial}{\partial \vartheta} = \frac{\partial}{\partial \vartheta}.$$

From the construction it follows that on γ

$$\psi^* \alpha = \alpha_0,$$
$$\psi^* d\alpha = d\alpha_0.$$

The proof of the theorem can now be completed with the same argument as before. Take the homotopy

$$\alpha_t = (1 - t)\alpha_0 + t\alpha_1$$

between α_0 and $\alpha_1 = \psi^* \alpha$ and use the fact that α_t and $d\alpha_t$ are constant along γ to construct an isotopy φ_t in a neighbourhood of γ such that

$$\varphi_t^* \alpha_t = \alpha_0 \quad \text{and} \quad \varphi_t|_\gamma = id \qquad t \in [0, 1].$$

The mapping $\psi \circ \varphi_1$ is then the diffeomorphism of the theorem. $\qquad\square$

8.2   Symplectification

Example The orientable *projectivization* PT^*V of the cotangent bundle of an n-dimensional manifold V becomes a contact manifold in a natural way. The elements in PT^*V are the equivalence classes of 1-forms ϑ at a point $p \in V$

$$(p, [\vartheta_p]) \in PT^*V$$

with $\vartheta_p \sim \vartheta_p'$ if $\vartheta_p' = \lambda\vartheta_p$ for some $\lambda > 0$. As such, the points in PT^*V are the "contact elements" of V. These are the oriented hyperplanes in T_pV represented as the kernel of the 1-forms ϑ_p in the equivalence class $[\vartheta_p]$. There exists a canonical plane field ξ in the tangent space of the manifold $M = PT^*V$. A vector X is in ξ if it describes a deformation of the field of contact planes whereby the point of contact moves in a direction lying in the contact element.

The situation is the same as in skating. On the icerink (manifold V) the skates are the contact elements. They are always moved in the direction of the skates.

Local coordinates on $M = PT^*V$ are e.g.

$$(q_1, \ldots, q_n, p_2', \ldots, p_n') = (q_1, \ldots, q_n, \frac{p_2}{p_1}, \ldots, \frac{p_n}{p_1})$$

with $dq_1 + \sum_{j=2}^{n} p'_j dq_j$ being a representative for the equivalence class $[\sum_{j=1}^{n} p_j dq_j]$. A tangent vector $X \in TM$ has the form

$$X = \sum_{j=1}^{n} a_j \frac{\partial}{\partial q_j} + \sum_{j=2}^{n} b_j \frac{\partial}{\partial p'_j}$$

and $X \in \xi$ if $\pi_* X = \sum a_j \frac{\partial}{\partial q_j} \in TV$ lies in the contact element determined by (q, p'). The condition for $X \in \xi$ therefore is

$$(dq_1 + \sum_{j=2}^{n} p'_j dq_j)(\sum_{j=1}^{n} a_j \frac{\partial}{\partial q_j}) = a_1 + \sum_{j=2}^{n} p'_j a_j = 0.$$

The differential form

$$\alpha = dq_1 + \sum_{j=2}^{n} p'_j dq_j$$

represents ξ and

$$d\alpha = \sum_{j=2}^{n} dp'_j \wedge dq_j$$

restricted to ξ is nondegenerate.

We will consider hypersurfaces in symplectic manifolds (P, ω).

Definition 8.4 $S \subset P$ *is a* hypersurface *if there exists a function* $f : P \to \mathbf{R}$ *such that*

$$S = \{f = 0\} \quad and \quad df|_S \neq 0.$$

The characteristic line field $\mathcal{L}_S$ *on a hypersurface in a symplectic manifold is determined by the Hamiltonian vector field X_f to the function f:*

$$\mathcal{L}_S = \{cX_f : c \in \mathbf{R}\}.$$

In different terms, a vector $X \in TS$ is in $\mathcal{L}_S$ if

$$\omega(X, Y) = 0 \quad for\ all\ Y \in TS.$$

(Note that the Hamiltonian vector field X_f, which is defined by $X_f \lrcorner \omega = df$, is tangent to S.)

If f and g are two functions which define the same hypersurface $S = \{f = 0\} = \{g = 0\}$, then $g = \lambda f$ for some scalar valued function $\lambda \neq 0$. The characteristic line field $\mathcal{L}_S$ remains the same and $X_g = \lambda X_f$ on S.

Definition 8.5 (Weinstein) *A* hypersurface S *in a symplectic manifold* (P, ω) *is of* contact type *if there exists a 1-form α on S such that*

(i) $d\alpha = i^*\omega$, *where* $i : S \hookrightarrow P$ *is the embedding,*

(ii) $\alpha(X) \neq 0$ *for all* $X \in \mathcal{L}_S \setminus \{0\}$.

Observe that for $\dim S > 3$ condition (i) can be replaced by

(i') $d\alpha|_{\alpha=0} = i^*\omega|_{\alpha=0}$.

This condition will then imply

$$i^*\omega = d\alpha + \alpha \wedge \vartheta$$

and $\alpha \wedge \vartheta$ must be closed since both $i^*\omega$ and $d\alpha$ are closed

$$d(\alpha \wedge \vartheta) = d\alpha \wedge \vartheta - \alpha \wedge d\vartheta = 0.$$

When restricted to $\{\alpha = 0\}$ this gives

$$d\alpha \wedge \vartheta|_{\alpha=0} = i^*\omega \wedge \vartheta|_{\alpha=0} = 0.$$

But if $\dim S > 3$, then $\vartheta|_{\alpha=0} \mapsto i^*\omega \wedge \vartheta|_{\alpha=0}$ is injective so that $\vartheta|_{\alpha=0} = 0$. Therefore $\vartheta = f\alpha$ and $\alpha \wedge \vartheta = \alpha \wedge f\alpha = 0$. It follows that $i^*\omega = d\alpha$.

Generally, the contact structure on a hypersurface S of contact type is not uniquely determined. The 1-form α which appears in the definition defines a contact structure. But there will in general exist closed 1-forms β on S such that one still has $(\alpha+\beta)(X) \neq 0$ for $X \in \mathcal{L}_S \setminus \{0\}$ and because $d(\alpha+\beta) = d\alpha = i^*\omega$, the form $\alpha + \beta$ will then define a different contact structure on S.

Definition 8.6 *The vector field* X *on the symplectic manifold* (P,ω) *is a sym-* plectic dilation *(or as a synonym a* homothetic vector field*) if* $\mathcal{L}_X\omega = \omega$.

Lemma 8.7 (Weinstein, [109]) *A hypersurface* $S \subset P$ *is of contact type if and only if there exists a symplectic dilation* X *(defined in a neighbourhood of* S*) which is transversal to* S.

Proof. If X is a transversal symplectic dilation, set $\bar{\alpha} = X \lrcorner \omega$. Then

$$\begin{aligned}
\mathcal{L}_X\omega &= d(X \lrcorner \omega) + X \lrcorner d\omega = d\bar{\alpha} \\
&= \omega
\end{aligned}$$

Define $\alpha = i^*\bar{\alpha}$ to obtain $d\alpha = i^*d\bar{\alpha} = i^*\omega$. The Hamiltonian vector field X_f then satisfies

$$\begin{aligned}
\alpha(X_f) &= \bar{\alpha}(X_f) = (X \lrcorner \omega)(X_f) = \omega(X, X_f) \\
&= -\omega(X_f, X) = -df(X) \neq 0
\end{aligned}$$

since X is transversal to S.

Conversely, if α is a 1-form on S such that $d\alpha = i^*\omega$ and $\alpha(X_f) \neq 0$, then there exists an extension $\overline{\alpha}$ of α defined in a neighbourhood U of S such that

$$\alpha = i^*\overline{\alpha} \quad \text{and} \quad d\overline{\alpha} = \omega \text{ in } U.$$

A proof of this fact will follow. If the vector field X is then defined by

$$X \lrcorner \omega = \overline{\alpha},$$

then X is a symplectic dilation:

$$\mathcal{L}_X \omega = d(X \lrcorner \omega) + X \lrcorner d\omega = d\overline{\alpha} = \omega.$$

The vector field X is transversal to S, since

$$-df(X) = \alpha(X_f) \neq 0.$$

For the proof of the extension result we need the following lemma (a variant of the Poincaré lemma).

Lemma 8.8 *Let S be a hypersurface in P and $i : S \to P$ the embedding map. If λ is a closed p-form near S such that $i^*\lambda = 0$, then there exists a $p-1$-form ϑ defined in a neighbourhood of S such that $\lambda = d\vartheta$ and $\vartheta|_S = 0$.*

Proof. Choose a Riemannian metric on P and consider the flow ψ_s of the vector field grad f (with f the defining function for S). On the neighbourhood U of S which consists of all points that can be reached from S within time $|s| < 1$

$$U = \{\psi_s(x) : x \in S, \, |s| < 1\}$$

we define the retraction $\varphi_t : U \to U$, $t \in [0, 1]$:

$$\varphi_t(z) = \psi_{ts}(x) \quad \text{if} \quad z = \psi_s(x), \, x \in S.$$

Clearly $\varphi_t|_S = \mathrm{id}_S$. $\varphi_1 = \mathrm{id}_U$ and $\varphi_0 : U \to S$. It can be assumed that λ is defined on U. Then

$$\begin{aligned}
\lambda - \varphi_0^*\lambda &= \int_0^1 \frac{d}{dt}(\varphi_t^*\lambda)dt \\
&= \int_0^1 \varphi_t^*(v \lrcorner d\lambda)dt + \int_0^1 d\varphi_t^*(v \lrcorner \lambda)\, dt,
\end{aligned}$$

where v is the vector field $\frac{d\varphi_t}{dt}$ (see the fundamental formula in section 8.1). Set then $I(\beta) = \int_0^1 \varphi_t^*(v \lrcorner \beta)dt$, so that

$$\lambda - \varphi_0^*\lambda = I(d\lambda) + dI(\lambda).$$

Now $i^*\lambda = 0$ if and only if $\varphi_0^*\lambda = 0$. Under this assumption one then has for a closed p-form λ:

$$\lambda = dI(\lambda) = d\vartheta \quad \text{with} \quad \vartheta = I(\lambda).$$

Furthermore, the construction gives $v(x) = 0$ at points $x \in S$ and this implies

$$\vartheta_{(x)} = \int_0^1 \varphi_{t\,(x)}^*(v \lrcorner \lambda)dt = 0$$

for all $x \in S$. $\qquad\square$

In order to complete the proof of Weinstein's lemma, take any extension $\tilde{\alpha}$ of α with $i^*\tilde{\alpha} = \alpha$ and set $\overline{\alpha} = \tilde{\alpha} + \vartheta$. The requirements

$$\begin{aligned}
d\overline{\alpha} &= d\tilde{\alpha} + d\vartheta = \omega \\
i^*\overline{\alpha} &= i^*\tilde{\alpha} + i^*\vartheta = \alpha
\end{aligned}$$

lead to the conditions

$$\begin{aligned}
d\vartheta &= \omega - d\tilde{\alpha} \\
i^*\vartheta &= 0
\end{aligned}$$

The form $\lambda = \omega - d\tilde{\alpha}$ is closed and

$$i^*\lambda = i^*(\omega - d\tilde{\alpha}) = d\alpha - i^*d\tilde{\alpha} = 0$$

Lemma 8.8 then shows that a form ϑ satisfying the above conditions exists. $\quad\square$

Contact manifolds with orientable contact structure ξ can always be realized as hypersurfaces of contact type in a symplectic manifold.

Definition 8.9 *Let (M, α) be a contact manifold with orientable contact structure ξ determined by the 1-form α and let $P \subset T^*M$ be the "positive" line bundle generated by α*

$$P = \{t\alpha : t > 0\} \cong M \times \mathbf{R}^+.$$

The natural symplectic form on T^*M restricts as a symplectic form $\omega = d(t\alpha)$ to P. (P, ω) is the symplectification of (M, α). M is identified with the graph of α considered as a hypersurface in P.

Clearly, M is a surface of contact type in P, since

$$i^*\omega = i^*d(t\alpha) = d\alpha$$

and $\alpha(X) \neq 0$ for all $X \in \mathcal{L}_M \setminus \{0\}$, because the generating vector field for $\mathcal{L}_M$ satisfies

$$dt = X \lrcorner \omega = X \lrcorner d(t\alpha) = dt(X) \cdot \alpha - \alpha(X)dt + tX \lrcorner d\alpha.$$

A contact transformation $f : M \to M$ lifts to an $\mathbf{R}^+$ equivariant symplectic transformation $F : P \to P$. If $f^*\alpha = \lambda\alpha$, set

$$F(p, t) = (f(p), \frac{t}{\lambda})$$

($\mathbf{R}^+$ acts on $M \times \mathbf{R}^+$ by multiplication in the second variable). F is symplectic:

$$F^*\omega = F^*d(t\alpha) = dF^*(t\alpha) = d(\frac{1}{\lambda}f^*(t\alpha)) = d(t\alpha) = \omega.$$

Conversely, any $\mathbf{R}^+$ equivariant symplectic mapping F is of the above form with f a contact transformation.

In the example $M = PT^*V$ the symplectification P of M will be the cotangent space T^*V less its zero section 0_V

$$P = T^*V \setminus 0_V.$$

8.3   Strictly pseudoconvex surfaces

The tangent space T_pS at any point p of a real hypersurface S in $\mathbf{C}^n$ contains a uniquely determined hyperplane ξ_p which is invariant under the complex structure J. If the plane field ξ is completely nonintegrable, then it defines a contact structure on S. Such is the situation for strictly pseudoconvex surfaces.

Consider a domain D given by a defining function ϱ

$$D = \{z \in \mathbf{C}^n : \varrho(z) > 0\}$$

with $d\varrho|_{\partial D} \neq 0$. The differential form

$$\vartheta \;\; = \;\; \frac{\partial\varrho - \bar{\partial}\varrho}{i} \;\; = \;\; -i\sum_{j=1}^{n}(\frac{\partial\varrho}{\partial z_j}dz_j - \frac{\partial\varrho}{\partial \bar{z}_j}d\bar{z}_j)$$

$$= \;\; \sum_{j=1}^{n}(-\frac{\partial\varrho}{\partial y_j}dx_j + \frac{\partial\varrho}{dx_j}dy_j)$$

defines the J-invariant hypersurfaces ξ_p in $T_p\partial D$

$$\xi_p = \{X \in T_p\partial D : \vartheta(X) = 0\}.$$

Clearly, $\vartheta \neq 0$ if $d\varrho \neq 0$. The form

$$L(X, Y) = -d\vartheta(X, JY)$$

is the *Levi form* for the domain D.

Definition 8.10 *The domain* $D = \{\varrho > 0\}$ *is* strictly pseudoconvex, *if the Levi form* L *restricted to* ξ *is positive definite at every point* $p \in \partial D$. *The surface* $S = \partial D$ *will be called a strictly pseudoconvex surface.*

The Levi form L on ξ is only determined up to a conformal factor. The defining function ϱ can be replaced by $f \cdot \varrho$ with $f \neq 0$. This will then change the Levi form on ξ by the factor f.

Pseudoconvexity is a biholomorphically invariant notion. If $f : G \to G'$ is a biholomorphic mapping and S a pseudoconvex surface in the domain G, then its image $f(S)$ is again pseudoconvex.

Observe that the surfaces of contact type and the pseudoconvex surfaces are of a different nature. One of the definitions is symplectically invariant, the other holomorphically. The only mappings in $\mathbf{C}^n$ which are both symplectic and holomorphic with respect to the standard structures are the unitary mappings, the translations and the compositions thereof.

Domains in $\mathbf{C}^n$ are Kähler manifolds. They come with an intrinsic symplectic structure which "blows up" at the boundary. The Kähler metric h is naturally defined through the *Bergman kernel* $K(z.\zeta)$ by the formula

$$h = \sum_{j,k=1}^{n} \frac{\partial^2}{\partial z_j \partial \bar{z}_k} \log K(z,z)\, dz_j\, d\bar{z}_k.$$

The *Kähler form* $\Omega = 2\,\mathrm{Re}(ih)$ is the associated real (symplectic) form

$$\Omega = i\partial\bar{\partial}\log K = i \sum_{j,k=1}^{n} \frac{\partial^2}{\partial z_j \partial \bar{z}_k} \log K(z.z) dz_j \wedge d\bar{z}_k$$

and $g = 2\,\mathrm{Re}\, h$ is the *Bergman metric*.

The complex structure J remains the same as on the underlying space $\mathbf{C}^n$. In real notation the form Ω can be written as

$$\Omega = -\frac{1}{2} dJd\log K.$$

A mapping $\varphi : D \to G$ between domains in $\mathbf{C}^n$, which is symplectic with respect to their Kähler forms Ω_D and Ω_G is a diffeomorphism satisfying $\varphi^* \Omega_G = \Omega_D$.

Proposition 8.11 [63] *A diffeomorphism $\varphi : D \to G$ between strictly pseudoconvex domains which extends smoothly to the boundaries and which is symplectic with respect to the Kähler forms Ω_D, Ω_G will extend as a contact transform to the boundaries.*

There is a partial converse to this statement which at least holds for mappings of the unit ball $B = \{|z| < 1\} \subset \mathbf{C}^n$.

Proposition 8.12 [63] *If $\psi : \partial B \to \partial B$ is a contact transform which embeds into a contact isotopy ψ_t ($\psi_0 = id$, $\psi_1 = \psi$), then ψ extends to a diffeomorphism $\varphi : B \to B$ which is symplectic with respect to the Kähler form: $\varphi^* \Omega_B = \Omega_B$.*

If mappings are considered which are symplectic with respect to the standard underlying symplectic structure ω_0 of $\mathbf{C}^n$, then neither of these statements will be true. In this context, the intrinsic symplectic structure given by the Kähler forms is much more appropriate than the underlying symplectic structure ω_0. Pseudoconvexity can be defined in a more general context.

Definition 8.13 *If (M, J) is a manifold with a complex structure J (not necessarily integrable) and S is a surface which is given by a defining function ϱ, $S = \{\varrho = 0\}$, $d\varrho|_S \neq 0$, then a J-invariant field ξ of hyperplanes in TS will be defined by the form*

$$\alpha = -i(\partial\varrho - \overline{\partial}\varrho) = -Jd\varrho,$$

where $\overline{\partial}\varrho = \frac{1}{2}(d\varrho + iJd\varrho)$. The surface S will be called J-convex, if the quadratic form $-d\alpha(X, JX)$ restricted to

$$\xi = \{\alpha = 0\}$$

is positive definite (at all points of S).

Note that in contrast to the case of strictly pseudoconvex surfaces in complex manifolds, the form $d\alpha(X, JY)$ will not necessarily be symmetric. The Levi form should then be defined by

$$L(X, Y) = -\frac{1}{2}(d\alpha(X, JY) + d\alpha(Y, JX)).$$

Proposition 8.14 *If S is a hypersurface of contact type in a symplectic manifold (P, ω), then there exists a complex structure J tamed by ω such that S is J-convex.*

Proof. We first construct J in a neighbourhood of S. Let X be the symplectic dilation transversal to S which exists in a neighbourhood of S (Lemma 8.7). The form $\alpha = X \lrcorner \omega$ then satisfies $d\alpha = \omega$. A new defining function σ for S can be constructed in a neighbourhood of S as follows: If φ_t is the flow generated by X, set

$$\sigma(x) = t \quad \text{if} \quad \varphi_t(x) \in S,$$

then $TS = \{d\sigma = 0\}|_S$ and $d\sigma(X) = X(\sigma) \equiv -1$ (see Figure 3).

Since ω is non-degenerate, there exists a unique vector field Y (defined in a neighbourhood of S) such that $d\sigma = Y \lrcorner \omega$. It follows that

$$\omega(X, Y) = -(Y \lrcorner \omega)(X) = -d\sigma(X) = 1.$$

At every point p in a neighbourhood of S, T_pP decomposes into the subspace $\{X, Y\}$ spanned by X and Y and its symplectic orthogonal $\{X, Y\}^\perp$. The complex structure J is then defined such that $\{X, Y\}$ and $\{X, Y\}^\perp$ are J-invariant and such that

$$JX = Y, \quad JY = -X.$$

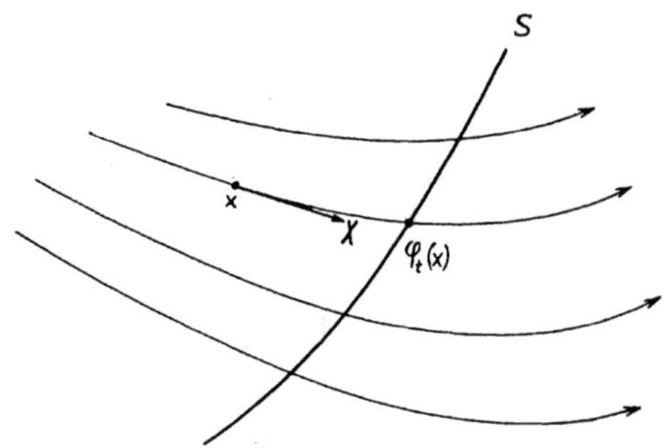

Figure 3

With this choice, the contact structure defined by $\alpha = X \lrcorner \omega$ coincides with the contact structure induced by the complex structure J:

$$J d\sigma = \alpha.$$

Indeed, both $J d\varrho$ and α vanish on

$$\{X, Y\}^{\perp} = \{Z : \alpha(Z) = 0 \text{ and } d\sigma(Z) = 0\}.$$

On the other hand,

$$
\begin{aligned}
J d\sigma(X) &= d\sigma(JX) = d\sigma(Y) = (Y \lrcorner \omega)(Y) = 0 \\
\alpha(X) &= \omega(X, X) = 0 \\
J d\sigma(Y) &= d\sigma(JY) = -d\sigma(X) = -(Y \lrcorner \omega)(X) = \omega(X, Y) = 1 \\
\alpha(Y) &= (X \lrcorner \omega)(Y) = \omega(X, Y) = 1
\end{aligned}
$$

We now have to construct J globally. In a neighbourhood U of a point $p \in S$ a metric μ_U can be chosen such that X and Y are orthonormal and such that the symplectic orthogonal $\{X, Y\}^{\perp}$ is also orthogonal to $\{X, Y\}$ with respect to μ_U. For neighbourhoods off the surface S any metric will do. By using a partition of unity we then arrive at a metric μ which is defined globally. The orthogonality properties near S are preserved. Since ω is non-degenerate, there exists a unique automorphism A of TP such that

$$\omega(Z, Z') = \mu(AZ, Z') \qquad \text{for all } Z, Z' \in TP.$$

Observe that $-A^2$ is positive definite with respect to μ:

$$
\begin{aligned}
\mu(-A^2 Z, Z') &= \omega(-AZ, Z') = \omega(Z'. AZ) \\
&= \mu(AZ', AZ) = \mu(AZ, AZ').
\end{aligned}
$$

At the same time this computation shows that A is skew-symmetric with respect to μ. Set therefore $J = (-A^2)^{-\frac{1}{2}} A$ and define a modified metric

$$\nu(Z, Z') = \mu(Z, (-A^2)^{\frac{1}{2}} Z').$$

Then J is compatible with ω, since

$$
\begin{aligned}
\omega(Z, JZ') &= \omega(Z, (-A^2)^{-\frac{1}{2}} AZ') \\
&= \mu(AZ, (-A^2)^{-\frac{1}{2}} AZ') = \mu(Z, -A^2(-A^2)^{-\frac{1}{2}} Z') \\
&= \nu(Z, Z')
\end{aligned}
$$

It must be verified that $JX = Y$ for the special vector fields defined above. But the metric μ in a neighbourhood of S is chosen such that X and Y are orthonormal and orhogonal to the symplectic orthogonal $\{X.Y\}^{\perp}$. The automorphism A therefore leaves invariant $\{X, Y\}$ as well as $\{X.Y\}^{\perp}$ and on $\{X, Y\}$ it reduces to

$$AX = Y, \qquad AY = -X,$$

since $1 = \omega(X, Y) = \mu(AX, Y) = -\mu(X, AY)$. The complex structure $J = (-A^2)^{-\frac{1}{2}} A$ therefore has the desired properties. $\qquad\qquad\square$

8.4   Contact structures on 3-manifolds

8.4.1   Plane fields on 3-manifolds

A contact structure ξ on a 3-dimensional manifold is locally determined by a differential 1-form. If in local coordinates $\varphi_j : U_j \to \mathbf{R}^3$ ξ is given by α_j:

$$\xi|_{U_j} = \{\alpha_j = 0\},$$

then there exist non-vanishing functions $f_{ij} : U_i \cap U_j \to \mathbf{R} \setminus \{0\}$ such that $\alpha_j = f_{ij}\alpha_i$ on $U_i \cap U_j$. The volume forms $\alpha_j \wedge d\alpha_j$ satisfy

$$\alpha_j \wedge d\alpha_j = f_{ij}^2 \, \alpha_i \wedge d\alpha_i \qquad f_{ij}^2 > 0$$

and therefore determine an orientation of the manifold M. If M is already oriented, then the contact structure ξ is called *positive*, if the orientation given by ξ coincides with the orientation of M. Otherwise ξ is a negative contact structure. A distribution ξ of 2-dimensional plane fields (and in particular a contact structure ξ) is *transversally oriented* if the local 1-forms α_j can be determined such that all transition functions f_{ij} are positive. It is then possible to construct a globally defined 1-form α such that

$$\xi = \{\alpha = 0\}$$

(just use a partition of unity for this construction).

Two distributions ξ_0 and ξ_1 are homotopic if there exists a continuous family ξ_t. $t \in [0, 1]$, of distributions connecting ξ_0 to ξ_1. On the sphere S^3, the homotopy classes of distributions ξ are classified by $\pi_3(\mathbf{R}P^2)$ in the following way:

S^3 identifies with $SU(2)$. Set $z_1 = x_1 + ix_2$. $z_2 = x_3 + ix_4$ and represent the elements of $SU(2)$ by

$$\begin{pmatrix} z_1 & -\overline{z_2} \\ z_2 & \overline{z_1} \end{pmatrix} \qquad |z_1|^2 + |z_2|^2 = 1.$$

The points x on S^3 are obtained as

$$x = \begin{pmatrix} z_1 & -\overline{z_2} \\ z_2 & \overline{z_1} \end{pmatrix} \begin{pmatrix} 1 \\ 0 \end{pmatrix}$$

and the group acts by left (or right) translation on S^3.

The tangent bundle TS^3 is therefore isomorphic to a product

$$TS^3 \cong S^3 \times \mathbf{g}$$

of S^3 with the Lie algebra $\mathbf{g}$ of $SU(2)$. As a vector space, $\mathbf{g} \cong \mathbf{R}^3$. The elements in $\mathbf{g}$ are the left invariant vector fields. A distribution ξ on S^3 determines now a mapping from S^3 to the space of 2-dimensional subspaces of $\mathbf{R}^3$. This space in turn is isomorphic to $\mathbf{R}P^2$. The homotopy classes of distributions ξ therefore are the elements in $\pi_3(\mathbf{R}P^2)$.

If transversally oriented distributions are classified up to homotopies, then these give rise to mappings from S^3 to the space of *oriented* two-dimensional real subspaces of $\mathbf{R}^3$ and this space is isomorphic to S^2. The homotopy classes of transversally oriented distributions are thus given by the elements of $\pi_3(S^2)$, i.e. by their Hopf invariant. For a discussion of the isomorphism $\pi_3(S^2) \cong \mathbf{Z}$ and the geometric ideas behind the Hopf invariant we refer to [15, p. 227ff].

The standard contact structure ξ_0 on S^3 consists of the complex invariant plane field. which can be described by the differential form

$$\begin{aligned} \vartheta &= -i(\partial - \bar\partial)\left(|z_1|^2 + |z_2|^2 - 1\right) \\ &= 2i(-\overline{z_1}\, dz_1 + z_1\, d\overline{z_1} - \overline{z_2}\, dz_2 + z_2\, d\overline{z_2}) \\ &= 4(-x_2\, dx_1 + x_1\, dx_2 - x_4\, dx_3 + x_3\, dx_4) \end{aligned}$$

The Lie algebra of $SU(2)$ is generated (as a vector space) by the elements

$$Y_2 = \begin{pmatrix} i & \\ & -i \end{pmatrix}, \quad Y_3 = \begin{pmatrix} & -1 \\ 1 & \end{pmatrix}, \quad Y_4 = \begin{pmatrix} & i \\ i & \end{pmatrix}.$$

At the point $x = (1, 0, 0, 0) \in S^3$ these correspond to the tangential vector fields

$$Y_2 \cong \frac{\partial}{\partial x_2}, \quad Y_3 \cong \frac{\partial}{\partial x_3}, \quad Y_4 \cong \frac{\partial}{\partial x_4}.$$

The action of $SU(2)$ extends to a holomorphic action on $\mathbf{C}^2$. Since ξ_0 is defined as the complex invariant plane field, it is invariant under the action of $SU(2)$. Therefore, the mapping $S^3 \to S^2$ associated to ξ_0 maps every point of S^3 onto the same two-dimensional oriented plane (spanned by Y_3 and Y_4) in $\mathbf{g} \cong \mathbf{R}^3$: this mapping is therefore constant. The Hopf invariant associated to ξ_0 is zero.

The homotopy classification for contact structures differs from the homotopy classification for plane distributions. In the case of the sphere there are two homotopy classes of contact structures which as plane fields both have Hopf invariant zero (hence are homotopic as plane fields) but which are not homotopic via contact structures. One of these structures is the standard structure ξ_0 (which is tight) and the other is the overtwisted contact structure with Hopf invariant zero.

The notions of tight and overtwisted contact structures will be discussed at length in the section 8.5. At this point we state Eliashberg's classification theorem for contact structures on the sphere.

Theorem 8.15 *On S^3 there are exactly two homotopy classes of transversally oriented contact structures with Hopf invariant zero, the tight and the overtwisted. There is exactly one homotopy class (the overtwisted) of transversally oriented contact structures in any homotopy class of plane fields with prescribed Hopf invariant different from zero.*

8.4.2 Invariant contact structures on the solid torus

Definition 8.16 *On $S^1 \times \mathbf{R}^2$ we will denote the unit tangent vector field in the direction of the circles $S^1 \times \{(x,y)\}$ by $\frac{\partial}{\partial\vartheta} = u$. A differential form ω on $S^1 \times \mathbf{R}^2$ is S^1-invariant, if $L_u\,\omega = 0$.*

The S^1-invariant one-forms $\omega = a\,dx + b\,dy + c\,d\vartheta$ satisfy thus

$$
\begin{aligned}
L_u\,\omega &= u\lrcorner d\omega + d(u\lrcorner\omega)\\
&= a_\vartheta\,dx + b_\vartheta\,dy + c_\vartheta\,d\vartheta = 0
\end{aligned}
$$

and can therefore be represented as

$$
\omega = \alpha + f\,d\vartheta
$$

with α a differential one-form and f a function on $\mathbf{R}^2$. The form ω defines a contact structure on $S^1 \times \mathbf{R}^2$ if $\omega \wedge d\omega \neq 0$. For S^1-*invariant* forms the condition

$$
\omega \wedge d\omega = (\alpha + f\,d\vartheta) \wedge (d\alpha + df \wedge d\vartheta) \neq 0
$$

reduces to

$$
\alpha \wedge df + f\,d\alpha \neq 0.
$$

At points where $f \neq 0$

$$\alpha \wedge df + f\, d\alpha = f^2 d(\tfrac{\alpha}{f}) \neq 0.$$

For such forms, the set

$$\gamma = \{(x,y) \in \mathbf{R}^2 : f(x,y) = 0\}$$

is a curve (or a collection of curves) without double points. Observe that on γ one has $\alpha \wedge df \neq 0$ and hence $df \neq 0$.

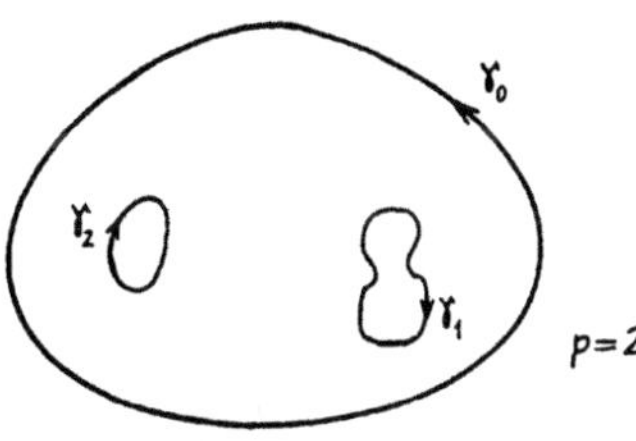

Figure 4

Lemma 8.17 *Let Δ be a bounded domain in $\mathbf{R}^2$ which is bounded by a finite number of Jordan curves. The positively oriented boundary is denoted by*

$$\partial\Delta = \sum_{j=0}^{p} \gamma_j$$

(see Figure 4).
Assume that the one-form μ is defined in a neighbourhood of $\partial\Delta$ and satisfies

$$d\mu = h\, dx \wedge dy \qquad with \qquad h|_{\partial\Delta} > 0.$$

If furthermore $\int_{\partial\Delta} \mu > 0$, then there exists a continuation $\tilde{\mu}$ of μ such that

$$d\tilde{\mu} = \tilde{h}\, dx \wedge dy \ with \ \tilde{h} > 0 \ in \ all \ of \ \Delta.$$

($\tilde{\mu}$ is called a continuation of μ if $\tilde{\mu} = \mu$ in a neighbourhood of $\partial\Delta$ and if $\tilde{\mu}$ is defined in all of Δ).

If such a continuation $\tilde{\mu}$ exists, then by Stokes' theorem

$$\int_{\partial\Delta} \mu = \int_{\partial\Delta} \tilde{\mu} = \int_{\Delta} d\tilde{\mu} > 0.$$

The condition $\int_{\partial\Delta} \mu > 0$ must therefore be satisfied if an extension $\tilde{\mu}$ of the above type exists.

The proof of the lemma starts with an extension $\tilde{h}$ of h such that $\tilde{h} > 0$ and

$$\int_{\partial\Delta} \mu = \int_{\Delta} \tilde{h}\, dx \wedge dy.$$

The form $\beta = \tilde{h}\, dx \wedge dy$ is closed and by de Rham's theorem there exists a form α on Δ such that $d\alpha = \beta$ and such that the periods can be adjusted:

$$\int_{\gamma_i} \alpha = \int_{\gamma_i} \mu \qquad i = 0, \ldots, p.$$

The form $\mu - \alpha$ is closed in a neighbourhood of $\partial\Delta$

$$d(\mu - \alpha) = d\mu - \beta = 0$$

and exact in neighbourhoods of the curves γ_i. Therefore

$$\mu = \alpha + df_i$$

in such neighbourhoods, and there exists then a function f on Δ which is a continuation of the functions f_i. The differential form $\tilde{\mu} = \alpha + df$ then responds to all the requirements of the lemma.

S^1-invariant contact structures on the torus

$$T = S^1 \times D^2 \subset S^1 \times \mathbf{R}^2, \qquad D^2 = \{(x,y) : x^2 + y^2 \leq 1\}$$

can be constructed with the help of the following proposition, which is due to Lutz (see [69, p. 155]).

Proposition 8.18 *Let* $\omega = \alpha + f\, d\vartheta$ *be a* S^1*-invariant contact form defined in a neighbourhood of* ∂T *and such that* $f \neq 0$ *on* ∂D*. Then there exists an extension* $\tilde{\omega}$ *of* ω *to a* S^1*-invariant contact form on* T*.*

Proof. It can be assumed that $f|_{\partial D} > 0$. Since ω is a contact form.

$$\alpha \wedge df + f d\alpha = f^2 d\left(\frac{\alpha}{f}\right) \neq 0 \text{ on } \partial D.$$

Set $d(\frac{\alpha}{f}) = h\, dx \wedge dy$. With the orientation of $\mathbf{R}^2$ chosen adequately, it can be assumed that $h > 0$ on ∂D. In the case that $\int_{\partial D} \frac{\alpha}{f} > 0$, the lemma provides an extension $\tilde{\mu}$ of $\mu = \frac{\alpha}{f}$ with $d\tilde{\mu} = \tilde{h}\, dx \wedge dy$, $\tilde{h} > 0$. Set $\tilde{\alpha} = \tilde{f}\, \tilde{\mu}$, where $\tilde{f}$ is a positive extension of f. Then $\tilde{\omega} = \tilde{\alpha} + \tilde{f}\, d\vartheta$ is the postulated extension of ω:

$$\tilde{\omega} \wedge d\tilde{\omega} = \tilde{f}^2 d\left(\frac{\tilde{\alpha}}{\tilde{f}}\right) \wedge d\vartheta = \tilde{f}^2 d\tilde{\mu} \wedge d\vartheta \neq 0.$$

In the case $\int_{\partial D} \frac{\alpha}{f} \leq 0$ consider the form

$$\omega_1 = \alpha_1 + f_1 d\vartheta = -x\,dy + y\,dx + (r^2 - \frac{1}{4})d\vartheta.$$

with $\omega_1 \wedge d\omega_1 = \frac{1}{2}dx \wedge dy \wedge d\vartheta$.

For $\varepsilon > 0$ denote by γ_ε the circle with center 0 and radius $\frac{1}{2} + \varepsilon$ and by Δ_ε the ring domain bounded by γ_ε and ∂D. Since $\lim_{\varepsilon \to 0} \int_{\gamma_\varepsilon} \frac{\alpha_1}{f_1} = -\infty$, it follows that

$$\int_{\partial D} \frac{\alpha}{f} - \int_{\gamma_\varepsilon} \frac{\alpha_1}{f_1} > 0$$

for sufficiently small ε. Determine first a positive function $\bar{f}$ defined on Δ_ε which interpolates between $f|_{\partial D}$ and $f_1|_{\gamma_\varepsilon}$. According to the lemma there exists then a form $\tilde{\mu} = \tilde{h}dx \wedge dy$, $\tilde{h} > 0$, which coincides with $\alpha/\bar{f}$ near ∂D and with $\alpha_1/\bar{f}_1$ near γ_ε. The invariant extension of ω is now

$$\tilde{\omega} = \begin{cases} \bar{f}\tilde{\mu} + \bar{f}\,d\vartheta & \text{on} \quad S^1 \times \Delta_\varepsilon \\ \omega_1 & \text{on} \quad S^1 \times (D^2 \setminus \Delta_\varepsilon). \end{cases}$$

This completes the proof of the proposition. $\qquad\qquad\square$

Consider the manifold $N = S^1 \times \mathbf{R}^2 \setminus \overset{\circ}{T}$. In polar coordinates r, ϑ' the standard contact form on $S^1 \times \mathbf{R}^2$ is given by

$$\omega_0 = d\vartheta + r^2 d\vartheta'.$$

The solid torus T can be glued back in via a unimodular automorphism $\varphi_A : S^1 \times S^1 \to S^1 \times S^1$

$$\begin{array}{ll} \vartheta & = a\bar{\vartheta} + b\bar{\vartheta}' \\ \vartheta' & = c\bar{\vartheta} + d\bar{\vartheta}' \end{array} \qquad A = \begin{pmatrix} a & b \\ c & d \end{pmatrix} \in GL(2, \mathbf{Z}).$$

The resulting surface is denoted by $T \underset{\varphi_A}{\bigcup} N$. On this surface there exists now a contact form $\tilde{\omega}_0$ which on N coincides with ω_0. Observe first that

$$\omega_A = d(a\bar{\vartheta} + b\bar{\vartheta}') + r^2 d(c\bar{\vartheta} + d\bar{\vartheta}')$$

is a continuation of ω_0 to

$$S^1 \times (D^2 \setminus \{0\})$$

and

$$\omega_A = (b + dr^2)d\bar{\vartheta}' + (a + cr^2)d\bar{\vartheta} = \alpha + f d\bar{\vartheta}$$

is S^1-invariant with respect to the action in the $\bar{\vartheta}$-variable. Choose then $r_0 \in (0,1)$ such that $a + cr_0^2 \neq 0$ and apply the proposition in order to extend ω_A from a neighbourhood of $S^1 \times \partial D_{r_0}^2$ to the solid torus $S^1 \times D_{r_0}^2$.

8.4.3 Martinet's theorem

Theorem 8.19 ([69]) *On every compact orientable 3-manifold there exists a contact structure.*

The proof is based on a structure theorem for compact orientable 3-manifolds which is due to Likorish [67]. We will first sketch the background for this result.

Start with a *self indexing* Morse function on M, i.e. a Morse function $f : M \to [0,3]$ with

$$\{p \in M : p \text{ critical point of } f \text{ with index } i\} \subseteq f^{-1}(i)$$

$i = 0,1,2,3$ and such that there exists only one relative maximum and one relative minimum. The level set $f^{-1}(\frac{3}{2})$ is a surface of some genus g and the sets $M_- = f^{-1}[0, \frac{3}{2}]$ and $M_+ = f^{-1}[\frac{3}{2}, 3]$ each consist of a 3-ball with g handles attached.

As an example take the sphere

$$S^3 = \{(z_1, z_2) \in \mathbf{C}^2 : |z_1|^2 + |z_2|^2 = 1\}$$

and the Morse function $f(z_1, z_2) = \frac{3}{2}(x_1 + 1)$, where $z_1 = x_1 + iy_1$. The surface $f^{-1}(\frac{3}{2}) = \{x_1 = 0\}$ is a sphere S^2 and the sets $f^{-1}[0, \frac{3}{2}]$ and $f^{-1}[\frac{3}{2}, 3]$ are 3-balls. The orientation preserving diffeomorphisms of S^2 are isotopic to the identity [95].

In the general situation, any orientation preserving diffeomorphism $f : \Sigma \to \Sigma$ of the compact surface of genus g is isotopic to a product of Dehn twists along a basic system of $3g - 1$ curves. Given such a diffeomorphism, there exists a finite number of disjoint solid tori V_i in $\Sigma \times (0,1)$ and a diffeomorphism

$$F : \Sigma \times [0,1] \setminus (\cup_i V_i) \to \Sigma \times [0,1] \setminus (\cup_i V_i)$$

such that

$$F|_{\Sigma \times \{0\}} = f \qquad F|_{\Sigma \times \{1\}} = id.$$

The tori V_i should be thought of as tori with cores, along which the Dehn twists operate.

Compact oriented 3-manifolds can then be constructed from the solid torus $T = S^1 \times D^2$ and a system T_i, $i = 1, \ldots, r$, of solid tori around a family of curves $\gamma_i \subset T$. These curves can be realized as graphs of differentiable mappings from S^1 into $\overset{\circ}{D}{}^2$. Given a system Φ of diffeomorphisms

$$\begin{aligned} \varphi_0 : \quad &\partial T \;\to\; \partial T \\ \varphi_i : \quad &\partial T \;\to\; \partial T_i \qquad i = 1, \ldots, r \end{aligned}$$

$r + 1$ copies of the torus T can then be glued in to give a manifold

$$M_\Phi = (T \setminus \cup_{i=1}^r T_i) \cup_{\varphi_0} T \cup_{\varphi_1} \ldots \cup_{\varphi_r} T.$$

Theorem 8.20 ([67]) *Any compact orientable 3-manifold is diffeomorphic to a manifold M_Φ of the above type.*

Martinet observes that the mappings describing the system of curves γ_i can be chosen in an arbitrary C^1-neighbourhood of the zero mapping and that the tori T_i can be chosen within arbitrarily small tubular neighbourhoods of the curves γ_i.

The proof of Martinet's result proceeds now as follows: M is diffeomorphic to some M_Φ. On the torus $T = S^1 \times D^2$ start with the canonical contact structure ξ_0 given by the form

$$\omega_0 = d\vartheta + x\,dy - y\,dx.$$

This structure is transverse to all curves γ which are graphs of mappings $S^1 \to D^2$ contained in a C^1-neighbourhood of the zero mapping. In particular, the curves γ_i which appear in the construction for M_Φ are transversal to the canonical contact structure. Theorem 8.3 then establishes the existence of diffeomorphisms ψ_i of T_i onto T such that

$$\psi_{i\,*}\,\xi_{0,i} = \xi_0,$$

where $\xi_{0,i}$ is the restriction of ξ_0 to T_i.

Since isotopic mappings $\varphi_i \sim \varphi_i'$ $(i = 0,\ldots,r)$ give diffeomorphic manifolds and since every automorphism of $S^1 \times S^1$ is isotopic to a unimodular transformation $\varphi_A \in GL(2, \mathbf{Z})$, it can be assumed that

$$\psi_i \circ \varphi_i : \partial T \to \partial T$$

is a unimodular transformation. The result from section 8.4.2 shows that the contact structure ξ_0 on $T \setminus \cup_{i=1}^r T_i$ extends to

$$M = (T \setminus \cup_{i=1}^r T_i) \cup_{\varphi_0} T \cup_{\varphi_1} \ldots \cup_{\varphi_r} T.$$

This gives the desired contact structure on M.

Recently Giroux [49] presented a new proof for Martinet's theorem. The constructive part is again based on Morse theory. A 2-dimensional surface C embedded in a 3-dimensional manifold M is called *essential* for the proper Morse function $f : M \to [0, \infty)$ if

(i) $f|_C$ is a proper Morse function.

(ii) every critical point of $f|_C$ is critical for f.

(iii) local extremal points of $f|_C$ are local extremal points of f.

Theorem 8.21 ([49]) *Let M be a (not necessarily compact) oriented 3-manifold.*

(A) *There exists a proper Morse function $f : M \to [0, \infty)$ which admits an essential surface.*

(B) *If $f : M \to [0, \infty)$ is a proper Morse function admitting an essential surface, then there exists a contact structure on M which induces the given orientation. Furthermore there exists a contact vector field X which is a pseudogradient for f, i.e. there exist a Riemannian metric and a positive function s such that*

$$Xf \geq s \, \|df\|^2 \quad \text{on } M.$$

Such contact structures are called convex *[36].*

Corollary 8.22 ([49, J]) *On every oriented 3-dimensional manifold there exists a convex contact structure which induces the given orientation.*

8.5 Two-dimensional surfaces in contact manifolds

A surface S of real dimension two embedded in a contact 3-manifold M carries a singular foliation, which is called the *characteristic foliation*. At the singular points $p \in S$ the tangent plane T_pS coincides with the plane ξ_p from the contact field ξ on M. At the other points, $TS \cap \xi$ defines a line field. The foliation is determined by this line field. If S is oriented and if ξ is transversally oriented, this line field will also be oriented. In the generic case, the singularities are isolated and non-degenerate. These singular points are called *elliptic* if the index of the line field is 1 and *hyperbolic* if the index is -1. In the presence of orientations, singular points p are *positive* if the orientations of T_pS and ξ_p coincide. They are *negative* if the orientations are opposite. A vector field Y directs the characteristic foliation, if $Y \in TS \cap \xi$, $Y = 0$ at singular points, and Y points in the "positive direction" (in the presence of orientation). The sign convention is such that a positive elliptic point is a source for the vector field. The leaves of the characteristic foliation are the orbits of Y.

8.5.1 Germs of contact structures on 2-dimensional surfaces.

Proposition 8.23 (Giroux) *Assume that ξ_0 and ξ_1 are contact structures defined in a neighbourhood of the oriented closed surface $S \subset M$ which induce the same characteristic foliation of S. Then there exists an isotopy of diffeomorphisms φ_t, $t \in [0, 1]$, defined in a neighbourhood of S such that $\varphi_0 = id, \varphi_{1*}\,\xi_0 = \xi_1$ and such that the characteristic foliation is preserved: $\varphi_t S = S$ and all contact structures $\varphi_{t*}\,\xi_0$ induce the same characteristic foliation.*

Proof. The normal bundle of S, which is isomorphic to $\Lambda^2 TS$, can be mapped diffeomorphically onto a tubular neighbourhood U of S. The surface S being oriented, this gives a diffeomorphism

$$S \times \mathbf{R} \cong \Lambda^2 TS \to U \subset M$$

and $S \times \{0\}$ is identified with the surface $S \subset M$. A germ ξ of a contact structure is transferred by this diffeomorphism to a contact structure on $S \times \mathbf{R}$ which is defined near $S \times \{0\}$ and can be given by a one-form

$$\alpha = \beta_t + u_t \, dt.$$

In this notation u_t is a function and β_t is a one-form on S, both depending on a real parameter t. The characteristic foliation induced on S by ξ is completely determined by $\{\beta_0 = 0\}$ and at its singularities $\beta_0 = 0$.

The nondegeneracy condition for contact forms is

$$\alpha \wedge d\alpha = dt \wedge \left[u_t d\beta_t + \beta_t \wedge \left(du_t - \frac{\partial \beta_t}{\partial t} \right) \right] \neq 0.$$

If both forms α_0 and α_1 induce the same characteristic foliation on S, then there is a nonvanishing function f, defined in a neighbourhood of S, such that the restrictions of α_0 and $f\alpha_1$ to S coincide. Therefore we can assume from the beginning that α_0 and α_1 restrict to the same form β_0 on S. If both forms α_0 and α_1 satisfy the nondegeneracy condition, then near S this will also be the case for their convex linear combinations

$$\alpha_s = (1 - s)\alpha_0 + s\alpha_1 \qquad s \in [0, 1].$$

Construct now an isotopy φ_s such that near S

$$\varphi_s^* \alpha_s = \lambda_s \alpha_0$$

for some nonvanishing function λ_s. The vector field $v = v_s$ generating φ_s has to satisfy (see section 8.1).

$$\frac{d}{ds}(\varphi_s^* \alpha_s) = \varphi_s^*\left(\tfrac{d}{ds} \log \lambda_s \circ \varphi_s^{-1} \, \alpha_s \right)$$

$$\frac{d\alpha_s}{ds} + v \lrcorner d\alpha_s + d(v \lrcorner \alpha_s) = \mu \alpha_s$$

with $\mu \circ \varphi_s = \frac{d}{ds} \log \alpha_s$. Take then the uniquely defined vector field v such that $v \lrcorner \alpha_s = 0$ and

$$v \lrcorner d\alpha_s = -\frac{d\alpha_s}{ds} + \mu \alpha_s$$

with $\mu = \frac{d\alpha_s}{ds}(T_s)$. For any vector $Y \in TS \cap \{\alpha_s = 0\} = TS \cap \{\alpha_0 = 0\}$ one has $\frac{d\alpha_s}{ds}(Y) = 0$ and this shows that $d\alpha_s(v, Y) = 0$. Consequently, v is tangent to S along S. $\qquad\square$

Definition 8.24 *A contact structure on a 3-manifold M is overtwisted, if there exists an embedded 2-disk $D \subset M$ such that ∂D is a limit cycle of the characteristic foliation and such that D contains exactly one singular point (which necessarily is elliptic). The contact structure is called* tight *if it is not overtwisted.*

Proposition 8.25 [33, Thm. 1.4.1] *If ξ is a tight contact structure on M, then for any embedded disc $D \subset M$, the characteristic foliation contains no limit cycle.*

This proposition is based on the Elimination Lemma (Giroux, Fuchs, Eliashberg). A complete proof for the elimination lemma is given in [32, p. 6]. The reduction of the proposition to the elimination lemma is presented in [33].

Lemma 8.26 (Elimination lemma) ([32, p. 6]) *Let S be an orientable embedded surface in a contact manifold (V, ξ) and $p, q \in S$ elliptic and hyperbolic points of the characteristic foliation F of the same sign. Let γ be a leaf of F joining p to q (γ is the separatrix of the hyperbolic point q which goes to p). Assume there are no other singularities in a neighbourhood of γ. Then there exists a $\mathbf{C}^0$-small isotopy $h_t : S \to V, t \in [0, 1]$, such that*

$$
\begin{aligned}
h_0 &= \; id \\
h_t &= \; id \text{ on } \gamma \text{ and outside a fixed neighbourhood } U \text{ of } \gamma \\
&(h_t \text{ is an embedding for all } t)
\end{aligned}
$$

and such that the characteristic foliation F_1 on $h_1 U$ has no singularities.

Proof. Choose a one-form α representing the contact structure ξ in a neighbourhood of γ and a surface element ω on S. This determines a vector field Y, $Y \lrcorner \omega = \alpha|_S$, directing the foliation, whose divergence at the singular points is positive:

$$(\operatorname{div}_\omega Y)\,\omega = d(Y \lrcorner \omega) = d\alpha|_S = k\omega.$$

In fact, at the singularities the function k is positive since the orientation of S given by ω and the orientation of ξ given by $d\alpha$ coincide.

The surface element ω can now be changed into an element $g\omega, g > 0$, such that the divergence of Y with respect to $g\omega$ is positive:

$$(\operatorname{div}_{g\omega} Y)\,g\omega = (\operatorname{div}_\omega Y + \frac{Yg}{g})\,g\omega.$$

Near the singular points g can be taken to be constant. Otherwise adjust g along the trajectories of the field such that

$$Y(\ln g) > -\operatorname{div}_\omega Y.$$

Set then $\beta = Y \lrcorner g\omega = g\alpha|_S$. The form $\beta + dt$ defines a contact structure on $U \times \mathbf{R}$. where $U \subset S$ is the neighbourhood above:

$$
\begin{aligned}
(\beta + dt) \wedge d(\beta + dt) &= (\beta + dt) \wedge d(Y \lrcorner g\omega) \\
&= (\beta + dt) \wedge (\mathrm{div}_{g\omega} Y) g\omega = (\mathrm{div}_{g\omega} Y) g\,\omega \wedge dt \\
&\neq 0
\end{aligned}
$$

A neighbourhood $W \subset U \times \mathbf{R}$ of the trajectory γ connecting the elliptic with the hyperbolic singularity can now be mapped by a contact transformation $h : W \to W'$ onto a neighbourhood W' of γ in the contact manifold (V, α):

$$
h^* \alpha = \lambda(\beta + dt)
$$

for some function $\lambda > 0$.

The contact vector field $\frac{\partial}{\partial t}$ on $U \times \mathbf{R}$ is mapped onto the contact vector field $X = h_* \frac{\partial}{\partial t}$ on $W' \subset V$ and X is transversal to S in a neighbourhood of γ.

Near γ the surface S will now be deformed as a surface in the contact manifold $U \times \mathbf{R}$, and the deformed surface will be mapped back to V by the contact mapping h. Within the manifold $U \times \mathbf{R}$ the deformed piece of a surface appears as the graph over U of a function $f : U \to \mathbf{R}$ with support in a neighbourhood of γ. The restriction of $\beta + dt$ to the graph of f is given by $\beta + df$. To be precise, $\beta + df$ is the pullback of $\beta + dt$ under the mapping $x \mapsto (x, f(x)). x \in U$. It suffices therefore to construct f with support near γ such that $\beta + df \neq 0$ on U. The form $\beta + dt$ restricted to the graph of f will then never vanish, hence the deformed surface will have no singular points. The function f can be constructed such that $f|_\gamma = 0$ and such that $\sup_{x \in U} f(x)$ is arbitrarily small.

Choose local coordinates $x : U \to \mathbf{R}^2$ such that the elliptic point is mapped to $(-1, 0)$. the hyperbolic point to $(1, 0)$ and the connecting trajectory onto the segment $\{(t. 0) : -1 \leq t \leq 1\}$. Furthermore, the local coordinates can be chosen such that for all x_2 in a neighbourhood of 0 but different from 0 the foliation is transversal to the lines $x_2 = \mathrm{const}$. For the surface element ω we choose the standard element $dx_1 \wedge dx_2$. The vector field Y' is defined by

$$
Y' \lrcorner \omega = \beta
$$

(see Figure 5).

Consider the rectangles

$$
R_c = \{(x_1, x_2) : |x_2| \leq c, \ |x_1| \leq 1 + c\}.
$$

For $c > 0$ sufficiently small, Y' is directed inward on the face $\{x_1 = 1 + c\} \cap R_c$ and directed outward on all other faces. There is a lower bound m for $|Y'|$ in $U \setminus R_c$ and an upper bound M for $|Y'|$ in R_c.

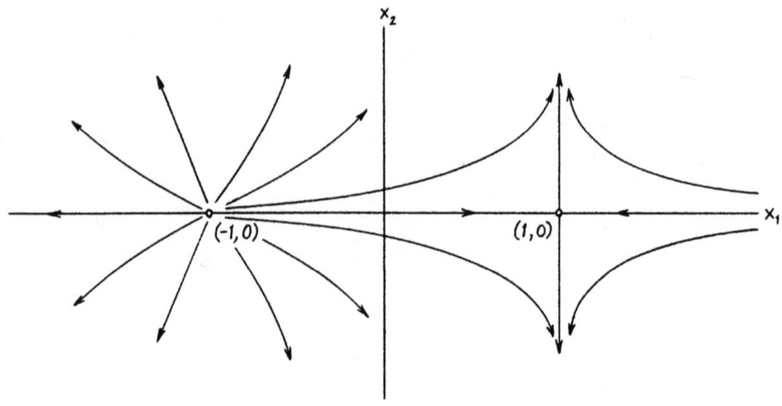

Figure 5

Given $d > c > 0$ choose auxiliary functions φ and $\psi : \mathbf{R} \to \mathbf{R}$ (see Figure 6) such that

$$
\begin{aligned}
\varphi &= \varphi_{c,d} \text{ is odd}, \quad \operatorname{supp}\varphi \subset [-d,d], \\
\varphi' &> M \quad \text{near } 0, \\
0 &\geq \varphi' > -\frac{m}{2} \quad \text{on } [c,d].
\end{aligned}
$$

It can be arranged that $\max |\varphi|$ is arbitrarily small.

$$
\begin{aligned}
\psi &= \psi_{c,d} \text{ is even}, \ \operatorname{supp}\psi \subset [-d-1,d+1], \\
\psi &= 1 \quad \text{on } [-c-1,c+1], \\
-\frac{2}{d-c} &< \psi' < 0 \quad \text{on } [c+1,d+1].
\end{aligned}
$$

Fix $d > c > 0$ such that $R_d \supset U$ and arrange for φ and ψ such that

$$
|\varphi\psi'| < \frac{m}{2}.
$$

We claim that the function $f(x_1,x_2) = \varphi(x_2) \cdot \psi(x_1)$ has the property that $\beta + df \neq 0$. We shall in fact show that the vector field $Y' + Y''$ given by

$$
(Y' + Y'')\lrcorner\omega = \beta + df
$$

does not vanish in U. In a neighbourhood of the separatrix γ

$$
|Y''| > M \geq |Y'|
$$

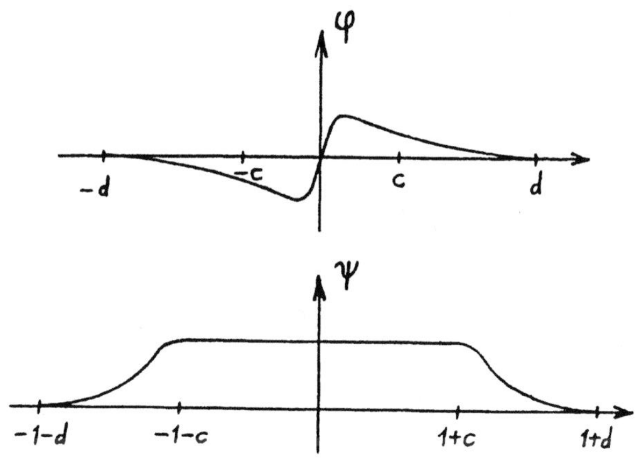

Figure 6

and therefore $Y' + Y'' \neq 0$. To the lines $\{x_2 = \text{const} \neq 0\} \cap R_c$, the field Y'' is parallel and the field Y' is transversal, hence $Y'' + Y' \neq 0$. Outside the rectangle R_c but still within $\{(x_1, x_2) : |x_1| \leq 1 + d, |x_2| \leq c\}$, the vector fields are still at an angle $< \pi$ and for $c \leq |x_2| \leq d$ one has

$$Y'' \lrcorner \omega \;=\; d(\varphi \cdot \psi) = \varphi \psi' dx_1 + \varphi' \psi dx_2$$
$$|Y''|^2 \;\leq\; \left(\frac{m}{2}\right)^2 + \max |\varphi' \psi|^2$$
$$\;<\; m^2 < |Y'|^2.$$

This completes the proof. $\qquad\square$

8.5.2 Invariants for curves in contact manifolds

We assume that the contact structure ξ on the oriented manifold M is described by the differential form α and $\alpha \wedge d\alpha > 0$. By a curve we will understand a closed immersed oriented curve $\gamma : S^1 \to M$. A *Legendre curve* λ (also called a horizontal curve) is a curve $\lambda : S^1 \to M$ which is tangent to ξ at all points ($\dot\gamma \in \xi$). A curve γ is *transversal,* if the tangent vectors never lie in ξ. It is positive transversal, if $\alpha(\dot\gamma) > 0$ (respectively negative transversal if $\alpha(\dot\gamma) < 0$).

Definition 8.27 *Let λ be a Legendre curve homologous to zero in M. Fix a relative homology class $\beta \in H_2(M, \lambda)$ and an oriented surface S with boundary λ which represents β. A new curve λ' is obtained by pushing λ slightly along a vector field X which is transversal to ξ. The* Thurston-Bennequin invariant *$tb(\lambda, \beta)$ is the homological intersection number of λ' with the surface S.*

The Thurston-Bennequin invariant remains unchanged if the orientation of λ is reversed. In the original definition given by Bennequin [11, p. 130] the curve λ is deformed in the direction of the vector field normal to the curve λ, but contained in ξ. This gives the same result. In fact, $tb(\lambda, \beta)$ depends only on ξ restricted to λ and on β.

In the case of the standard structure on $\mathbf{R}^3$ given by $\omega_0 = dz + y\,dx - x\,dy$, or more generally $\omega = dz + u\,dx + v\,dy$, the invariant tb can be read off from a count of the self intersection points of the projected curve $\pi \circ \lambda$ in the (x, y)-plane. The intersections must be counted with the appropriate sign (see Figure 7).

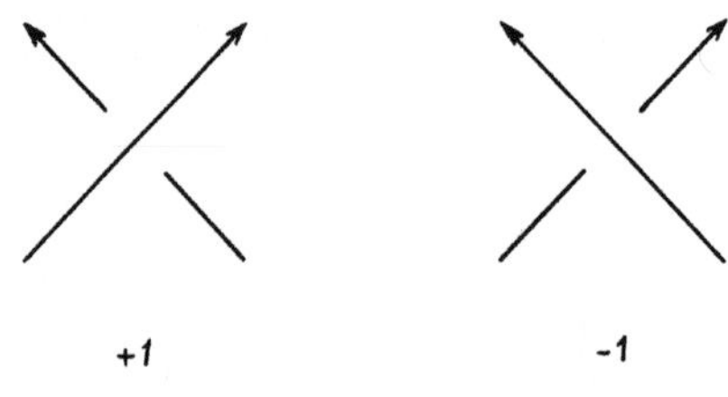

Figure 7

Definition 8.28 *Assume that λ is a Legendre curve homologous to zero. Let T be a tangent vector field along λ and S an oriented surface with boundary λ, which represents $\beta \in H_2(M, \lambda)$. Choose a trivilization of the bundle $\xi|_S$. Then the degree of T with respect to the trivilization depends only on β. It is called the* rotation number $r(\lambda, \beta)$ *of the curve λ.*

The rotation number is the total variation of the angle of the tangent T with respect to a frame defining the trivialization of ξ. If the orientation of λ is changed, then the rotation number also changes sign.

Definition 8.29 *Assume that γ is a transversal curve homologous to zero and let S be an oriented surface with boundary γ which represents $\beta \in H_2(M, \gamma)$. Take a nonvanishing horizontal vector field X on S (i.e. $X \in \xi|_S$, $X \neq 0$) and push γ slightly along X. The intersection number $l(\gamma', S)$ of the resulting curve γ' with the surface S is independent of the vector field X chosen. It is called the* self-linking number $l(\gamma, \beta)$ *of γ with respect to β.*

In fact, if X and Y are two horizontal nonvanishing vector fields, then the difference of the linking numbers of γ' with S and of γ'' (the curved obtained by pushing along Y) with S is the degree of the mapping φ which associates to every point on γ the angle between the vectors X and Y in the plane ξ. But φ extends to a mapping $M \to S^1$ and therefore the degree of φ is zero.

In manifolds M with a tight contact structure (Definition 8.24) the invariants characterize trivial knots (i.e. curves which bound an embedded disc) up to isotopies. We state without proof the following result of Eliashberg.

Theorem 8.30 ([32, Thm. 5.1.1 and 5.7.1]) *Assume that the contact structure ξ on M is tight.*

(a) *If γ and γ' are transversal trivial knots with $l(\gamma) = l(\gamma')$, then γ and γ' are isotopic as transversal knots.*

(b) *If λ and λ' are trivial Legendre knots with $r(\lambda) = r(\lambda')$ and $tb(\lambda) = tb(\lambda')$, then λ and λ' are isotopic as transversal knots.*

The invariants are related to each other. We will now formulate and prove Bennequin's equality.

If a Legendre curve λ is slightly pushed in direction of its oriented normal N within ξ, then the resulting curve λ^+ is negatively transversal to ξ (see Figure 8).

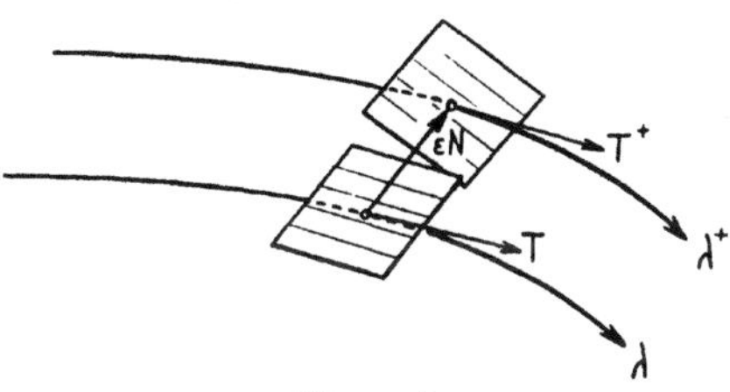

Figure 8

Apply Stokes' formula to a rectangle R with one side on λ, one on λ^+ and the two remaining sides given by ε times the normal vectors N. Denote by T and T^+ the tangent vectors to the curves λ and λ^+. Then up to second order terms in ε

$$\alpha(T^+) \approx \int_{\lambda^+ \cap \partial R} \alpha \approx \int_{\partial R} \alpha = \int_R d\alpha$$
$$\approx d\alpha(\varepsilon N, T) = -\varepsilon \, d\alpha(T, N).$$

The orientation of the normal vector N is defined by $d\alpha(T, N) > 0$. It follows that $\alpha(T^+) < 0$. In this way, Legendre curves can be approximated by negatively (or positively) transversal curves.

Theorem 8.31 (Bennequin) *If λ is a Legendre curve homologous to zero in M, then*

$$l(\lambda^\pm) = tb(\lambda) \pm r(\lambda).$$

Proof. We follow Bennequin [11, Prop. 6]. Choose a surface S with $\partial S = \lambda$ which represents the homology class $\beta \in H_2(M, \lambda)$ and extend the horizontal normal vector field N to a vector field on all of S. Pushing S by ε along this field gives a new surface S^+ with $\partial S^+ = \lambda^+ = \lambda + \varepsilon N$ representing an element $\beta^+ \in H_2(M, \lambda^+)$. Let X be a nonvanishing transversal vector field defined near S and choose sufficiently small positive numbers ε and ε'. The invariant $tb(\lambda, \beta)$ counts how many times $\lambda + \varepsilon' X$ winds around λ with respect to the surface S, or what amounts to the same, how many times $\lambda + \varepsilon' X + \varepsilon N$ winds around $\lambda + \varepsilon N$ with respect to S^+. On the other hand, $l(\lambda^+, \beta^+)$ counts how often $\lambda + \varepsilon N + \varepsilon' N$ winds around $\lambda^+ = \lambda + \varepsilon N$ with respect to S^+. The difference $l(\lambda^+, \beta^+) - tb(\lambda, \beta)$ is therefore the total variation of the angle of X with N along λ^+, or what is the same, along λ. But this is the rotation number

$$l(\lambda^+, \beta^+) - tb(\lambda, \beta) = r(\lambda, \beta).$$

The equality for the curve λ^- is obtained by changing α into $-\alpha$. Then the orientation of S is changed, the normal N becomes $-N$ and thus λ^+ turns into λ^-. Both l and tb remain unchanged but r changes sign. $\square$

Let us come back to the situation of an oriented compact surface embedded in a manifold M with transversally oriented contact structure ξ. Assume that the singularities of the characteristic foliation are isolated elliptic or hyperbolic points. Denote by e_+, h_+ the number of positive elliptic and hyperbolic points, respectively, and use e_-, h_- for negative singularities. Set

$$d_\pm = e_\pm - h_\pm.$$

Take a vector field N normal to S and push the boundary curve $\gamma = \partial S$ slighty in the direction of N. The intersection number of the resulting curve $\gamma' = \gamma + \varepsilon N$ with S is zero, if the surface is embedded.

Proposition 8.32 ([11, Thm. 7]) *For an embedded oriented surface S with a transversal boundary $\gamma = \partial S$*

$$-l(\gamma) = d_+ - d_-.$$

Consider a vector field Y directing the characteristic foliation. Since γ is transversal, the intersection number of $\gamma + \varepsilon Y$ with S is the same as the intersection number of $\gamma + \varepsilon N$ with S. It therefore vanishes. Take then a nonvanishing vector field X on M. The difference of the intersection numbers of $\gamma + \varepsilon X$ with S and of $\gamma + \varepsilon Y$ with S measures the total variation of the angle φ between X and Y along the curve γ. The total variation of φ along the boundary of a small disc centered at p is 2π for positive elliptic and negative hyperbolic points p and -2π otherwise. This proves the formula.

Remark. Bennequin's theorem is more precise. It gives an expression for $l(\gamma)$ in the case of an immersed and not only of an embedded surface.

8.5.3 Index theorems for 2-dimensional surfaces

A point p on an oriented 2-dimensional manifold S embedded in a 4-dimensional manifold M with (not necessarily integrable) complex structure J is a *point of complex tangency* if T_pS is a complex subspace of T_pM

$$JT_p S = T_p S.$$

For generic embeddings, points of complex tangency are isolated. The tangencies are *positive* if the orientation of T_pS agrees with the complex orientation and they are negative otherwise. Assume that ω is a symplectic structure on M, compatible with J so that $g(X,Y) = \omega(X, JY)$ defines a positive definite J-invariant quadratic form on TM. In the neighbourhood of an isolated tangency p, choose a vector field N along S which is normal to S (with respect to g). The field of J-invariant 2-dimensional planes orthogonal to N and symplectically orthogonal to N intersects TS along a directed line field η which is singular at p. The tangency is called *elliptic* or *hyperbolic*, according to whether the line field η is elliptic or hyperbolic near p. Observe that different choices for the normal field are homotopic and will lead to the same result.

According to [13] a 2-dimensional surface embedded in $\mathbf{C}^2$ can be represented in normal form

$$z = w\overline{w} + 2\,\beta\,\mathrm{Re}\,w^2 + O(w^3), \quad \beta \geq 0$$

near an isolated tangency $p = (0,0)$ (see section 8.6.1). The normal vector field to the surface

$$z = w\overline{w} + 2\,\beta\,\mathrm{Re}\,w^2$$

can be chosen in the direction of the imaginary axis of z. The plane field is given by $z = \mathrm{const}$ and the integral curves for η project onto the level lines of the function

$$u^2 + v^2 + 2\beta(u^2 - v^2).$$

Hence the point of tangency $p = (0,0)$ of a surface in normal form is elliptic if $0 \leq \beta < \frac{1}{2}$ and hyperbolic if $\beta > \frac{1}{2}$.

Consider then a real 2-dimensional closed oriented surface S embedded in a complex 2-dimensional manifold M and assume that all points of tangency are isolated and either hyperbolic or elliptic. The number of positive elliptic tangencies is denoted by $e_+(S)$, the number of positive hyperbolic tangencies by $h_+(S)$ and similarly for negative complex tangencies. For embedded closed manifolds $S \hookrightarrow M$, let $\chi(S)$ denote the Euler characteristic, $\nu(S)$ the normal Euler number and $c(S)$ the first Chern class of M evaluated on the homology class realized by S.

Theorem 8.33 ([13], [64])

$$e_\pm - h_\pm = \frac{1}{2}\big(\chi(S) + \nu(S) \pm c(S)\big).$$

For spheres embedded in $\mathbf{C}^2$ this theorem was proved by E. Bishop [13]. In this case $\nu(S) = 0$ and $c(S) = 0$ so that

$$e_\pm - h_\pm = 1.$$

The general form of the theorem is due to Lai [64, Thm 4.10 and 5.11].

Following Bishop and Bennequin we will give a complete proof of this theorem for the case of an oriented closed surface S embedded in $\mathbf{C}^2$.

The *Gauss mapping* associates to the point $p \in S$ the tangent space T_pS, considered as an element of the Grassmann manifold $G_{2,2}$ of oriented 2-dimensional subspaces of $\mathbf{R}^4$. The *Plücker coordinates* a_{ij} on the Grassmannian $G_{2,2}$ are determined as follows: Elements $P \in G_{2,2}$ are given by ordered pairs v_1, v_2 of othonormal vectors. Set

$$v_1 \wedge v_2 = \sum_{i<j} a_{ij}\, e_i \wedge e_j.$$

Then the coefficients satisfy

$$a_{12}\, a_{34} - a_{13}\, a_{24} + a_{14}\, a_{23} = 0,$$
$$\sum_{i<j} a_{ij}^2 = 1.$$

Conversely $P \in G_{2,2}$ is completely determined by a set of coordinates a_{ij} satisfying these two relations. The linear coordinate change

$$\begin{aligned}
x &= (x_1, x_2, x_3) = (a_{12} + a_{34}, a_{23} + a_{14}, -a_{13} + a_{24}) \\
y &= (y_1, y_2, y_3) = (a_{12} - a_{34}, a_{23} - a_{14}, -a_{13} - a_{24})
\end{aligned}$$

will transform these relations into

$$|x| = |y| = 1,$$

so that $G_{2,2}$ is diffeomorphic to the product of two unit spheres S_1 and S_2 in $\mathbf{R}^3$

$$G_{2,2} = S_1 \times S_2.$$

The submanifold

$$S_2^+ = \{(1,0,0)\} \times S_2 \subset G_{2,2}$$

is the manifold of complex lines with their complex orientation, whereas

$$S_2^- = \{(-1,0,0,)\} \times S_2$$

is the manifold of complex lines with their orientation reversed.

The Gauss mapping has the components $g_j = pr_j \circ g$, where $pr_j : G_{2,2} \to S_j$ $(j = 1, 2)$, are the projections onto the 2-spheres S_j. The positive complex tangencies are given by $g^{-1}(S_2^+) = g_1^{-1}(1,0,0)$ and the negative tangencies by $g^{-1}(S_2^-) = g_1^{-1}(-1,0,0)$.

Lemma 8.34 *A point of complex tangency is elliptic or hyperbolic according to whether g_1 preserves or reverses the orientation.*

Proof. Assume that in local coordinates $w = u + iv$ the embedding is given in normal form

$$
\begin{aligned}
z_1 &= w \\
z_2 &= w\overline{w} + 2\beta \mathrm{Re}(w^2) + O(|w|^3) \qquad \beta \geq 0.
\end{aligned}
$$

At points q near $p = (0,0)$ the tangent space $T_q S$ is given by $v_1 \wedge v_2$ with

$$
\begin{aligned}
v_1 &= e_1 + (2 + 4\beta)ue_3 + O(|w|^2) \\
v_2 &= e_2 + (2 - 4\beta)ve_3 + O(|w|^2)
\end{aligned}
$$

(where e_1, e_2, e_3, e_4 is the standard basis in $\mathbf{R}^4$). Within second order approximation the vectors are orthogonal and the coordinates of $v_1 \wedge v_2 \in G_{2,2}$ are given by

$$
\begin{aligned}
x &= (1, -(2 + 4\beta)u, -(2 - 4\beta)v), \\
y &= (1, -(2 + 4\beta)u, -(2 - 4\beta)v).
\end{aligned}
$$

At $(1, 0.0) \in S_1$ the tangent space is spanned by $\frac{\partial}{\partial x_2}$ and $\frac{\partial}{\partial x_3}$. The tangent map to g_1 at $w = 0$ (in local coordinates) maps $\frac{\partial}{\partial u}$ onto $-(2 + 4\beta)\frac{\partial}{\partial x_2}$ and $\frac{\partial}{\partial v}$ onto $-(2 - 4\beta)\frac{\partial}{\partial x_3}$. It preserves the orientation if $0 \leq \beta < \frac{1}{2}$ and reverses it if $\beta > \frac{1}{2}$. For negative tangencies the calculation is similar. $\square$

The proof of the theorem can now be completed. The degree of the first component g_1 of the Gauss mapping calculated near $(1, 0, 0)$ is $e^+ - h^+$ and calculated near $(-1, 0, 0)$ it is $e^- - h^-$, therefore

$$
e^+ - h^+ - e^- + h^- = 0.
$$

If S is embedded in $\mathbf{C}^2$, then by a result of Whitney (see Chern and Spanier [18]) there exists a nonvanishing vector field N normal to S in $\mathbf{C}^2$. The complex lines orthogonal to N determine a line field on TS with singularities at the points of complex tangency. The index of this line field is $+1$ at elliptic and -1 at hyperbolic points. By the Euler-Poincaré theorem

$$
e^+ - h^+ + e^- - h^- = \chi(S).
$$

It follows that

$$
e_\pm - h_\pm = \frac{1}{2}\chi(S) = \mathrm{degree}(g_1).
$$

Obstruction theory provides an interpretation of the formula

$$d_{\pm} = \frac{1}{2}(\chi(S) + \nu(S) \pm c(S))$$

also in the case of bordered surfaces. Assume that S is an oriented compact embedded surface in a complex 2-dimensional manifold M such that the points of complex tangency are isolated elliptic or hyperbolic points, with no complex tangencies on ∂S. Consider a vector bundle $E \to S$ and a triangulation $S = \cup \Delta_j$ of S such that each 2-simplex Δ_j lies in an open set U over which the bundle is isomorphic to the product bundle $U \times \mathbf{R}^2$. The triangulation can be chosen such that the points of complex tangency are not on the 1-skeleton of the triangulation. Start now with a nonvanishing section of E on the 1-skeleton of the triangulation. This gives mappings

$$\alpha_j : \partial\Delta_j \to \mathbf{R}^2 \setminus \{0\}$$

$\partial\Delta_j$ is homeomorphic to S^1 and $\mathbf{R}^2 \setminus \{0\}$ retracts onto S^1. Thus α_j can be extended to Δ_j if the homotopy class $[\alpha_j]$ of α_j in $\pi_1(S^1)$ vanishes, i.e. if its winding number vanishes. The elements $[\alpha_j]$ define a cochain with values in $\mathbf{Z} = \pi_1(S^1)$. It is a well known fact that this cochain is in fact a cocycle. The obstruction to extending the nonvanishing vector field is the cohomology class of this cocycle in the relative cohomology $H^2(S, \partial S; \mathbf{Z})$.

If the vector bundle over S is the normal bundle, then the relative normal Euler number is the relative cohomology class $\nu \in H^2(S^2, \partial S; \mathbf{Z})$ evaluated on S. In the case $\partial S = \emptyset$ the homology class $\nu(S)$ is the obstruction to the existence of a nonvanishing vector field transversal to S.

The relative first Chern class $c(S)$ is defined similarly. Consider the problem of extending the tangent vector field T of ∂S and the vector field $N \in TS$ which is outward transversal to T (and tangent to the surface) as $\mathbf{C}$-linearly independent vector fields. Here the vector bundle is the complex tangent bundle of M pulled back to S via the embedding. Consider again a trivialization $U_j \times \mathbf{C}^2$ of the bundle in a neighbourhood of Δ_j. The mapping α_j from $\partial\Delta_j$ to the pairs of complex linearly independent vector fields can be considered as a mapping

$$\alpha_j : \partial\Delta_j \to GL(2, \mathbf{C}).$$

Since $GL(2, \mathbf{C})$ retracts onto $U(2)$, the homotopy class $[\alpha_j]$ of α_j in $\pi_1(U(2)) = \mathbf{Z}$ will have to vanish if an extension to Δ_j be possible. Again it can be shown that $[\alpha_j]$ is a cocycle with values in $\mathbf{Z}$.

By definition, the relative first Chern class c is the relative cohomology class of this cocycle. Thus c is an element of $H^2(S, \partial S; \mathbf{Z})$.

Theorem 8.35 ([52]) *Let S be an oriented compact surface embedded in a 4-manifold with a complex structure J such that the points of complex tangency* *are isolated elliptic or hyperbolic points, with no complex tangencies on ∂S. Then with the above definitions of ν and c,*

$$\nu(S) = -\chi(S) + e_+ + e_- - h_+ - h_- = -\chi(S) + d_+ + d_-$$

and

$$c(S) = d_+ - d_-.$$

The Bishop equality

$$d_\pm = \frac{1}{2}\left(\chi(S) + \nu(S) \pm c(S)\right)$$

is an immediate consequence of these formulas.

Proof. Extend T from ∂S to S as a tangent vector field (still called T) with singularities at isolated points q_j, which are not points of complex tangency. Then

$$\sum_j \text{index}_{q_j} T = \chi(S).$$

The vector field iT is transverse to S outside the points q_j and outside the complex tangencies p_j. The index of iT at points q_j is

$$\text{index}_{q_j} iT = -\,\text{index}_{\,q_j} T$$

and at the points of complex tangency

$$\text{index}_{p_j} iT = \pm 1$$

depending on whether p_j is an elliptic $(+1)$ or a hyperbolic (-1) point. Thus

$$
\begin{aligned}
\nu(S) &= -\chi(S) + e_+ + e_- - h_+ - h_- \\
&= -\chi(S) + d_+ + d_-.
\end{aligned}
$$

Let us prove that at an elliptic point of tangency p,

$$\text{index}_p iT = 1.$$

For this it suffices to take the model case

$$z = x + iy = w\overline{w} + 2\beta\,\text{Re}(w^2) = (1 + 2\beta)u^2 + (1 - 2\beta)v^2.$$

Consider the tangent vector field

$$T = \frac{\partial}{\partial u} + (2 + 4\beta)u\frac{\partial}{\partial x}$$

which is regular at 0. The projection of iT onto the normal bundle spanned by the vectors

$$
\begin{aligned}
V_1 &= \frac{\partial}{\partial x} - u(2+4\beta)\frac{\partial}{\partial u} - v(2-4\beta)\frac{\partial}{\partial v} \\
V_2 &= \frac{\partial}{\partial y}
\end{aligned}
$$

is given by

$$
\begin{aligned}
pr(iT) &= pr\left(\frac{\partial}{\partial v} + (2+4\beta)u\frac{\partial}{\partial y}\right) \\
&= -(2-4\beta)v\, V_1 + (2+4\beta)u V_2.
\end{aligned}
$$

The index of this vector field is the winding number of the image of the circle $|w| = 1$ under the mapping

$$
(u,v) \mapsto (-(2-4\beta)v, (2+4\beta)u).
$$

In complex notation, the mapping is $w \mapsto 2iw + 4i\beta\overline{w}$, thus

$$
\text{index}\,(pr\, iT) = 1
$$

if $|\beta| < \frac{1}{2}$.

In order to prove the formula $c(S) = d_+ - d_-$, extend the vector fields T and N to vector fields tangent to S with possible singularities at points $q_j \in S$ (which are not complex tangencies). The vector fields T and N are $\mathbf{C}$-independent outside the points q_j and the points p_j of complex tangency. It is no problem to extend the vector fields T and N as $\mathbf{C}$-independent vector fields in $TM|_S$ near the points q_j. Consider then the normal form

$$
\begin{aligned}
z_1 &= w, \qquad 0 \le \beta < \frac{1}{2} \\
z_2 &= w\overline{w} + 2\beta\,\mathrm{Re}\,w^2 + O(|w|^3)
\end{aligned}
$$

for a positive elliptic point p. Near p look at the Gauss mapping $S \to G_{2,2}$. The vector fields T and N determine real subspaces of $\mathbf{C}^2$ (considered as elements of $G_{2,2}$) outside the elliptic point. But they cannot be extended as $\mathbf{C}$-independent vector fields across the singularity. The index in $\pi_1(U(2))$ is determined by the considerations above. It is $+1$ for positive elliptic (and for negative hyperbolic) and -1 for positive hyperbolic (and negative elliptic) points. It follows that

$$
c(S) = d_+ - d_-.
$$

$\square$

8.5.4 Bennequin's inequality

Bennequin shows [11, p. 147] that for the standard structure ξ_0 on S^3 the inequality

$$l(\partial S) \leq -\chi(S)$$

holds whenever S is an immersed surface with an embedded transversal boudary ∂S. If ∂S is a Legendre boundary, then this result in combination with the equality

$$l(\lambda^{\pm}) = tb(\lambda) \pm r(\lambda)$$

(see section 8.5.2) shows that

$$tb(\partial S) \leq -\chi(S) \pm r(\partial S).$$

From this inequality Bennequin deduces that the standard structure ist tight, i.e. the characteristic foliation of embedded discs cannot have closed integral curves avoiding the singular points.

The argument is the following: Assume that S is an embedded disc with Legendrian boundary and such that there are no singular points on ∂S. There exists then a vector field X along ∂S which is transversal to ξ_0 and tangent to S. The intersection number of the curve $\lambda^+ = \partial S + \varepsilon X$ with S will then be zero, since λ^+ will not meet S, therefore $tb(\partial S) = 0$. On the other hand, the inequality shows that

$$tb(\partial S) \leq -\chi(S) - |r(\partial S)| < 0,$$

which clearly is a contradiction.

In [32] Eliashberg showed that Bennequin's inequality holds for arbitrary tight structures. We will give a sketch of the proof for Eliashberg's result. The fact that the standard structure ξ_0 on S^3 is tight (Bennequin's theorem) will be obtained as a corollary to the theorem of Eliashberg and Gromov, which states that all fillable structures are tight (see section 8.6.3).

Theorem 8.36 ([32]). *Assume that γ is a transversal curve homological to zero in a tight contact manifold (M, ξ). Then for every embedded surface S with boundary γ*

$$l(\gamma) \leq -\chi(S)$$

(S represents a homology class $\mu \in H_2(M, \gamma)$ and l depends on μ).

Corollary 8.37 *For an embedded disc S in a tight manifold with a Legendre boundary ∂S avoiding the singular points the inequality*

$$tb(\partial S) \leq -\chi(S) - |r(\partial S)|$$

holds.

Proof of the corollary. Take a vector field X along ∂S which is transversal to ξ and tangent to S. The curve $\gamma^{\pm} = \partial S \pm \varepsilon X$ is (negatively respectively positively) transversal and by Bennequin's theorem 8.31

$$tb(\partial S) \pm r(\partial S) = l(\gamma^{\pm}) \leq -\chi(S).$$

□

The proof of the theorem given by [31] consists in showing that negative elliptic points of S can be cancelled via a small perturbation with (negative) hyperbolic points. This shows that $d_- \leq 0$. From Bennequin's theorem we have

$$l(\partial S) = -d_+ + d_-$$

and consequently

$$l(\partial S) \leq -d_+ - d_- = -\chi(S).$$

The deformation argument, which permits cancellation. is based on the elimination lemma 8.26. Eliashberg introduces the notion of a Legendrian polygon (roughly an immersed surface with piecewise Legendrian boundary and singularities at the vertices). After a small deformation, the basin of attraction $B(p)$ of an elliptic point is a Legendrian polygon. There must be a negative hyperbolic point on $\partial B(p)$, otherwise all the positive hyperbolic points could be cancelled leaving an embedded disc with a Legendrian boundary. But this is impossible since the contact structure is tight.

Let us observe that $d_- \leq 0$ generally only holds for embedded and not for immersed surfaces. An example is given by Bennequin ([11. p. 135]). The present proof therefore cannot be used for immersed surfaces.

8.6   Holomorphic filling

8.6.1   Bishop's theorem

Consider a manifold S of real dimension two embedded in $\mathbf{C}^2$. A point $p \in S$ is a *complex tangency,* if $T_p S$ is a complex subspace of $\mathbf{C}^2$. The embedded manifold S is *completely real,* if there are no complex tangencies. Generically, complex tangencies are isolated, they are either elliptic or hyperbolic (see below). If the manifold S is oriented, then a complex tangency is positive or negative according to whether the orientation of $T_p S$ agrees or disagrees with the orientation induced from $\mathbf{C}^2$. Generically, in a neighbourhood of a complex tangency p holomorphic coordinates (w, z) can be introduced such that near p the manifold S is represented as the graph of

$$z = w\overline{w} + 2\beta \mathrm{Re}\, w^2 + O(w^3)$$

with $0 \leq \beta < \frac{1}{2}$ in the elliptic and $\frac{1}{2} < \beta < \infty$ in the hyperbolic case. In a generic situation. the parabolic case $\beta = \frac{1}{2}$ will not occur.

The representation of S is obtained as follows: By a $\mathbf{C}$-linear change of coordinates in $\mathbf{C}^2$ it may be achieved that $p = (0, 0)$ and $T_p = \{z = 0\}$, such that locally S is given by

$$z = \alpha w^2 + \beta \overline{w}^2 + \gamma w\overline{w} + O(w^3).$$

If we make the generic assumption $\gamma \neq 0$, we can replace z by z/γ to obtain

$$z = \alpha w^2 + \beta \overline{w}^2 + w\overline{w} + O(w^3).$$

Choose ϑ so that $\beta e^{-2i\vartheta} \geq 0$ and replace w by $we^{i\vartheta}$ to arrive at $\beta \geq 0$. Finally substitute z by $z + (\alpha - \beta)w^2$ to get

$$z = -(\alpha - \beta)w^2 + \alpha w^2 + \beta \overline{w}^2 + w\overline{w} + O(w^3) = \beta(w^2 + \overline{w}^2) + w\overline{w} + O(w^3).$$

This shows that locally S can be represented as the graph of

$$z = w\overline{w} + 2\beta \mathrm{Re}\, w^2 + O(w^3)$$

with $0 \leq \beta < \infty$.

Theorem 8.38 (Bishop [13]) *Let p be an elliptic tangency of a 2-dim$_\mathbf{R}$ manifold S embedded in $\mathbf{C}^2$. Then there exists a continuous family $(w_e, z_e) : D \to \mathbf{C}^2$ of holomorphic discs with boundaries in S and such that their boundaries fill a neighbourhood of p in S. (The family depends on the real parameter e.)*

Proof. Set $Q(w) = w\overline{w} + 2\beta \mathrm{Re}\, w^2$ and consider the normalized Riemannian mapping

$$e\tau : D \to \{Q < k\} \qquad e > 0$$

between the disc $D = \{|\zeta| < 1\}$ and the ellipse $\{Q < k\}$. We normalize by $\tau(0) = 0$, $\tau'(0) = 1$ (see Figure 9).

For sufficiently small parameters $e > 0$, a holomorphic disc $(w_e, z_e) : D \to \mathbf{C}$ with boundary in S will be constructed. In the first approximation $w_e = e\tau$.

Fix $e > 0$ and set $(w, z) = (w_e, z_e)$. The function w will be of the form

$$w(\zeta) = e\tau(\zeta) + \zeta f(\zeta)$$

with f holomorphic in D, continuous in $\overline{D}$ and $f(0) = f'(0) = 0$. If we let g be the local representation for S:

$$g(w) = w\overline{w} + 2\beta \mathrm{Re}\, w^2 + \lambda(w),$$

then the crucial condition for the determination of $z(\zeta)$ is that $g(w(\zeta))$ restricted to ∂D be the boundary value of a holomorphic function.

$$
\begin{aligned}
g(w(\zeta)) &= Q(e\tau + \zeta f) + \lambda(e\tau + \zeta f) \\
&= Q(e\tau) + 2\beta(e\tau\zeta f + e\overline{\tau}\,\overline{\zeta f}) + e\tau\overline{\zeta f} + e\overline{\tau}\zeta f + Q(\zeta f) + \lambda(e\tau + \zeta f) \\
&= Q(e\tau) + 2e\,\mathrm{Re}\,[f(2\beta\tau\zeta + \overline{\tau}\zeta)] + Q(\zeta f) + \lambda_1(e\tau + \zeta f) \\
&\qquad + i\lambda_2(e\tau + \zeta f),
\end{aligned}
$$

where $\lambda = \lambda_1 + i\lambda_2$.

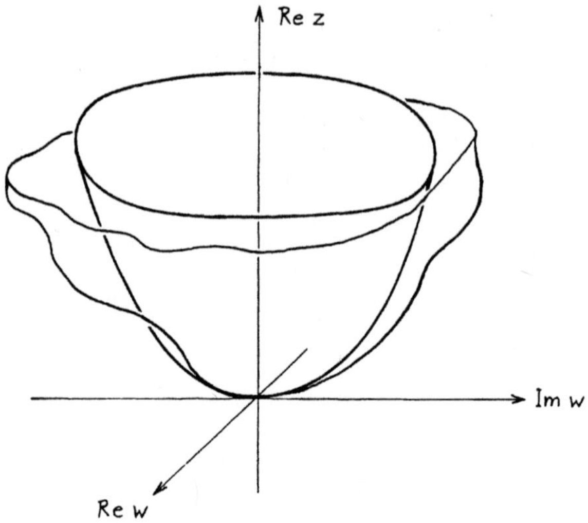

Figure 9

The *Hilbert transform* T defined for functions on ∂D maps the boundary values $u|_{\partial D}$ of the real part u of a holomorphic function $u + iv$ onto the boundary values $v|_{\partial D}$ of the imaginary part and is normalized by $T1 = 0$. The function

$$-T\lambda_2 + i\lambda_2$$

defined on ∂D is the boundary value of a holomorphic function on D. The condition for $g(w(\zeta))$ can therefore be restated as:

$$k\,e^2 + 2e\mathrm{Re}\,[f(2\beta\tau\zeta + \bar\tau\zeta)]|_{|\zeta|=1} + Q(\zeta f)|_{|\zeta|=1} + (\lambda_1 + T\lambda_2)(e\tau + \zeta f)|_{|\zeta|=1}$$

must be the boundary value of a holomorphic function. Since the right hand side is real, it must be constant. For the function restricted to ∂D this means

$$2e\mathrm{Re}\,[f(2\beta\tau\zeta + \bar\tau\zeta)] = C - Q(\zeta f) - (\lambda_1 + T\lambda_2).$$

The winding number of the image of ∂D under the mapping $2\beta\tau\zeta + \bar\tau\zeta$ is zero, since $2\beta < 1$. There exists therefore a holomorphic function γ with

$$\arg\gamma|_{\partial D} = \arg(2\beta\tau\zeta + \bar\tau\zeta)|_{\partial D}$$

and we can thus define a positive function δ on ∂D such that

$$(2\beta\tau\zeta + \bar\tau\zeta)|_{\partial D} = \delta\gamma|_{\partial D}.$$

The functional equation for f (restricted to ∂D) reduces to

$$2e\delta\,\mathrm{Re}\,(f\gamma) = C - Q(\zeta f) - \lambda_1 - T\lambda_2$$

and the constant C is implicitly defined by the condition $f(0) = 0$. The equation can formally be solved for $f\gamma$

$$f\gamma = (1 + iT)\frac{1}{2e\delta}(C - Q(\zeta f) - \lambda_1 - T\lambda_2)$$

and the condition for C translates into

$$\int_{|\zeta|=1} \frac{C}{\delta}\,d\vartheta = \int_{|\zeta|=1} \frac{Q(\zeta f) + \lambda_1 + T\lambda_2}{\delta}\,d\vartheta.$$

Choose now a suitable space of functions in which the functional equation can be solved, e.g. $H^{1,2}(\partial D)$ (this is the space chosen in Bishop's original treatment). In this space, the norm of the Hilbert transform is 1.

The functions λ_1 and λ_2 vanish to third order at the origin, therefore if $\|f_j\| \le a \le 1$ ($j = 1, 2$), then for $\lambda = \lambda_1$ or $\lambda = T\lambda_2$,

$$\|\lambda \circ (e\tau + \zeta f_1) - \lambda \circ (e\tau + \zeta f_2)\| = O((e + a)^2)\,\|f_1 - f_2\|.$$

The constants in the functional equation can be estimated as follows:

$$|C_1 - C_2|\int_{|\zeta|=1}\frac{d\vartheta}{\delta} \le \int_{|\zeta|=1}|Q(\zeta f_1) - Q(\zeta f_2)|\,d\vartheta +$$

$$\int_{|\zeta|=1}\frac{1}{\delta}|(\lambda_1 + T\lambda_2)\circ(e\tau + \zeta f_1) - (\lambda_1 + T\lambda_2)\circ(e\tau + \zeta f_2)|\,d\vartheta,$$

$$|C_1 - C_2| \le K_1\,a\,\|f_1 - f_2\| + K_2\,(e + a)^2\,\|f_1 - f_2\|$$

with constants K_j independent of $e \le 1$ and $a \le 1$. We are now in a position to prove that the mapping

$$f \mapsto \frac{1}{\gamma}(1 + iT)\frac{1}{2e\delta}(C - Q(\zeta f) - (\lambda_1 + T\lambda_2)\circ(e\tau + \zeta f))$$

is contracting in the domain $\{\|f\| \le a\}$, provided e and a are small enough. Taking into account that

$$\|Q(\zeta f_1) - Q(\zeta f_2)\| \le K_3\,\|f_1 - f_2\|\,a,$$

we obtain

$$\|\gamma f_1 - \gamma f_2\| \le \frac{K_4}{e}\left(a\,\|f_1 - f_2\| + (e + a)^2\,\|f_1 - f_2\|\right)$$

$$\|f_1 - f_2\| \le K_5\frac{a + (e + a)^2}{e}\,\|f_1 - f_2\|.$$

Choose e such that $K_5 e < 1/4$. Then for $a \le \frac{e}{3}K_5^{-1}$ ($\le \frac{e}{3}$) the mapping is contracting in the domain $\{\|f\| \le a\}$. The contraction factor is bounded by $\frac{1}{3} + K_5\frac{(4e/3)^2}{e} \le \frac{1}{3} + \frac{4}{9} = 1$. $\qquad\square$

8.6.2   One-parameter families of holomorphic discs

Let S be a sphere embedded in a symplectic manifold (M, ω) with a compatible complex structure J. We consider the problem of finding a 3-dimensional ball B such that B is foliated by holomorphic discs whose boundaries fill up $S^* = S \setminus \{\text{points of complex tangency}\}$.

The theorem of Bedford and Gaveau deals with the situation that a holomorphic disc $f : D \to M$ with $f(\partial D) \subset S^*$ is prescribed. The assertion then is that there is a whole family of holomorphic discs close to f whose boundaries foliate a neighbourhood of $f(\partial D)$ in S^*. Bedford and Gaveau [9] proved the theorem in the context of holomorphic embeddings into $\mathbf{C}^2$ (with standard structure). the surface S being a graph in a strictly pseudoconvex hypersurface in $\mathbf{C}^2$.

Hofer [55] extended this theorem to the present more general situation. Previously some general variants of the theorem were stated by [50] and [30]. In Hofer's version the surface S is still embedded in a contact manifold M which itself is naturally embedded in its symplectification $W = \mathbf{R} \times M$. Yet the complex structure which goes with it is in general not integrable and the surface S is not necessarily a graph in the contact manifold M. As it turns out, the prescribed surface S can be any compact orientable 2-dimensional completely real surface embedded in a 4-dimensional symplectic manifold (W, ω) with a compatible complex structure J, such that the boundary $\partial f D$ of the initial holomorphic disc will not meet the boundary of S. Note, however, that the index hypothesis formulated below will further restrict the situation.

Let $f : \bar{D} \to W$ be an embedded holomorphic disc with $f(\partial D)$ contained in the completely real compact surface $S \subset W$, $f(\partial D) \cap \partial S = \emptyset$. Then over the boundary ∂D the pullback bundle $(f|_{\partial D})^* TS$ is a totally real subbundle of $(f|_{\partial D})^* TW$. Via a trivialization $\psi : f^* TW \to \bar{D} \times \mathbf{C}^2$ the subbundle $(f|_{\partial D})^* TS$ determines a closed curve $L : S^1 \to G_{2,2}$ in the Grassmann manifold $G_{2,2}$ of 2-dimensional oriented real subspaces of $\mathbf{C}^2$. The *index* $\mathrm{ind}\,(f, S)$ of f with respect to S is defined to be the Maslov index of L. To be more explicit, take the tangent vector field T along $f(\partial D)$ and choose a vector field $V \in TS$, defined along $f(\partial D)$. such that $T \wedge V$ determines the orientation of S. Then via the trivialization ψ the pair (T, V) describes an element of $\pi_1(GL(2, \mathbf{C}))$ and the index of f is the element in $\mathbf{Z}$ under the isomorphism $\pi_1(GL(2, \mathbf{C})) \cong \mathbf{Z}$. The index $\mathrm{ind}\,(f, S)$ is independent of the choice of the trivialization ψ.

For $p \in (2, \infty)$ denote by $\mathcal{B}$ the Banach manifold of all mappings $f \in H^{1,p}(\bar{D}, W)$ with $f(\partial D) \subset S$. Let $L^p(f) = L^p(f^* TW)$ be the space of all L^p-sections along f. Then as in chapter 6, $\mathcal{E} = \cup_{f \in \mathcal{B}} \{f\} \times L^p(f)$ is a Banach space bundle over $\mathcal{B}$ and the $\bar{\partial}$-mapping associated with the complex structure J on W

$$\bar{\partial} : f \mapsto f_x + J(f) f_y$$

is a section of $\mathcal{E}$ over $\mathcal{B}$. (For the disc $D = \{|z| < 1\}$ the parameter $z = x + iy$ is used).

Theorem 8.39 ([55]) *Assume that $u_0 : \bar{D} \to W$ is an embedded holomorphic disc ($\bar{\partial} u_0 = 0$ on D) and $u_0(\partial D) \subset S \setminus \partial S$, where S is an embedded completely real compact surface. Assume further that $\mathrm{ind}(u_0. S) = 1$. Then there exists a smooth embedding $\varphi : (-\varepsilon, \varepsilon) \times \bar{D} \to W$ such that*

$$\begin{aligned}
\varphi(0, \cdot) &= u_0 \\
\varphi(\tau, \cdot) &\in \mathcal{B} \qquad \text{for all } \tau \in (-\varepsilon, \varepsilon) \\
\bar{\partial}\varphi(\tau, \cdot) &= 0 \qquad \text{on } D.
\end{aligned}$$

Remark. In [56] the definition of the index is based on the Grassmannian of unoriented 2-planes, whereas the index used here comes from the Grassmannian of oriented 2-planes. The two indices differ by a factor of 2.

Proof. There exists a smooth chart $\sigma : U \to \mathbf{C}^2$ defined in a neighbourhood of $u_0(\bar{D})$ such that

$$\sigma \circ u_0(z) = (z, 0), \quad z \in \bar{D}$$

and the tangent mapping $T\sigma$ satisfies

$$T\sigma \circ J = i \, T\sigma$$

at all points $u_0(z), z \in \bar{D}$. The second equation says that along u_0 the complex structure J pulls back to the standard structure on $\mathbf{C}^2$. Within the disc $\bar{D} = \bar{D} \times \{0\}$ this can be achieved, since u_0 is holomorphic. The mapping σ^{-1} has then to be adjusted such that the tangent vectors $(0. 1)$ and $(0, i) \in T_{(z,0)}\mathbf{C}^2$ at points $(z, 0) \in \bar{D} \times \{0\}$ are mapped onto vectors in $T_{u_0(z)}W$ satisfying

$$T\sigma^{-1}(0, i) = J T\sigma^{-1}(0. 1).$$

Denote by $\hat{J}$ the pull back under σ^{-1} of the complex structure J

$$\hat{J} = T\sigma \circ J \circ T\sigma^{-1}$$

and by $\hat{S}$ the image of $S \cap U$ under σ. The index of u_0 is then given as the index

$$\mathrm{ind}\,(u_0, S)$$

of the closed curve $e^{i\vartheta} \mapsto T_{(e^{i\vartheta}, 0)}\hat{S}$ of real planes in $\mathbf{C}^2$.

It must be shown that there exists a smooth family $\sigma \circ \varphi : (-\varepsilon, \varepsilon) \times \bar{D} \to \mathbf{C}^2$ of $\hat{J}$-holomorphic mappings with boundary values in $\hat{S}$. The mapping $u : D \to \mathbf{C}^2$ is $\hat{J}$-holomorphic if

$$u_x + \hat{J}(u)u_y = 0.$$

The linearization of this equation at $u = (id, 0) = \sigma \circ u_0$ is then obtained in the following standard fashion:

$$\begin{aligned}
u(z) &= (z, 0) &+& \;\varepsilon(h, k)(z) &+& \;O(\varepsilon^2), \\
\hat{J}(u) &= \hat{J}(z, 0) &+& \;\varepsilon D_2\hat{J}(z, 0)(0. k). &+& \;O(\varepsilon^2)
\end{aligned}$$

since $D_1 \hat{J}(z, 0) = 0$.

$$
\begin{aligned}
0 &= u_x + \hat{J}(u)u_y \\
&= \varepsilon\big((h, k)_x + \hat{J}(z,0)(h,k)_y + D_2\hat{J}(z,0)(0,k)(z,0)_y\big) + O(\varepsilon^2).
\end{aligned}
$$

This gives for the linearization

$$
(h, k)_x + i(h, k)_y + (Ak, Bk) = 0,
$$

where A and B are real linear mappings.

The linearization of the boundary condition

$$
(z, 0) + \varepsilon(h, k) + O(\varepsilon^2) \in \hat{S}, \quad z \in \partial D
$$

is given by

$$
(h, k)(z) \in T_{(z,0)}\hat{S}, \quad z \in \partial D.
$$

$\hat{S}$ is completely real and $T_{(z,0)}\hat{S}$ contains the tangent vector $X := (iz, 0)$. Suppose Y is a vector field along ∂D such that X and Y span $T_{(z,0)}\hat{S}$ at every point $z \in \partial D$. Since $T_{(z,0)}\hat{S}$ is completely real, the second component pr_2Y of the vector field Y never vanishes. The assumption $\mathrm{ind}\,(u_0, S) = 1$ is equivalent to the statement that the winding number of $pr_2Y : \partial D \to \mathbf{C}$ is zero.

Consider a closed path $z \mapsto \Gamma(z)$, $z \in \partial D$, of totally real oriented subspaces of $\mathbf{C}^2$ and set

$$
V_p^\Gamma = \{(h, k) \in H^{1,p}(\bar{D}, \mathbf{C}^2) : (h(z), k(z)) \in \Gamma(z) \text{ for almost all } z \in \partial D\}.
$$

For $2 < p < \infty$, V_p^Γ is a Banach space. To continue the proof, we now make use of the following theorem of Hofer [56].

Theorem 8.40 *Assume that* $(iz, 0) \in \Gamma(z)$ *for all* $z \in \partial D$ *and* $\mathrm{ind}\,\Gamma = 1$. *Then the solution set* $L \subset V_p^\Gamma$ *of the partial differential equation*

$$
\begin{aligned}
h_x + ih_y + A(z)k &= 0 & \text{on } D \\
k_x + ik_y + B(z)k &= 0 & \text{on } D \\
(h(z), k(z)) \in \Gamma(z) & & \text{on } \partial D
\end{aligned}
$$

is 4-dimensional. Moreover every solution for which $k(z_0) \neq 0$ *for some* $z_0 \in \bar{D}$ *satisfies* $k(z) \neq 0$ *for all* $z \in \bar{D}$.

Applying this theorem to the linearized equation we see that the solution set is given by

$$
\begin{aligned}
L &= \mathbf{R}(h_0, k_0) \oplus \{(c + irz - \bar{c}z^2, 0) : r \in \mathbf{R}, c \in \mathbf{C}\} \\
&= L_1 \oplus L_2.
\end{aligned}
$$

Note that L_2 describes the Lie algebra of the Möbius group $PSL(2, \mathbf{R})$ acting on the unit disc. According to the theorem, $k_0(z) \neq 0$ in all of $\bar{D}$.

We now identify vector fields in a neighbourhood U of the zero section in $T\mathbf{C}^2$ with mappings $\mathbf{C}^2 \to \mathbf{C}^2$. As an identification mapping $\psi : U \to \mathbf{C}^2$ choose a smooth mapping such that its restriction ψ_x to U_x (the fiber at x intersected with U) is an embedding with

$$\psi_x(0) = x, \quad T\psi_x(0) = id \qquad (x \in \mathbf{C}^2)$$

and moreover such that for $z \in \partial D$,

$$\psi_{(z,0)} T_{(z,0)} \hat{S} \subset \hat{S}.$$

Look then for solutions u_τ of the equation $\bar{\partial} u_\tau = 0$ with $u_\tau(\partial D) \subset \hat{S}$ which are of the form

$$u_\tau(z) = \psi(\tau(h_0(z), k_0(z)) + \Delta_\tau(z)).$$

Here τ is a real parameter and Δ_τ a "small" vector field. This vector field is required to be orthogonal to L with respect to a suitably chosen inner product on $\bigcap_{2 < p < \infty} V_p^\Gamma$. By the implicit function theorem there is an $\varepsilon > 0$ such that for all $|\tau| < \varepsilon$ there exists a unique $\Delta_\tau \in \bigcap_{2 < p < \infty} V_p^\Gamma$ such that

$$(u_\tau)_x + \hat{J}(u_\tau)(u_\tau)_y = 0.$$

In addition, Δ is a smooth map $(-\varepsilon, \varepsilon) \to V_p^\Gamma$ $(2 < p < \infty)$ with $\Delta_0 = 0$ and $\frac{\partial}{\partial \tau} \Delta_\tau|_{\tau=0} = 0$. By elliptic regularity theory, Δ is C^∞-smooth if, as we assume, the complex structure J and the surface S are C^∞-smooth. Since $k_0(z) \neq 0$ for all $z \in \bar{D}$ we conclude that $\varphi(\tau, z) = \sigma^{-1} u_\tau(z)$ is an embedding (possibly after restricting the parameter domain for τ), satisfies $\varphi(0, z) = u_0(z)$ and solves $\bar{\partial}\varphi(\tau, \cdot) = 0$. $\qquad\qquad\square$

The solution φ to this problem is unique up to reparametrizations (see [55, Thm. 8]).

8.6.3   Fillable contact structures

Definition 8.41 *A compact 3-manifold with oriented contact structure ξ is holomorphically fillable if it is the J-convex boundary $M = \partial\Omega$ of a compact symplectic 4-manifold (Ω, ω) with complex structure J tamed by ω.*

We shall usually assume that Ω is a domain contained in a bigger symplectic manifold (W, ω) and that J is defined in all of W. Recall that J is *tamed* by ω if $\omega(X, JX) > 0$ for all tangent vectors $X \neq 0$. It is not assumed that ω be J-invariant.

Boundaries of strictly pseudoconvex bounded domains in $\mathbf{C}^2$ or more generally in a Kähler manifold are typical examples of holomorphically fillable contact manifolds. According to proposition 8.14, the hypersurfaces of contact type $M = \partial\Omega$, with Ω a domain with compact closure in a symplectic manifold can be equipped with a holomorphically fillable contact structure. In contrast to these examples, note that though symplectification yields an embedding of (M, ξ) into a symplectic manifold $W = M \times \mathbf{R}$, the manifold M does not appear as the boundary of a bounded domain in W.

Throughout this section, (M, ξ) will be a holomorphically fillable contact manifold. Our main aim is to discuss the proof of the following theorem due to Gromov [50], see also Eliashberg [27].

Theorem 8.42 *Holomorphically fillable contact structures are tight.*

Corollary 8.43 *The standard structure ξ_0 on the sphere S^3 is tight.*

This is a fundamental result of Bennequin [11, Corollaire 2, p. 150]. The proof given by Bennequin uses his famous inequalities (see section 8.5.4 for a detailed discussion).

The scheme of the proof is the following: Assume that the manifold (M, ξ) is overtwisted. Then there exists an embedded disc D with an elliptic singular point $p \in D$ such that its boundary ∂D is an embedded Legendrian curve and such that there are no singular points on $\bar{D}$ apart from p. According to [27, p. 45] (see also [28, p.629]) a neighbourhood U of $\bar{D}$ can then be chosen such that $S = \partial U$ is homeomorphic to the sphere and its characteristic foliation has exactly two singular points $p_{\pm}$ which are elliptic and of opposite sign. Furthermore the foliation has two limit cycles.

Use then that (M, ξ) is holomorphicaly fillable in order to construct a family f_t, $t \in (-1, 1)$, of holomorphic discs $f_t : (\Delta, \partial\Delta) \to (\Omega, S)$, which fill S. (Here $\Delta \subset \mathbf{C}$ denotes the unit disk.) This means that

$$f : (-1, 1) \times \bar{\Delta} \to \bar{\Omega}$$

is an embedding, $f_t|_\Delta$ is holomorphic, $\lim_{t \to \pm 1} f_t = p_\pm$ and the curves $f_t(\partial D)$, $t \in (-1, 1)$, foliate $S \setminus \{p_+, p_-\}$.

Now M is a J-convex manifold. Holomorphic discs $f_t : (\Delta, \partial\Delta) \to (\Omega, S)$ will not touch M at an interior point nor will the tangent plane to the disc at a boundary point coincide with a contact plane ξ in M (the holomorphic discs intersect $M = \partial\Omega$ transversally). Therefore the characteristic foliation μ of S has to be transversal to the foliation $f_t(\partial\Delta)$, $t \in (-1, 1)$. But this is impossible if the characteristic foliation μ has limit cycles. This contradiction shows that there are no embedded overtwisted discs in M. The contact structure is tight.

The method of filling by holomorphic discs was developed by Bedford and Gaveau in the classical setting. They showed that a surface S which appears as a graph in the boundary of a strictly pseudoconvex domain in $\mathbf{C}^2$ can always be filled by holomorphic discs, provided it contains only two elliptic singular points of opposite sign. The compactness argument of Bedford and Gaveau was based on a priori estimates. In the general case of a nonintegrable complex structure, compactness will be a consequence of Gromov's theorem (see below). The method has been developped further by Bedford and Klingenberg [10], who showed that holomorphic filling is also possible in the presence of hyperbolic singularities.

In the nonintegrable case the theorem on filling by holomorphic discs and its applications are due to Gromov [50] and Eliashberg [31] and [35].

Theorem 8.44 *Assume that S is a 2-sphere embedded in a holomorphically fillable contact manifold $M = \partial\Omega$ such that S has exactly two singular points p_+ and p_- (which necessarily are elliptic and of opposite sign). If the complex structure J is integrable near the singular points, then S admits a filling*

$$f : (-1, 1) \times \bar{\Delta} \to \bar{\Omega}$$

by holomorphic discs: f is an embedding, $f_t = f(t, \cdot)$ is holomorphic in Δ for all t, the curves $f_t(\partial D)$ foliate S and $\lim_{t \to \pm 1} f_t = p_\pm$.

The theorem should be true without the additional hypothesis that J is integrable near the singular points. However, so far all efforts to generalize Bishop's theorem to the nonintegrable case failed. Up to some modifications our proof follows Eliashberg's reasoning in [35].

By Bishop's theorem (section 8.6.1) there exist families of embedded holomorphic discs f_t and $g_s : (\Delta, \partial\Delta) \to (\Omega, S)$, $t \in (-1, -1 + \varepsilon), s \in (1 - \varepsilon, 1)$ with $\lim_{s \to 1} g_s = p_+$ and $\lim_{t \to -1} f_t = p_-$. Observe that the index $\mathrm{ind}\,(f_t, S)$ for the discs in Bishop's family is equal to 1.

If $f_{t_0} : (\Delta, \partial\Delta) \to (\Omega, S)$ is an embedded holomorphic disc with $\mathrm{ind}\,(f_{t_0}, S) = 1$, then by the implicit function theorem in Hofer's setting (Theorem 8.39) there exists an open interval $I \ni t_0$ and an embedding $\varphi : I \times \bar{\Delta} \to \Omega$ such that for all $t \in I$, $\varphi_t(\Delta)$ is a holomorphic disc with boundary in S. Furthermore, φ is uniquely determined. This shows that the Bishop family starting at say -1 can be uniquely continued.

Next apply Gromov's compactness theorem for families of holomorphic discs with boundaries in $S \setminus \{p_+, p_-\}$. (See [112] for a thorough discussion). In the generic case (and the complex structure allows for slight deformations off the surface S) bubbling off cannot occur in the interior of the discs. That bubbles cannot split off the boundaries is due to the fact that $f_t(\partial\Delta)$ always has to be transversal to the characteristic foliation.

As a result, the family of holomorphic discs starting at -1 can be continued until at some point t_0 the holomorphic disc f_{t_0} hits a holomorphic disc, say g_s from the Bishop family starting at p_+. In this case, f_{t_0} will have to coincide with g_s and the family of holomorphic discs extends to a filling of S. The argument that upon touching, f_{t_0} and g_s have to coincide is of a subtle nature. That the surfaces cannot touch at an interior point follows from results of [75]. A proof for the fact that they cannot touch at boundary points is given by Eliashberg [35, prop. 2.2.A]. In the present situation, it should also be possible to adapt the proof of Bedford and Gaveau, since the touching would occur in a region where by hypothesis the complex structure is integrable.

Let us now show how the theorem of filling by holomorphic discs can be applied. Assume that p is an elliptic singular point on the 2-dimensional suface S embedded in the contact manifold $M = \partial\Omega$. Near p the complex structure J can be changed into an integrable one which is still tamed by ω and such that the characteristic foliation μ on S is unchanged. For the proof of this statement, consider the standard Heisenberg structure on the contact manifold $H = \partial D$

$$D = \{(z_1, z_2) \in \mathbf{C}^2 : \operatorname{Im} z_2 - |z_1|^2 > 0\}.$$

It is described by the form

$$\vartheta := (\overline{z_1}dz_1 - z_1 d\overline{z_1}) - dt \qquad (t = \operatorname{Re} z_2).$$

By Darboux' theorem there exists a contact mapping $\varphi : U \to H$ defined in a neighbourhood U of p with $\varphi(p) = 0$. The tangent mapping φ_* restricted to ξ at p will in general not be complex linear, but φ can be composed with a contact transform $\lambda : H \to H$ such that at p, $(\lambda \circ \varphi)_*|_\xi$ is complex linear. (λ can be chosen as a quasiconformal mapping with constant complex dilation). Choose next an extension φ of the contact mapping $\lambda \circ \varphi$ to a neighbourhood of p in Ω such that at the point p the tangent mapping is holomorphic.

$$J_0 \varphi_*(p) = \varphi_*(p)J,$$

where J_0 is the standard complex structure in $\mathbf{C}^2$. The pullback structure $\tilde{J} = \varphi_*^{-1} J_0 \varphi_*$ then defines an integrable structure in a neighbourhood of $p \in \bar{\Omega}$ with $\tilde{J}(p) = J(p)$ and the contact structure ξ induced by $\tilde{J}$ is unchanged. The symplectic form ω is tamed by J and therefore by all complex structures sufficiently C^0-close to J. The new complex structure can thus be chosen equal to $\tilde{J}$ in a neighbourhood of p and equal to J outside a slightly bigger neighbourhood. At intermediate points interpolate between the two complex structures such that ξ remains a complex invariant subspace.

The previous argument which showed that holomorphically fillable contact structures are tight used the method of filling by holomorphic discs. We just showed that the additional hypothesis on integrability of the complex structure which appears in the theorem on filling by holomorphic discs can be dealt with.

8.7 Eliashberg's classification of 3-dimensional contact structures

In 1989 Eliashberg [28] classified overtwisted contact structures on compact 3-manifolds and in 1992 he showed [33] that the standard structure is the only tight structure on S^3 (up to isotopies). For general compact 3-manifolds the classification problem for tight structures remains largely open. The classification for contact structures on $\mathbf{R}^3$ was completed in 1993 [34].

Let us state (without proof) the relevant theorems:

Theorem 8.45 ([28]) *The classification of overtwisted oriented contact structures on a compact 3-manifold up to isotopy (fixed near the boundary ∂M) coincides with the homotopy (rel ∂M) classification of oriented tangent plane fields.*

As we have observed in section 8.4.1, the homotopy classification of oriented tangent plane fields on S^3 is given by $\pi_3(S^2) \cong \mathbf{Z}$. In every homotopy class $[\alpha_k]$. $k \in \mathbf{Z}$, of oriented tangent plane fields on S^3. there exists thus up to isotopies exactly one positive (and one negative) overtwisted contact structure $\zeta_k, k \in \mathbf{Z}$.

Theorem 8.46 ([33]) *Any tight positive contact structure on S^3 is isotopic to the standard contact structure ξ_0.*

In the homotopy class $[\alpha_0]$ of oriented tangent plane fields on S^3 there exist therefore (up to isotopies) exactly two positive contact structures, the tight (ξ_0) and the overtwisted (ζ_0).

Theorem 8.47 ([33]) *The space of tight contact structures on S^3 fixed at a point $p \in S^3$ is contractible.*

From this it can be derived that any contact transformation $h : S^3 \to S^3$ (with respect to the standard structure ξ_0) arises as the time-one map h_1 of a flow h_t. $t \in [0,1]$, of contact transformations with $h_0 = id$.

According to a result of Cerf (see [17] and the discussion in [33]) a flow f_t of diffeomorphisms of S^3 exists with $f_0 = id$ and $f_1 = h$. It can be assumed that h and f_t, $t \in [0,1]$, fix the contact structure ξ_0 at a point $p \in S^3$. Apply then the contraction mapping φ_s from the theorem to the pullback structures $\xi_t = f_{t*}\xi_0$ (with $\xi_1 = \xi_0$). The contraction φ_s will then map ξ_t along a path $\varphi_s(\xi_t). 0 \leq s \leq s_0$, to the standard structure ξ_0 and $\varphi_s(\xi_t)$ depends smoothly on t and s. According to Gray's theorem 8.2 there exist mappings g_t with $g_{t*}\xi_0 = \xi_t$. Observe that the construction of these mappings can be normalized such that the mappings g_t themselves are uniquely determined by the paths $\varphi_s(\xi_t)$. They will depend smoothly on t and $g_t = id$ if $\varphi_s(\xi_t) \equiv \xi_0$. Therefore $h_t = g_t^{-1} \circ f_t$ is a contact flow with $h_0 = id$, $h_1 = h$.

The contact flow h_t extends to a symplectic flow of (B, ω). where ω is the $SU(2,1)$-invariant Kähler form on the ball B.

Eliashberg's proof of the previous theorem (and of Cerf's theorem) actually goes the other way around. Using the method of filling by holomorphic discs, he shows ([33, Lemma 6.2]) that any contact automorphism h of (S^3, ξ_0) extends as a diffeomorphism to all of B. In combination with Theorem 8.2 this then also gives a proof of Cerf's theorem.

A Generalities on Homology and Cohomology

This section should be skipped by the reader familiar with the subject. For reference we will just summarize the main properties without motivation and proof. The reader is referred to the extensive literature on algebraic topology, e.g. [23], [97]. Homology and cohomology are functors, ways of converting topological problems into algebraic ones. They assign abelian groups to topological spaces and homomorphisms to continuous maps of one space to another. The main properties of these functors are summarized by the *Eilenberg-Steenrod axioms*, which will be listed below.

A *pair of spaces* (X, A) is a topological space X together with a subset $A \subseteq X$ with the relative topology. A *map* $f : (X, A) \to (Y, B)$ is a continuous map $f : X \to Y$ such that $f(A) \subseteq B$. It is customary to abbreviate $(X, \emptyset)$ by (X) or just X.

Definition. A family of pairs of spaces and maps of such pairs is called an *admissible category for homology theory* if it satisfies the following five conditions.

(1) If $(X, A) \in A$ then all pairs and inclusion maps in the following diagram, called the *lattice* of (X, A)

$$
\begin{array}{ccccc}
 & & (X, \emptyset) & & \\
 & \nearrow & & \searrow & \\
(\emptyset, \emptyset) \to (A, \emptyset) & & & & (X, A) \to (X, X) \\
 & \searrow & & \nearrow & \\
 & & (A, A) & &
\end{array}
$$

are in A.

(2) If $f : (X, A) \to (Y, B)$ is in A, then (X, A) and (Y, B) are in A together with all maps that f defines of members of the lattice of (X, A) to corresponding members of the lattice of (Y, B).

(3) Whenever the composition of two maps in A is defined, it is in A.

(4) If $I = [0, 1]$ and $(X, A) \in A$, then the cartesian product $(X, A) \times I = (X \times I, A \times I)$ is in A and the maps

$$
g_0, g_1 : (X, A) \to (X, A) \times I
$$

defined by $g_0(x) = (x, 0)$, $g_1(x) = (x, 1)$ are in A.

(5) A contains a space P_0 consisting of a single point. If $X, P \in A$, $f : P \to X$ and P is a single point, then $f \in A$.

Examples of admissible categories are:

A_1 = all pairs (X, A) and all maps of such pairs. This is the largest admissible category.

$\mathcal{A}_c$ = all pairs (X, A) with X compact, A closed in X, and all maps of such pairs.

$\mathcal{A}_{lc}$ = all pairs (X, A) where X is a locally compact Hausdorff space, A is closed in X, and all proper maps of such pairs.

A.1   Axioms for homology

A homology theory on an admissible category $\mathcal{A}$ consists of three kinds of objects.

(1) an R-module $H_k(X, A)$ for every pair $(X, A) \in \mathcal{A}$ and every integer $k \geq 0$. It is called the *k-dimensional relative homology group of X modulo A.* Here, R is a commutative ring with a unit element. When one wants to specify R one speaks of homology *with coefficients in R.*

(2) For each map $f : (X, A) \to (Y, B)$ in $\mathcal{A}$ and each nonnegative integer k, a homomorphism

$$f_* : H_k(X, A) \to H_k(Y, B),$$

called the *homomorphism induced by f.*

(3) For each $(X, A) \in \mathcal{A}$ and each nonnegative integer k, a homomorphism

$$\partial : H_k(X, A) \to H_{k-1}(A),$$

called the *boundary operator,* where by definition $H_{-1}(A) = 0$.

The following are the Eilenberg-Steenrod axioms which must be fulfilled for a homology theory.

Axiom 1. If $f = $ identity, then $f_* = $ identity.
Axiom 2. $(gf)_* = g_* f_*$.
Axiom 3. $\partial f_* = (f|A)_* \partial$, i.e. the diagram

$$
\begin{array}{ccc}
H_k(X, A) & \xrightarrow{f_*} & H_k(Y, B) \\
\partial \downarrow & & \downarrow \partial \\
H_{k-1}(A) & \xrightarrow{(f|A)_*} & H_{k-1}(B)
\end{array}
$$

commutes.

Axiom 4. (Exactness axiom) If (X, A) is admissible and $i : A \hookrightarrow X$, $j : X \hookrightarrow (X, A)$ are inclusion maps, then the sequence, called the *homology sequence of (X, A),*

$$\ldots H_k(A) \xrightarrow{i_*} H_k(X) \xrightarrow{j_*} H_k(X, A) \xrightarrow{\partial} H_{k-1}(A) \xrightarrow{i_*} \ldots \xrightarrow{j_*} H_0(X. A) \xrightarrow{\partial} 0$$

is exact.

Axiom 5. (Homotopy axiom) If the admissible maps $f, g : (X, A) \to (Y, B)$ are homotopic in $\mathcal{A}$. then for all k the homomorphisms $f_*, g_* : H_k(X, A) \to H_k(Y, B)$ coincide.

Axiom 6. (Excision axiom) If $U \subset X$ is open, $\bar{U} \subset \operatorname{int} A$ and the inclusion map $(X \setminus U, A \setminus U) \hookrightarrow (X, A)$ is admissible, then it induces isomorphisms $H_k(X \setminus U, A \setminus U) \cong H_k(X, A)$ for all k.
Axiom 7. (Dimension axiom) If P is an admissible space consisting of a single point, then $H_k(P) = 0$ for $k \neq 0$.

The axioms are consistent since they are satisfied if $H_k(X, A) = 0 \, \forall \, k$. To proof the existence of nontrivial homologies one has to actually construct some homology, like the singular, Čech, etc.

Axioms 5 and 6 have the following equivalent versions.

Axiom 5'. If (X, A) is admissible and $f, g : (X, A) \to (X, A) \times I$ are defined by $f(x) = (x, 0)$, $g(x) = (x, 1)$, then $f_* = g_*$.
Axiom 6'. If $X = \operatorname{int} X_1 \cup \operatorname{int} X_2$ for subsets X_1 and X_2 of X, X_1 closed and the inclusion $i : (X_1, X_1 \cap X_2) \hookrightarrow (X, X_2)$ is admissible, then it induces isomorphisms $i_* : H_k(X_1, X_1 \cap X_2) \cong H_k(X, X_2)$ for each k.

A.2   Axioms for cohomology

In going from homology to cohomology, $H_k(X, A)$ is replaced by $H^k(X, A)$ which is again an R-module, $f_* : H_k(X, A) \to H_k(Y, B)$ is replaced by the homomorphism $f^* : H^k(Y, B) \to H^k(X, A)$ and ∂ is replaced by the homomorphism $\delta : H^{k-1}(A) \to H^k(X, A)$, the *coboundary operator*.

Axiom 1c. If $f = \text{identity}$, then $f^* = \text{identity}$.
Axiom 2c. $(gf)^* = f^* g^*$.
Axiom 3c. $\delta(f|A)^* = f^* \delta$.
Axiom 4c. If $i : A \hookrightarrow X$ and $j : X \hookrightarrow (X, A)$ are admissible maps, then the *cohomology sequence of* (X, A)

$$0 \xrightarrow{\delta} H^0(X, A) \ldots \xrightarrow{\delta} H^k(X, A) \xrightarrow{j^*} H^k(X) \xrightarrow{i^*} H^k(A) \xrightarrow{\delta} H^{k+1}(X, A) \ldots$$

is exact.
Axiom 5c. If $f, g : (X, A) \to (Y, B)$ are homotopic (and admissible), then $f^* = g^*$.
Axiom 6c. If U is open in X, $\bar{U} \subset \operatorname{int} A$ and the inclusion map $i : (X \setminus U, A \setminus U) \hookrightarrow (X, A)$ is admissible, then it induces isomorphisms $H^k(X, A) \cong H^k(X \setminus U, A \setminus U)$.
Axiom 7c. If the admissible space P consists of a single point, then $H^k(P) = 0$ for $k \neq 0$.

If the ring R equals $\mathbf{Z}$, then homology and cohomology are dual: Letting $H_k(X, A)$ be the character group of $H^k(X, A)$ and f_*, ∂ the homomorphisms dual to f^*, δ yields a system satisfying axioms 1 to 7. The dual of each theorem about $\{H_k, f_*, \partial\}$ is a true theorem about $\{H^k, f^*, \delta\}$. Passing from a theorem to its dual reverses arrows, replaces subgroups by factor groups and vice versa.

For a general ring with unit there is only a partial duality. The dual of each theorem on homology is true for cohomology in general only as far as the additive structure of the R-modules $H^k(X, A)$ is concerned.

A.3   Homomorphisms of (co)homology sequences

A map $f : (X, A) \to (Y, B)$ defines maps

$$f_1 : X \to Y, \quad f_2 = (f_1|A) : A \to B.$$

The collection of induced homomorphisms f_*, f_{1*}, f_{2*} will be called f_{**}. It forms a homomorphism of the homology sequence of (X, A) into that of (Y, B), i.e. the following diagram commutes:

$$
\begin{array}{ccccccccc}
\to & H_{k+1}(X.A) & \xrightarrow{\partial} & H_k(A) & \xrightarrow{i_*} & H_k(X) & \xrightarrow{j_*} & H_k(X.A) & \to \\
 & \downarrow f_* & & \downarrow f_{2*} & & \downarrow f_{1*} & & \downarrow f_* & \\
\to & H_{k+1}(Y.B) & \xrightarrow{\partial} & H_k(B) & \xrightarrow{i'_*} & H_k(Y) & \xrightarrow{j'_*} & H_k(Y,B) & \to
\end{array}
$$

This follows immediately from axioms 2 and 3. For cohomology there is an analogous homomorphism f^{**} of cohomology sequences.

Theorem A.1 ([23. Thm. 4.2]) *Let $f : (X, A) \to (Y, B)$ be admissible. If $f_{1*} : H_k(X) \to H_k(Y)$ and $f_{2*} : H_k(A) \to H_k(B)$ are isomorphisms for all k, then so are $f_* : H_k(X, A) \to H_k(Y, B)$ and f_{**} is an isomorphism.*

The dual theorem for cohomology is true, too.

A.4   The (co)homology sequence of a triple

Suppose $B \subseteq A \subseteq X$ and the inclusions

$$i : (A, B) \hookrightarrow (X, B), \quad j : (X, B) \hookrightarrow (X, A)$$

are admissible. Then (X, A, B) is called an *admissible triple*. Let $j' : A \hookrightarrow (A, B)$ be the inclusion, $\partial : H_k(X, A) \to H_{k-1}(A)$ the boundary operator and $\delta : H^{k-1}(A) \to H^k(X.A)$ the coboundary operator.

Definition. The *boundary operator of the triple* (X, A, B) is the homomorphism

$$\bar{\partial} : H_k(X, A) \to H_{k-1}(A, B)$$

defined by $\bar{\partial} = j'_* \partial$. The *coboundary operator for the triple* $(X.A, B)$ is the homomorphism

$$\bar{\delta} : H^{k-1}(A, B) \to H^k(X, A)$$

defined by $\bar{\delta} = \delta \, j'^*$.

Theorem A.2 ([23, Thms. 10.2, 10.2c]) *The* homomology sequence *of the triple* $(X.A,B)$

$$\ldots \to H_k(A,B) \xrightarrow{i_*} H_k(X,B) \xrightarrow{j_*} H_k(X,A) \xrightarrow{\bar{\partial}} H_{k-1}(A,B) \to \ldots$$

and the cohomology sequence *of the triple* $(X.A.B)$

$$\ldots \to H^{k-1}(A,B) \xrightarrow{\bar{\delta}} H^k(X,A) \xrightarrow{j^*} H^k(X.B) \xrightarrow{i^*} H^k(A,B) \to \ldots$$

are both exact.

Note that when $B = \emptyset$ these sequences are just those for the pair (X,A) from axioms 4 and 4c.

A *map* $f : (X',A',B') \to (X,A,B)$ of one triple into another is a map of X' into X carrying A' into A and B' into B. It is called *admissible* if the maps

$$f_1 : (X',A') \to (X,A), \quad f_2 : (A',B') \to (A,B), \quad f_3 : (X',B') \to (X,B)$$

defined by f are admissible.

By axioms 2 and 3, an admissible map $f : (X',A',B') \to (X,A,B)$ induces a homomorphism of the homology sequence of (X',A',B') into that of (X,A,B), i.e. the diagram

$$
\begin{array}{ccccccccc}
\to & H_k(A',B') & \xrightarrow{i'_*} & H_k(X',B') & \xrightarrow{j'_*} & H_k(X',A') & \xrightarrow{\bar{\partial}'} & H_{k-1}(A',B') & \to \\
 & \downarrow{f_{2*}} & & \downarrow{f_{3*}} & & \downarrow{f_{1*}} & & \downarrow{f_{2*}} & \\
\to & H_k(A,B) & \xrightarrow{i_*} & H_k(X,B) & \xrightarrow{j_*} & H_k(X,A) & \xrightarrow{\bar{\partial}} & H_{k-1}(A,B) & \to
\end{array}
$$

commutes. Analogously, f induces a homomorphism of the cohomology sequence of (X,A,B) into that of (X',A',B').

Proposition A.3 ([23, Thm. 10.4]) *If* $B \neq \emptyset$ *and* (X,A,B) *is an admissible triple, then the following equivalences hold.*

$$
\begin{array}{ccc}
H_k(X,A) = 0 \ \forall k & \Longleftrightarrow & i_* : H_k(A,B) \cong H_k(X,B) \ \forall k \\
H_k(X,B) = 0 \ \forall k & \Longleftrightarrow & \bar{\partial} : H_k(X,A) \cong H_{k-1}(A,B) \ \forall k \\
H_k(A,B) = 0 \ \forall k & \Longleftrightarrow & j_* : H_k(X,B) \cong H_k(X,A) \ \forall k
\end{array}
$$

Corresponding equivalences hold for cohomology.

Proposition A.4 ([23, Thm. 10.5]) *Let* (X,A,B) *be an admissible triple. If* $B \hookrightarrow A$ *induces isomorphisms* $H_k(B) \cong H_k(A) \ \forall k$, *then* $(X,B) \hookrightarrow (X,A)$ *induces isomorphisms* $H_k(X,B) \cong H_k(X,A) \ \forall k$. *If* $A \hookrightarrow X$ *induces isomorphisms* $H_k(A) \cong H_k(X) \ \forall k$, *then* $(A,B) \hookrightarrow (X,B)$ *induces isomorphisms* $H_k(A,B) \cong H_k(X,B) \ \forall k$. *The analogs for cohomology hold.*

A.5   Homotopy equivalence and contractibility

All pairs, maps and homotopies will be assumed to be admissible.

Definition. Two pairs (X, A) and (Y, B) are *homotopically equivalent* (in $\mathcal{A}$) if there exist maps

$$f : (X, A) \to (Y, B), \qquad g : (Y, B) \to (X, A)$$

such that gf is homotopic to the identity map of (X, A) and fg is homotopic to the identity map of (Y, B). The pair of maps f, g (or just one of them) is called a *homotopy equivalence*. f and g are called *homotopy inverses* of each other.

By the homotopy axiom, homotopically equivalent spaces have the same homology and cohomology:

Proposition A.5 *A homotopy equivalence* $f : (X, A) \to (Y, B)$ *induces isomorphisms* $f_* : H_k(X, A) \cong H_k(Y, B)$ *for all* k *and an isomorphism of the homology sequences of* (X, A) *and* (Y, B). *The analog for cohomology holds.*

Recall that by definition of an admissible category there exists an admissible space P_0 consisting of a single point. If the unique map $f : X \to P_0$ is admissible, then the kernel of $f_* : H_0(X) \to H_0(P_0)$ is called the *reduced 0-dimensional homology group of* X, and is denoted by $\tilde{H}_0(X)$. The factor group of $H^0(X)$ modulo the image of $f^* : H^0(P^0) \to H^0(X)$ is called the *reduced 0-dimensional cohomology group of* X.

If P consists of a single point, then $\tilde{H}_0(P) = \tilde{H}^0(P) = 0$.

Definition. A space X is *homologically trivial* if $H_k(X) = 0$ for $k \neq 0$ and $\tilde{H}_0(X) = 0$. If $A \neq \emptyset$, the pair (X, A) is *homologically trivial* if $H_k(X, A) = 0$ $\forall\, k$. The definitions for cohomology are analogous.

Proposition A.6 *Every space contractible to a point over itself is (co)homologically trivial.*

Definition. A pair $(X', A') \subseteq (X, A)$ is called a *retract* of (X, A) if there is a map $f : (X, A) \to (X', A')$ such that $f(x) = x \; \forall\, x \in X'$. It is a *deformation retract* if the composition of f with the inclusion $(X', A') \hookrightarrow (X. A)$ is homotopic to the identity map of (X, A). (Note that this definition depends on the category $\mathcal{A}$.)

Every deformation retract of (X, A) is homotopy equivalent to (X, A), since the retraction map and the inclusion map are homotopy inverses.

Corollary A.7 *If* (X', A') *is a deformation retract of* (X, A), *then the inclusion map* $(X', A') \hookrightarrow (X, A)$ *induces isomorphisms of the (co)homology sequences.*

A.6 Direct sums

Assume $X = X_1 \cup \ldots \cup X_n$ is the disjoint union of closed (hence open) subsets of X and $A = A_1 \cup \ldots \cup A_n$, where $A_m \subseteq X_m$ for $m = 1, \ldots, n$. Assume further all pairs and inclusions formed of the sets X_m, A_m and their unions are admissible. Let $i_m : (X_m, A_m) \hookrightarrow (X, A)$ be the inclusion.

Theorem A.8 ([23, Thm. 13.2]) *The homomorphisms $i_{m*} : H_k(X_m, A_m) \to H_k(X, A)$ yield an injective representation of $H_k(X, A)$ as a direct sum, i.e. each $\alpha \in H_k(X, A)$ can be represented uniquely as $\alpha = \sum_{m=1}^{n} i_{m*}\alpha_m$, where $\alpha_m \in H_k(X_m, A_m)$.*

Theorem A.9 ([23, Thm. 13.2c]) *The homomorphisms*

$$i_m^* : H^k(X, A) \to H_k(X_m, A_m)$$

yield a projective representation of $H^k(X, A)$ as a direct sum, i.e. for each sequence $\alpha_m \in H^k(X_m, A_m)$ $(m = 1, \ldots, n)$ there is a unique element $\alpha \in H^k(X, A)$ with $i_m^* \alpha = \alpha_m$, for $m = 1, \ldots, n$.

A.7 Triads

A triad is a generalization of an admissible triple.

Definition. A *triad* $(X; X_1, X_2)$ consists of a space X and two subsets X_1, X_2 such that $X, X_1, X_2, X_1 \cup X_2, X_1 \cap X_2$ and all pairs and inclusion maps formed from these are admissible. The triad $(X; X_1, X_2)$ is called *proper* with respect to (co)homology if the inclusions

$$k_1 : (X_2, X_1 \cap X_2) \hookrightarrow (X_1 \cup X_2, X_1), \quad k_2 : (X_1, X_1 \cap X_2) \hookrightarrow (X_1 \cup X_2, X_2)$$

induce isomorphisms of the (co)homology groups in all dimensions.

Note that the last condition depends on the homology theory. It is automatically satisfied if $X_1 \supseteq X_2$, i.e. every admissible triple is a proper triad.

Theorem A.10 ([23, Thms. 14.2, 14.2c]) *A triad $(X; X_1, X_2)$ is proper with respect to homology if and only if the inclusions $i_m : (X_m, X_1 \cap X_2) \hookrightarrow (X_1 \cup X_2, X_1 \cap X_2)$ $(m = 1, 2)$ yield for each k an injective representation of $H_k(X_1 \cup X_2, X_1 \cap X_2)$ as a direct sum. The analog for cohomology holds with 'injective' replaced by 'projective'.*

Definition. The *boundary operator* $\tilde{\partial} : H_k(X, X_1 \cup X_2) \to H_{k-1}(X_1, X_1 \cap X_2)$ of a proper triad $(X; X_1, X_2)$ is defined by $\tilde{\partial} = k_{2*}^{-1} l_{2*} \partial$, where $l_2 : X_1 \cup X_2 \hookrightarrow (X_1 \cup X_2, X_2)$ is the inclusion. (Note that $\tilde{\partial}$ is the boundary operator of the triple $(X, X_1 \cup X_2, X_2)$ followed by an excision map.) The *coboundary operator* $\tilde{\delta} : H^{k-1}(X_1, X_1 \cap X_2) \to H^k(X, X_1 \cup X_2)$ is defined by $\tilde{\delta} = \delta \, l_2^* \, (k_2^*)^{-1}$.

Theorem A.11 ([23, Thms. 14.4, 14.4c]) *The homology sequence of a proper triad* $(X; X_1, X_2)$

$$\xrightarrow{\overset{\circ}{\partial}} H_k(X_1, X_1 \cap X_2) \to H_k(X, X_2) \to H_k(X, X_1 \cup X_2) \xrightarrow{\overset{\circ}{\partial}} H_{k-1}(X_1, X_1 \cap X_2) \to$$

and its cohomology sequence

$$\to H^{k-1}(X_1, X_1 \cap X_2) \xrightarrow{\overset{\circ}{\delta}} H^k(X, X_1 \cup X_2) \to H^k(X, X_2) \to H^k(X_1, X_1 \cap X_2) \xrightarrow{\overset{\circ}{\delta}}$$

are exact, where the unmarked arrows are induced by inclusions.

Note that if $X_1 \supseteq X_2$, then the (co)homology sequence of the triad $(X; X_1, X_2)$ reduces to that of the triple (X, X_1, X_2). As for triples, a map between proper triads induces a homomorphism between their corresponding (co)homology sequences.

Theorem A.12 ([23, Thms. 14.6, 14.6c]) *Let* $(X; X_1, X_2)$ *be a proper triad with* $X = X_1 \cup X_2$. *If* f, f_1, f_2 *are maps* $(X, X_1 \cap X_2) \to (Y, B)$ *such that*

$$f_1|X_1 = f|X_1, \quad f_1(X_2) \subseteq B, \qquad f_2|X_2 = f|X_2, \quad f_2(X_1) \subseteq B,$$

then $f_* = f_{1*} + f_{2*}$ *and* $f^* = f_1^* + f_2^*$.

A.8   Mayer-Vietoris sequence of a triad

Theorem A.13 ([23, Thm. 15.3]) *The* Mayer-Vietoris *sequence of a proper triad* $(X; X_1, X_2)$ *with* $X = X_1 \cup X_2$ *and* $A = X_1 \cap X_2$,

$$\ldots \to H_k(A) \xrightarrow{\psi} H_k(X_1) + H_k(X_2) \xrightarrow{\varphi} H_k(X) \xrightarrow{\Delta} H_{k-1}(A) \to \ldots$$

is exact. The maps ψ, φ, Δ *are defined as follows*
$$\psi = (h_{1*}, -h_{2*}), \quad \varphi(\alpha_1, \alpha_2) = m_{1*}\alpha_1 + m_{2*}\alpha_2, \quad \Delta = \partial_1 k_{2*}^{-1} l_{2*} = -\partial_2 k_{1*}^{-1} l_{1*},$$
where $h_i : A \hookrightarrow X_i$, $m_i : X_i \hookrightarrow X$, $k_1 : (X_2, A) \hookrightarrow (X, X_1)$, $k_2 : (X_1, A) \hookrightarrow (X, X_2)$ *and* $l_i : X \hookrightarrow (X, X_i)$ *are the inclusion maps, and* $\partial_i : H_k(X_i, A) \to H_{k-1}(A)$ *are the boundary operators.*

Theorem A.14 ([23, Thm. 15.3c]) *The* Mayer-Vietoris *cohomology sequence of a proper triad as above,*

$$\ldots \to H^{k-1}(A) \xrightarrow{\Delta} H^k(X) \xrightarrow{\varphi} H^k(X_1) + H^k(X_2) \xrightarrow{\psi} H^k(A) \to \ldots$$

is exact. Here, the maps are defined by

$$\begin{aligned}
\Delta &= l_2^*(k_2^*)^{-1}\delta_1 = -l_1^*(k_1^*)^{-1}\delta_2, \\
\varphi &= (m_1^*, m_2^*), \\
\psi(\alpha_1, \alpha_2) &= h_1^*\alpha_1 - h_2^*\alpha_2.
\end{aligned}$$

Proposition A.15 ([23, Thms. 15.4, 15.4c]) *If* $f : (X; X_1, X_2) \to (Y; Y_1, Y_2)$ *is a map of one proper triad into another, and* $X = X_1 \cup X_2$, $Y = Y_1 \cup Y_2$, *then* f *induces a homomorphism of the Mayer-Vietoris sequence of the first triad into that of the second and a homomorphism of the Mayer-Vietoris cohomology sequence of the second triad into that of the first.*

In the general case $X \supset X_1 \cup X_2$ there are relative Mayer-Vietoris sequences.

Theorem A.16 ([23, Thm. 15.7]) *The relative Mayer-Vietoris sequence of a proper triad* $(X; X_1, X_2)$,

$$\ldots \to H_k(X, X_1 \cap X_2) \overset{\psi}{\to} H_k(X, X_1) + H_k(X, X_2) \overset{\varphi}{\to}$$
$$H_k(X, X_1 \cup X_2) \overset{\Delta}{\to} H_{k-1}(X, X_1 \cap X_2) \to \ldots$$

is exact. The maps are defined as follows

$$\begin{aligned}
\psi &= (h_{1*}, -h_{2*}), \\
\varphi(\alpha_1, \alpha_2) &= m_{1*}\alpha_1 + m_{2*}\alpha_2, \\
\Delta &= n_{1*}k_{2*}^{-1}\partial_2 = -n_{2*}k_{1*}^{-1}\partial_1,
\end{aligned}$$

where $h_i : (X, X_1 \cap X_2) \hookrightarrow (X, X_i)$, $m_i : (X, X_i) \hookrightarrow (X, X_1 \cup X_2)$, $n_i : (X_i, X_1 \cap X_2) \hookrightarrow (X, X_1 \cap X_2)$, $k_1 : (X_2, X_1 \cap X_2) \hookrightarrow (X_1 \cup X_2, X_1)$. $k_2 : (X_1, X_1 \cap X_2) \hookrightarrow (X_1 \cup X_2, X_2)$ *are the inclusion maps, and* $\partial_i : H_k(X, X_1 \cup X_2) \to H_{k-1}(X_1 \cup X_2, X_i)$ *is the boundary operator.*

Theorem A.17 ([23, Thm. 15.7c]) *The relative Mayer-Vietoris cohomology sequence of a proper triad* $(X; X_1, X_2)$,

$$\ldots \to H^{k-1}(X, X_1 \cap X_2) \overset{\Delta}{\to} H^k(X, X_1 \cup X_2) \overset{\varphi}{\to}$$
$$H^k(X, X_1) + H^k(X, X_2) \overset{\psi}{\to} H^k(X, X_1 \cap X_2) \to \ldots$$

is exact. Here, the maps are defined by

$$\begin{aligned}
\Delta &= \delta_2(k_2^*)^{-1}n_1^* = -\delta_1(k_1^*)^{-1}n_2^*, \\
\varphi &= (m_1^*, m_2^*), \\
\psi(\alpha_1, \alpha_2) &= h_1^*\alpha_1 - h_2^*\alpha_2.
\end{aligned}$$

Proposition A.18 ([23, Thms. 15.8, 15.8c]) *If* $f : (X; X_1, X_2) \to (Y; Y_1, Y_2)$ *is a mapping of proper triads, then it induces homomorphisms of the relative Mayer Vietoris sequences in homology and in cohomology.*

References

[1] W. Abikoff, *The Real Analytic Theory of Teichmueller Spaces*, Springer Lecture Notes in Math. **820**, 1980.

[2] R. Abraham and J. Marsden, *Foundations of Mechanics*, Benjamin-Cummings, Menlo Park, California, 1978.

[3] R. A. Adams, *Sobolev Spaces*, Academic Press. 1975.

[4] H. Amann, E. Zehnder, *Periodic solutions of asymptotically linear Hamiltonian systems*, Manuscripta Math. **32** (1980). 149–189.

[5] V. Arnold and A. Givental, *Symplectic Geometry*, Encyclopaedia of Mathematical Sciences, vol. 4, *Dynamical Systems IV*, Springer Verlag, Berlin Heidelberg New York, 1990.

[6] V. I. Arnol'd. *Sur une propriété topologique des applications globalement canoniques de la mécanique classique*, C. R. Acad. Sci. Paris **261** (1965), 3719–3722.

[7] V. I. Arnol'd, *Mathematical Methods of Classical Mechanics*, Springer 1978. Appendix 9.

[8] A. Banyaga, *Sur le groupe des difféomorphismes qui préservent une forme sympléctique*, Comment. Math. Helv. **53** (1978). 174–227.

[9] E. Bedford and B. Gaveau, *Envelopes of holomorphy of certain 2-spheres in $\mathbf{C}^2$*, Amer. J. Math. **105** (1983), 975–1009.

[10] E. Bedford and W. Klingenberg, *On the envelope of holomorphy of a 2-sphere in $\mathbf{C}^2$*, J. Amer. Math. Soc. 4 (1991), 623–646.

[11] D. Bennequin, *Entrelacements et équations de Pfaff*, Astérisque **107–108** (1983), 83–161.

[12] I. Berstein and T. Ganea, *Homotopical nilpotency*, Illinois J. Math. **5** (1961), 99–130.

[13] E. Bishop, *Differentiable manifolds in complex euclidean space*, Duke Math. J. **32** (1965). 1–22.

[14] R. Bott, *Lectures on Morse theory, old and new*. Bull. Amer. Math. Soc. **7** (1982), 331–358.

[15] R. Bott and L. W. Tu, *Differential Forms in Algebraic Topology*, GTM 82. Springer-Verlag, 1982.

[16] P. Buser, *Geometry and Spectra of Compact Riemann Surfaces*, Birkhäuser. Boston Basel Stuttgart, 1992.

[17] J. Cerf, *Sur les difféomorphismes de la sphère de dimension trois* ($\Gamma_4 = 0$), Springer Lecture Notes in Math. **53**, 1968.

[18] S. S. Chern and E. Spanier, *A theorem on orientable surfaces in four-dimensional space*, Comm. Math. Helv. **25** (1951), 205–209.

[19] Y. Choquet-Bruhat and C. DeWitt-Morette, *Analysis, manifolds and Physics*, North-Holland, revised edition, 1982.

[20] C. C. Conley, *Isolated invariant sets and the Morse index*, CBMS Regional Conf. Series Math. **38**, Amer. Math. Soc, Providence. RI, 1978.

[21] C. Conley and E. Zehnder, *The Birkhoff-Lewis fixed point theorem and a conjecture of V. I. Arnold,* Invent. Math. **73** (1983). 33–49.

[22] B. Doubrovine. S. Novikov, A. Fomenko, *Géometrie contemporaine,* Vol II, Mir, Moscou, 1982.

[23] S. Eilenberg and N. Steenrod, *Foundations of Algebraic Topology,* Princeton University Press, Princeton, NJ, 1952.

[24] I. Ekeland and H. Hofer, *Symplectic topology and Hamiltonian dynamics,* Math. Z. **200** (1989), 355–378.

[25] I. Ekeland and H. Hofer, *Symplectic topology and Hamiltonian dynamics II,* Math. Z. **203** (1990), 553–567.

[26] Y. Eliashberg. *The complexification of contact structures on a 3-manifold,* Usp. Math. Nauk. 6(40), (1985), 161–162.

[27] Y. Eliashberg. *Three Lectures on Symplectic Topology in Cala Gonone, Basic Notions. Problems and Some Methods,* Rendiconti Seminari Facoltà Scienze, Univ. Cagliari, Supplemento al Vol. **58** (1988).

[28] Y. Eliashberg. *Classification of overtwisted contact structures on 3-manifolds,* Invent. Math. **98** (1989), 623–637.

[29] Y. Eliashberg. *Topological characterization of Stein manifolds of dimension* > 2, Int. J. of Math. **1** (1990), 29–46.

[30] Y. Eliashberg. *On symplectic manifolds with some contact properties,* J. Diff. Geom. **33** (1991), 233–238.

[31] Y. Eliashberg. *Filling by holomorphic discs and its applications,* London Math. Soc. Lect. Notes Ser. **151** (1991), 45–67.

[32] Y. Eliashberg. *Legendrian and transversal knots in tight contact manifolds,* Topological Methods in Modern Mathematics ?? (1991), 171–193.

References   231

[33] Y. Eliashberg, *Contact 3-manifolds, twenty years since J. Martinet's work*. Ann. Inst. Fourier **92** (1992), 165–192.

[34] Y. Eliashberg, *Classification of contact structures on* $\mathbf{R}^3$, International Mathematics Research Notices **3** (1993).

[35] Y. Eliashberg, *Topology of 2-knots in* $\mathbf{R}^4$ *and symplectic geometry*, preprint.

[36] Y. Eliashberg and M. Gromov, *Convex symplectic manifolds*. to appear in Proc. of Symposia in Pure Math., 1991.

[37] A. Fathi, F. Laudenbach and V. Poénaru, *Traveaux de Thurston sur les surfaces*, Astérisque **66-67**, Société Mathématique de France, Paris, 1979.

[38] A. Floer, *The unregularized gradient flow of the symplectic action*, Commun. Pure Appl. Math **41** (1988), 775–813.

[39] A. Floer, *A relative Morse index for the symplectic action*, Commun. Pure Appl. Math **41** (1988), 393–407.

[40] A. Floer, *Symplectic fixed points and holomorphic spheres*, Comm. Math. Phys. **120** (1989), 575–611.

[41] A. Floer, *Witten's complex and infinite dimensional Morse theory*, J. Diff. Geom. **30** (1989), 207–221.

[42] A. Floer, *Cuplength estimates on Lagrangian intersections*, Comm. Pure Appl. Math. **42** (1989), 335–357.

[43] A. Floer and H. Hofer, *Coherent orientations for periodic orbit problems in symplectic geometry*, preprint, 1991.

[44] A. Floer and H. Hofer, *Symplectic homology I: Preliminaries and recollections*, preprint, Ruhr Universität Bochum, 1991.

[45] A. Floer and H. Hofer, *Symplectic homology I: open sets in* $\mathbf{C}^n$, preprint.

[46] A. Floer, H. Hofer, C. Viterbo, *The Weinstein conjecture in* $P \times \mathbf{C}^l$, Math. Z. **203** (1990), 469–482.

[47] A. Floer, H. Hofer, K. Wysocki, *Applications of symplectic homology I*, preprint.

[48] D. Gilbarg and N. Trudinger, *Elliptic Partial Differential Equations of the Second Order*, Berlin 1984.

[49] E. Giroux, *Convexité en topologie de contact*, Comm. Math. Helv. **66** (1991), 637–677.

[50] M. Gromov, *Pseudoholomorphic curves in symplectic manifolds*, Invent. Math. **82** (1985), 307–347.

[51] V. Guillemin and S. Sternberg. *Symplectic techniques in physics*, Cambridge University Press, 1984.

[52] V. Harlamov and Y. Eliashberg, *On the number of complex points of a real surface in a complex surface*, Proc. LITC-82, 1982, 143–148.

[53] S. Helgason, *Differential Geometry and Symmetric Spaces*, Academic Press, New York, 1962.

[54] M. W. Hirsch. *Differential Topology*, GTM 33, Springer, 1976.

[55] H. Hofer, *Lusternik-Schnirelman theory for Lagrangian intersections*, Ann. Inst. Poincaré **5** (1988), 465–499.

[56] H. Hofer, *Pseudoholomorphic curves in symplectizations with applications to the Weinstein conjecture in dimension three*, preprint

[57] H. Hofer and E. Zehnder, *Periodic solutions on hypersurfaces and a result by C. Viterbo.* Invent. math. **90** (1987), 1–9.

[58] H. Hofer and E. Zehnder, *A new capacity for symplectic manifolds*, in: Analysis et cetera, Ed. P. Rabinowitz and E. Zehnder, Academic Press, 1990, pp. 405–428.

[59] H. Hofer and E. Zehnder, *Symplectic invariants and Hamiltonian dynamics*, preprint, May 1993.

[60] L. Hörmander. *The Analysis of Linear Partial Differential Operators*, Vol III, Springer, Berlin, 1983.

[61] C. Hummel, *Geometrische Eigenschaften pseudoholomorpher Kurven*, Diplomarbeit, Universität Freiburg im Breisgau, 1992.

[62] S. Kobayashi and K. Nomizu, *Foundations of Differential Geometry*, vol. I and II, Interscience Publ., New York, 1969.

[63] A. Korányi, H. M. Reimann, *Contact transformations as limits of symplectomorphisms*, preprint.

[64] H. F. Lai, *Characteristic classes of real manifolds immersed in complex manifolds*, Trans. Amer. Math. Soc. **172** (1972), 1–33.

[65] S. Lang, *Differential Manifolds*, Springer, 1985.

[66] H.B. Lawson Jr., *Lectures on Minimal Submanifolds*, vol. I. Publish or Perish, Inc., Berkeley, 1980.

[67] W. B. Likorish, *A foliation for 3-manifolds*, Ann. Math. **82** (1965), 414–420.

[68] R. Lutz, *Structures de contact sur les fibrés principaux en cercles de dimension 3*, Ann. Inst. Fourier **3** (1977), 1–15.

[69] J. Martinet, *Formes de contact sur les variétés de dimension 3*, Springer Lecture Notes in Math. **209** (1971), 142–163.

[70] D. McDuff, *Examples of symplectic structures*, Invent. Math. **89** (1987), 13–36.

[71] D. McDuff, *Elliptic methods in symplectic geometry*, Bull. Amer. Math. Soc. **23** (1990), 311–358.

[72] D. McDuff, *The structure of rational and ruled symplectic 4-manifolds*, J. Amer. Math. Soc., **3** (1990), 679–712.

[73] D. McDuff, *Blow ups and symplectic embeddings in dimension 4*, Topology **30** (1991), 409–421.

[74] D. McDuff, *Rational and ruled symplectic 4 manifolds*, J. Am. Math. Soc. **3** (1991), 679-712.

[75] D. McDuff, *Lectures on symplectic 4 manifolds*. CIMPA Summer school, Nice 1992.

[76] D. McDuff, *Singularities of J-holomorphic curves in almost complex 4-manifolds*, The Journal of Geometric Analysis. **3** (1992), 249–266.

[77] J. W. Milnor, *Morse Theory*, Princeton University Press, Princeton, NJ, 1963.

[78] J. W. Milnor and J. D. Stasheff, *Characteristic classes*, Ann. of Math. Studies **67**, Princeton University Press, Princeton, NJ, 1974

[79] C. Morrey, *Multiple integrals in the calculus of variations*, Berlin 1966.

[80] J. Moser, *On the volume elements of a manifold*, Trans. Amer. Math. Soc. **120** (1965), 286–294.

[81] D. Mumford, *Algebraic Geometry I, Complex Projective Varieties*, Springer Verlag, New York, 1976.

[82] R. S. Palais, *Lusternik-Schnirelman theory on Banach manifolds*, Topology **5** (1966), 115–132.

[83] R. Palais, *Foundations of global nonlinear analysis*, Benjamin, New York, 1968.

[84] R. S. Palais, *Critical point theory and minimax principle*. Proc. Sympos. Pure Math., vol. 15 (1970), Amer. Math. Soc., pp. 185–212.

[85] P. Pansu, *Notes sur les pages 316 à 323 de l'article de M. Gromov: Pseudoholomorphic curves in symplectic manifolds,* preprint. Ecole Polytechnique, Palaiseau, 1986.

[86] T. Parker and J. Wolfson, *A compactness theorem for Gromov's moduli space,* preprint.

[87] P. Rabinowitz, *Periodic solutions of Hamiltonian systems.* Commun. Pure Appl. Math. **31** (1978), 157–184.

[88] J. W. Robbin and D. Salamon, *Dynamical systems, shape theory and the Conley index,* Ergodic Theory Dynamical Systems **8** (1988). 375–393.

[89] J. W. Robbin and D. Salamon, *The Maslov Index for Paths,* preprint 1993.

[90] D. Salamon, *Morse theory, the Conley index and Floer homology,* Bull. London Math. Soc. **22** (1990), 113–140.

[91] D. Salamon and E. Zehnder, *Morse theory for periodic solutions of Hamiltonian systems and the Maslov index,* Comm. Pure Appl. Math. **45** (1992), 1303–1360.

[92] J. T. Schwartz, *Nonlinear Functional Analysis,* Courant Institute of Mathematical Sciences, New York, 1965.

[93] M. Schwarz, *Morse Homology,* Birkhäuser, 1993.

[94] J.-C. Sikorav, *Problème d'intersection et de points fixes en géometrie Hamiltonienne,* Comment. Math. Helv. **62** (1987), 61–72.

[95] S. Smale, *Diffeomorphisms of the 2-sphere,* Proc. Amer. Math. Soc. **10** (1959), 621–626.

[96] S. Smale, *On gradient dynamical systems,* Ann. of Math. **74** (1961), 199–206.

[97] E. H. Spanier. *Algebraic Topology,* McGraw-Hill, New York. 1966.

[98] S. Sternberg. *Lectures on Differential Geometry,* 2nd ed.. Chelsea, New York, 1983

[99] M. Struwe. *Existence of periodic solutions of Hamiltonian systems on almost every energy surface,* Bol. Soc. Bras. Mat. **20** (1990). 49–58.

[100] C. Viterbo. *A proof of Weinstein's conjecture in* $\mathbf{R}^{2n}$, Annales Inst. Poincaré, Anal. nonlinéaire **4** (1987), 337–356.

[101] C. Viterbo, *Intersection de sous-variétés Lagragiennes, fonctionelles d'action et indice des systèmes Hamiltoniens,* Bull. Soc. Math. France **115** (1987), 361–390.

[102] C. Viterbo, *A new obstruction to embedding Lagrangian tori,* Invent. math. **100** (1990) 301–320.

[103] C. Viterbo, *Symplectic topology as the geometry of generating functions,* Math. Ann. **292** (1992), 685–710.

[104] C. Viterbo, *Recent progress in periodic orbits of autonomous Hamiltonian systems and applications to symplectic geometry.*

[105] A. Weil, *Variétés Kaehleriennes,* Hermann, Paris. 1957.

[106] A. Weinstein, *Symplectic manifolds and their Lagrangian submanifolds,* Adv. in Math. **6** (1971), 329–346.

[107] A. Weinstein, *Lectures on symplectic manifolds.* CBMS Regional Conf. Series in Math. **29** (1977).

[108] A. Weinstein, *Periodic orbits for convex Hamiltonian systems,* Ann. Math. **108** (1978), 507–518.

[109] A. Weinstein, *On the hypotheses of Rabinowitz' periodic orbit theorems,* J. Diff. Equations **33** (1979) 353–358.

[110] E. Witten, *Supersymmetry and Morse theory,* J. Diff. Geom. **17** (1982), 661–692.

[111] J. Wolfson, *Gromov's compactness of pseudoholomorphic curves and symplectic geometry,* J. Diff. Geom. **28** (1988), 383–405.

[112] R. Ye, *Gromov's compactness theorem for pseudoholomorphic curves,* preprint 1992.

[113] K. Yosida, *Functional Analysis,* Springer-Verlag. Berlin 1980.

[114] A. Zygmund, *Trigonometric Series,* Cambridge University Press, 1959.

Index

Progress in Mathematics

Edited by:

J. Oesterlé
Départment de Mathématiques
Université de Paris VI
4, Place Jussieu
75230 Paris Cedex 05, France

A. Weinstein
Department of Mathematics
University of California
Berkeley, CA 94720
U.S.A.

Progress in Mathematics is a series of books intended for professional mathematicians and scientists, encompassing all areas of pure mathematics. This distinguished series, which began in 1979, includes authored monographs, and edited collections of papers on important research developments as well as expositions of particular subject areas.

We encourage preparation of manuscripts in such form of TeX for delivery in camera-ready copy which leads to rapid publication, or in electronic form for interfacing with laser printers or typesetters.

Proposals should be sent directly to the editors or to: Birkhäuser Boston, 675 Massachusetts Avenue, Cambridge, MA 02139, U.S.A.

MIX
Papier aus verantwortungsvollen Quellen
Paper from responsible sources
FSC® C105338

If you have any concerns about our products,
you can contact us on
ProductSafety@springernature.com

In case Publisher is established outside the EU,
the EU authorized representative is:
Springer Nature Customer Service Center GmbH
Europaplatz 3, 69115 Heidelberg, Germany

Printed by Libri Plureos GmbH
in Hamburg, Germany